T0340362

RANDOM DYNAMICAL SYSTEMS IN FINANCE

RANDOM DYNAMICAL SYSTEMS IN FINANCE

ANATOLIY SWISHCHUK

UNIVERSITY OF CALGARY

SHAFIQUL ISLAM

UNIVERSITY OF PRINCE EDWARD ISLAND

CRC Press is an imprint of the
Taylor & Francis Group, an **informa** business
A CHAPMAN & HALL BOOK

CRC Press
Taylor & Francis Group
6000 Broken Sound Parkway NW, Suite 300
Boca Raton, FL 33487-2742

First issued in paperback 2019

ISBN-13: 978-1-4398-6718-1 (hbk)
ISBN-13: 978-0-367-38014-4 (pbk)

Visit the Taylor & Francis Web site at
http://www.taylorandfrancis.com

and the CRC Press Web site at
http://www.crcpress.com

Contents

List of Figures

Preface

The theory and applications of random dynamical systems (RDS) are at the cutting edge of research in both mathematics and economics. There are many papers on RDS and also some books on RDS. As excellent examples we would like to mention *Random Dynamical Systems* by Ludwig Arnold (Springer, 2003) and *Random Dynamical Systems: Theory and Applications* by Rabi Bhattacharya and Mukul Majumdar (Cambridge, 2007).

Random dynamical systems have especially been studied in many contexts in economics, particularly in modeling long run evolution of economic systems subject to exogenous random shocks.

There are some papers on applications of RDS in economics, and a few papers on RDS in finance. However, there is no book containing any consideration of RDS in finance. Thus, this is the right time to publish a book on this topic.

Finance modeling with RDS is in its infancy. Our book is the first book that contains applications of random dynamical systems in finance.

In this way, the book is useful not only for researchers and academic people, but also for practitioners who work in the financial industry and for graduate students specializing in RDS and finance.

Anatoliy Swishchuk
Calgary, AB, Canada

Shafiqul Islam
Charlottetown, PEI, Canada

Acknowledgment

We would like to thank our colleagues and students for many useful discussions and unforgettable collaborations.

Anatoliy Swishchuk thanks Jianhong Wu, Anatoliy Ivanov, Yulia Mishura, Yuriy Kazmerchuk, Anna Kalemanova-Schlösser, Edson Alberto Coayla Teran, and LiFeng Zhang.

Shafiqul Islam wishes to give special recognition and appreciation to two individuals at Concordia University, Montreal, Canada. Pawel Góra and Abraham Boyarsky of Concordia University introduced and taught Shafiqul Islam the subject: random dynamical systems.

Anatoliy Swishchuk and Shafiqul Islam wish to recognize the patient encouragement and understanding of their families: wives Maryna Swishchuk and Monzu Ara Begum, and daughters Julia Swishchuk and Anika Tabassum, respectively. Without their dedication and support this book would remain just a dream. Anatoliy Swishchuk also thanks his son Victor Swishchuk for continuous support and inspiration.

We would also like to thank the anonymous referee for many remarks and comments. Our great appreciation and many thanks go to Matt Davison, who helped us to improve the book with many valuable and important comments, suggestions, and additions, especially on the matter of financial applications issues.

We both appreciate enormous help from David Grubbs (Chapman & Hall) and Shashi Kumar (Chapman & Hall) during preparation of this book.

Anatoliy Swishchuk
Calgary, AB, Canada

Shafiqul Islam
Charlottetown, PEI, Canada

Chapter 1

Introduction

This book is devoted to the study of random dynamical systems (RDS) and their applications in finance. The theory of RDS, developed by L. Arnold and co-workers, can be used to describe the asymptotic and qualitative behavior of systems of random and stochastic differential/difference equation in terms of stability, invariant manifolds, attractors, etc. Usually, a RDS consists of two parts: the first part is a model for the noise path, leading to a RDS, and the second part is the dynamics of a model.

In this book, we present many models of RDS and develop techniques in the RDS which can be implemented in finance.

Let us present just a few of many examples that can be used in finance or/and economics.

One of the examples of a model of RDS that can be used in finance is a geometric Markov renewal process (GMRP) for a stock price, which is defined as follows (see Chapter 6 for details):

$$S_t := S_0 \prod_{k=1}^{\nu(t)} (1+\rho(x_k)), \; t \in \mathbf{R}_+,$$

where function $\rho(x) > -1$ is continuous and bounded on phase space X of a Markov chain $x_n, \quad n \in \mathbf{Z}_+$, $\nu(t)$ is a counting process. This model is a generalization of Cox-Ross-Rubinstein binomial model for stock price (see [4], Chapter 6) and Aase's geometric compound Poisson process (see [1], Chapter 6).

The second example of a model of RDS that can be used in economics is a Ramsey (see [10], Chapter 13) stochastic model for capital that takes into account the delay and randomness in the production cycle (see Chapter 13 for details):

$$dK(t) = [AK(t-T) - u(K(t))C(t)]\, dt + \sigma(K(t-T))dw(t)$$

where K is the capital, C is the production rate, u is a control process, A is a positive constant, σ is a standard deviation of the "noise" $w(t)$. The "initial capital"

$$K(t) = \phi(t), \quad t \in [-T, 0],$$

is a continuous bounded positive function and depends not only on current t, but also on the past before t.

One more example is associated with a model for a stock price $S(t)$ that includes regime switching, delay, noise and Poisson jumps (see Chapter 12 for details):

$$\begin{aligned} dS(t) &= [a(r(t))S(t) + \mu(r(t))S(t-\tau)]dt + \sigma(r(t))S(t-\rho)dW(t) \\ &+ \int_{-1}^{\infty} yS(t)\nu(dy,dt). \end{aligned}$$

This model includes not only the current state of the stock price $S(t)$, but also, e.g., histories, $S(t-\tau)$ and $S(t-\rho)$, where ρ and τ are delayed parameters, and sudden shocks (Poisson jumps).

Dynamical systems are mathematical models of real-world problems and they provide a useful framework for analyzing various physical (see [7] and [9] in Chapter 3), engineering, social, and economic phenomena (see [37] in Chapter 3). A random dynamical system is a measure-theoretic formulation of a dynamical system with an element of randomness. A deterministic dynamical system is a system in which no randomness is involved in the development of future states of the system. The fundamental problem in the ergodic theory of deterministic dynamical systems is to describe the asymptotic behavior of trajectories defined by a deterministic dynamical system. In general, the long-time behavior of trajectories of a chaotic deterministic dynamical system is unpredictable (see [2] in Chapter 2). Therefore, it is natural to describe the behavior of the system as a whole by statistical means. In this approach, one attempts to describe the dynamics by proving the existence of an invariant measure and determining its ergodic properties (see [2] in Chapter 2). In particular, the existence of invariant measures which are absolutely continuous with respect to Lebesgue measure is very important from a physical point of view, because computer simulations of orbits of the system reveal only invariant measures which are absolutely continuous with respect to Lebesgue measure (see [18] in Chapter 3). The Birkhoff Ergodic Theorem (see [2] in Chapter 2) states that if $\tau : (X, \mathcal{B}, \mu) \to (X, \mathcal{B}, \mu)$ is ergodic and $\mu-$invariant and E is a measurable subset of X then the orbit of almost every point of X occurs in the set E with asymptotic frequency $\mu(E)$.

The Frobenius–Perron operator P_τ is the main tool for proving the existence of absolutely continuous invariant measures (acim) of a transformation τ. It is well known that f is the density of an acim μ under a transformation τ if and only if $P_\tau f = f$. In 1940, Ulam and von Neumann found examples of transformations having absolutely continuous invariant measures. In 1957, Rényi (see [35] in Chapter 3) defined a class of transformations that have an acim. Rényi's key idea of using distortion estimates has been used in the more general proofs of Adler and Flatto (see [2] in Chapter 3). In 1973, Lasota and Yorke (see [10] in Chapter 2) proved a general sufficient condition for the existence of an absolutely continuous invariant measure for piecewise expanding C^2 transformations. Their result was an important generalization of Rényi's (see [35] in Chapter 3) result using the theory of bounded variation and their essential observation was that, for piecewise expanding transformations, the Frobenius–Perron operator is a contraction. The bounded variation technique has been generalized in a number of directions (see [27] in Chapter 3). In Chapter 2 of this book, we briefly review deterministic dynamical systems, ergodic theory, the Frobenious–Perron operator, invariant measures, and stochastic perturbations. Many of these fundamental results in Chapter 2 of this book will be useful for Chapters 3–4. For more detailed results on the existence, properties, and approximations of invariant measures for deterministic dynamical systems, see the book by Boyarsky and Góra (see [2] in Chapter 2). The book by Ding and Zhou, (see [4] in Chapter 2) is another good reference for deterministic dynamical systems.

Random dynamical systems provide a useful framework for modeling and analyzing various physical, social, and economic phenomena (see [9], [37], and [38] in Chapter 3). A random dynamical system of special interest is a random map where the process switches from one map to another according to fixed probabilities (see [34] in Chapter 3) or, more generally, position dependent probabilities (see [3–6] and [16] in Chapter 3]. The existence and properties of invariant measures for random maps reflect their long-time behavior and play an important role in understanding their chaotic nature. Random maps have applications in the study of fractals (see [7] in Chapter 3), in modeling interference effects in quantum mechanics (see [9] in Chapter 3), in computing metric entropy (see [38] in Chapter 3), and in forecasting the financial markets (see [3] in Chapter 3). In 1984, Pelikan (see [34] in Chapter 3) proved sufficient conditions for the existence of acim for random maps with constant probabilities. Morita (see [32] in Chapter 3) proved a spectral decomposition theorem. In Chapter 3 of this book, we first present a general setup for a random dynamical system from Arnold's sense (see [1] in Chapter 3). Then we present skew product and random maps with constant probabilities. Some fundamental results on the properties of the Frobenius–Perron operator for random maps with constant probabilities are also presented in Chapter 3. We present necessary and sufficient conditions for the existence of absolutely continuous invariant measures for random maps. Moreover, we present two important properties of invariant measures for random maps with constant probabilities. At the end of Chapter 3, we present some applications of random maps in finance.

Position dependent random maps are more general random maps where the probabilities of choosing component maps are position dependent. Góra and Boyarsky (see [14] in Chapter 4) proved sufficient conditions for the existence of acim for random maps with position dependent probabilities. Bahsoun and Góra proved sufficient average expanding conditions for the existence of acim for position dependent random maps in one and higher dimensions (see [2] in Chapter 4), weakly convex and concave position dependent random maps (see [5] in Chapter 3). Bahsoun, Góra, and Boyarsky proved the sufficient condition for the existence of Markov switching random map with position dependent switching matrix (see [3] in chapter 4). In Chapter 4 of this book, we first present position dependent random maps and properties of the Frobenius–Perron operator. Then we present the existence of invariant measures for random maps, Markov switching random maps in one and higher dimensions.

Froyland (see [14] in Chapter 3) extended Ulam's method for a single transformation to random maps with constant probabilities (see [34] in Chapter 3). Góra and Boyarsky proved the convergence of Ulam's approximation for position dependent random maps (see [14] in Chapter 4). For Markov switching random maps, Froyland (see [14] in Chapter 3) considered the constant stochastic irreducible matrix W and proved the existence and convergence of Ulam's approximation of invariant measures. In Chapter 4 of this book, we also present numerical schemes for the approximation of invariant measures for position dependent random maps. Applications of position dependent random maps in finance are presented at the end of Chapter 4 of this book.

Chapter 5 is devoted to the study of random evolutions (REs). In mathematical language, a RE is a solution of stochastic operator integral equation in a Banach space. The operator coefficients of such equations depend on random parameters. The random evolution (RE), in physical language, is a model for a dynamical system whose state of evolution is subject to random variations. Such systems arise in many branches of science, e.g., random Hamiltonian and Shroedinger's equations with random potential in quantum mechanics, Maxwell's equation with a random reflective index in electrodynamics, transport equation, storage equation, etc. There are a lot of applications of REs in financial and insurance mathematics (see [11], Chapter 5). One of the recent applications of RE is associated with geometric Markov renewal processes which are regime-switching models for a stock price in financial mathematics, which will be studied intensively in the next chapters. Another recent application of RE is a semi-Markov risk process in insurance mathematics (see [11], Chapter 5). The REs are also examples of more general mathematical objects such as multiplicative operator functional (MOFs) (see [7, 10], Chapter 5), which are random dynamical systems in Banach space. The REs can be described by two objects: 1) operator dynamical system $V(t)$ and 2) random process $x(t)$. Depending on structure of $V(t)$ and properties of the stochastic process $x(t)$ we have different kinds of REs: continuous, discrete, Markov, semi-Markov, etc. In this chapter we deal with various problems for REs, including martingale property, asymptotical behavior of REs, such as averaging, merging, diffusion approximation, normal deviations, averaging, and diffusion approximation in reducible phase space for $x(t)$ rate of convergence for limit theorems for REs.

Chapters 6–9 deal with geometric Markov renewal processes (GMRP) as a special case of REs. We study approximation of GMRP in ergodic, merged, double averaged, diffusion, normal deviation, and Poisson cases. In all these cases we present applications of the obtained results to finance, including option pricing formulas.

In Chapter 6 we introduce the geometric Markov renewal processes as a model for a security market and study these processes in a series scheme. We consider its approximations in the form of averaged, merged, and double averaged geometric Markov renewal processes. Weak convergence analysis and rates of convergence of ergodic geometric Markov renewal processes, are presented. Martingale properties, infinitesimal operators of geometric Markov renewal processes are presented and a Markov renewal equation for expectation is derived. As an application, we consider the case of two ergodic classes. Moreover, we consider a generalized binomial model for a security market induced by a position dependent random map as a special case of a geometric Markov renewal process.

In Chapter 7 we study the geometric Markov renewal processes in a diffusion approximation scheme. Weak convergence analysis and rates of convergence of ergodic geometric Markov renewal processes in a diffusion scheme are presented. We present European call option pricing formulas in the case of ergodic, double averaged, and merged diffusion geometric Markov renewal processes.

Chapter 8 is devoted to the normal deviations of the geometric Markov renewal processes for ergodic averaging and double averaging schemes. Algorithms of averaging define the averaged systems (or models) which may be considered as the first approximation. Algorithms of diffusion under balance condition define diffusion models which may be considered as the second approximation. In this chapter we consider the algorithms of construction of the first and second approximation in the case when the balance condition is not fulfilled. Some applications in finance are presented; in particular, option pricing formulas in this case are derived.

In Chapter 9, we introduce the Poisson averaging scheme for the geometric Markov renewal processes. Financial applications in Poisson approximation schemes of the geometric Markov renewal processes are presented, including option pricing formulas.

Chapter 10 considers the stochastic stability of fractional (B,S)-security markets, that is, financial markets with a stochastic behavior that is caused by a random process with long-range dependence, fractional Brownian motion. Three financial models are considered. They arose as a result of different approaches to the definition of the stochastic integral with respect to fractional Brownian motion. The stochastic stability of fractional Brownian markets with jumps is also considered. In Appendix, we give some definitions of stability, Lyapunov indices, and some results on rates of convergence of fractional Brownian motion, which we use in our development of stochastic stability.

In Chapter 11, we study the stochastic stability of random dynamical systems arising in the interest rate theory. We introduce different definitions of stochastic stability. Then, the stochastic stability of interest rates for the Black-Scholes, Vasicek, Cox-Ingersoll-Ross models and their generalizations for the case of random jump changes are studied.

The subject of Chapter 12 is the stability of trivial solution of stochastic differential delay in Ito's equations with Markovian switchings and with Poisson bifurcations. Throughout the work stochastic analogue of second Lyapunov method is used.

Some applications in finance are considered as well.

RDS in the form of stochastic differential delay equations and their optimal control have received much attention in recent years. Delayed problems often appear in applications in physics, biology, engineering, and finance. Optimal controls of delayed RDS in finance in some specific and general settings are considered in Chapters 13 and 14, respectively.

Chapter 13 is devoted to the study of optimal control of random delayed dynamical systems and their applications. By using the Dynkin formula and solution of the Dirichlet-Poisson problem developed in Chapter 5, the Hamilton-Jacobi-Bellman (HJB) equation and the inverse HJB equation are derived. Application is given to a stochastic model in economics (stochastic Ramsey's model).

In Chapter 14 the problem of RDS arising in optimal control theory for vector stochastic differential delay equations (SDDEs) and its applications in mathematical finance and economics is studied. By using the Dynkin formula and solution of the Dirichlet-Poisson problem developed in Chapter 5, the Hamilton-Jacobi-Bellman (HJB) equation and the converse HJB equation are derived. Furthermore, applications are given to an optimal portfolio selection problem and a stochastic Ramsey model in economics.

The analogue of the Black-Scholes formula for vanilla call option price in conditions of (B,S)-securities market with delayed/past-dependent information is derived in Chapter 15. A special case of a continuous version of GARCH is considered. The results are compared with the results of the Black and Scholes (1973) formula.

All references are provided at the end of each chapter.

Thus, the book contains a variety of RDS which are used for approximations of financial models, studies of their stability and control, and presents many option pricing formulas for these models.

The book will be useful for researchers and academics who work in RDS and mathematical finance, and also for practitioners working in the financial industry. It will also be useful for graduate students specializing in the areas of RDS and mathematical finance.

Chapter 2

Deterministic Dynamical Systems and Stochastic Perturbations

2.1 Chapter overview

In this chapter we review deterministic dynamical systems and their invariant measures. Deterministic dynamical systems are special cases of random dynamical systems, and theories of deterministic dynamical systems play an important role for the study of random dynamical systems. The existence and properties of absolutely continuous invariant measures for deterministic dynamical systems reflect their long-time behavior and play an important role in understanding their chaotic nature. The Frobenius–Perron operator for deterministic dynamical systems is one of the key tools for the study of invariant measures for deterministic dynamical systems. In Chapter 3 and Chapter 4 we will see that the Frobenius–Perron operator for random dynamical systems is a combination of the Frobenius–Perron operator of the individual component systems which are deterministic dynamical systems. In this chapter we focus our special attention on the class of piecewise monotonic and expanding deterministic dynamical systems. Moreover, we present stochastic perturbations of deterministic dynamical systems. For the Frobenius–Perron operator and existence of invariant measures we closely follow [2, 4, 9, 10] and the references therein. For the stochastic perturbations we closely follow [7, 8, 9, 11] and the references therein.

2.2 Deterministic dynamical systems

Let $(X,\mathcal{B},\mu)$ be a normalized measure space where X is a set, $\mathcal{B}$ is a σ-algebra of subsets of X and μ is a measure such that $\mu(X)=1$. Let ν be another measure on $(X,\mathcal{B})$. The measure μ is absolutely continuous with respect to ν if for any $A\in\mathcal{B}$ with $\nu(A)=0$, we have $\mu(A)=0$. Let $I=[a,b]$ be an interval of the real line $\mathbb{R}$. Throughout this chapter, we consider $X=I=[0,1]$ and we denote by $V_I(\cdot)$ the standard one dimensional variation of a function on $[0,1]$ and let $BV(I)$ be the space of functions of bounded variations on I equipped with the norm $\|\cdot\|_{BV}=V_I(\cdot)+\|\cdot\|_1$, where $\|\cdot\|_1$ denotes the L^1 norm on $L^1(I,\mathcal{B},\mu)$.

Definition 2.1 *Let $\tau: I\to I$ be a transformation such that for any initial $x\in I$, the nth iteration of x under τ is defined by $\tau^n(x)=\tau\circ\tau\circ\ldots\circ\tau(x)$* n times. *The transforma-*

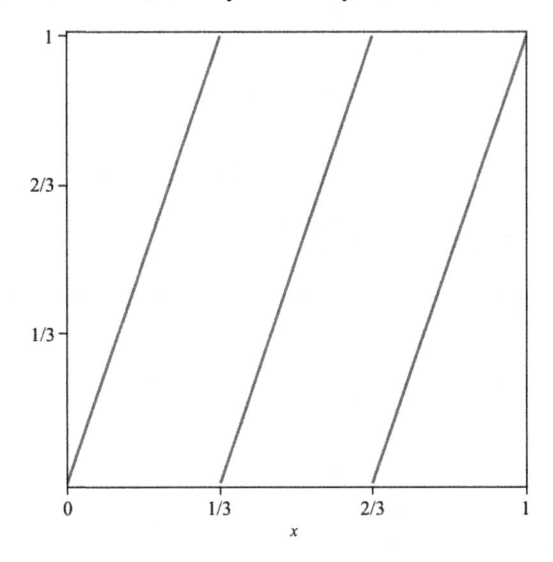

Figure 2.1 *The piecewise expanding map τ.*

tion $\tau : I \to I$ is **non-singular** *if for any $A \in \mathcal{B}$ with $\mu(A) = 0$, we have $\mu(\tau^{-1}(A)) = 0$. The transformation τ* **preserves the measure** *μ or* **the measure μ is τ-invariant** *if $\mu(\tau^{-1}(A)) = \mu(A)$ for all $A \in \mathcal{B}$. In this case the quadruple $(X, \mathcal{B}, \mu, \tau)$ is called a* **deterministic dynamical system**. *A family $\mathcal{B}^*$ of subsets of I is a* **π-system** *if and only if $\mathcal{B}^*$ is closed under intersections.*

The following Theorem (Theorem 3.1.1 in [2]) is useful for checking whether a transformation preserves a measure:

Theorem 2.2 *[2] Let $(I, \mathcal{B}, \mu)$ be a normalized measure space and $\tau : I \to I$ be a measurable transformation. Let $\mathcal{B}^*$ be a $\pi-$ system that generates $\mathcal{B}$. Then μ is τ-invariant if $\mu(\tau^{-1}(A)) = \mu(A)$ for any $A \in \mathcal{B}^*$.*

Example 2.1 *Consider the measure space $([0,1], \mathcal{B}, \lambda)$, where $\mathcal{B}$ is σ-algebra on $[0,1]$ and λ the Lebesgue measure on $[0,1]$. Let $\tau : [0,1] \to [0,1]$ be a map (see Figure 2.1) defined by*

$$\tau(x) = \begin{cases} 3x, & 0 \leq x < \frac{1}{3}, \\ 3x - 1, & \frac{1}{3} \leq x < \frac{2}{3}, \\ 3x - 2, & \frac{2}{3} \leq x \leq 1, \end{cases}$$

For any interval $[x,y] \subset [0,1]$, $\tau^{-1}([x,y]) = [\frac{x}{3}, \frac{y}{3}] \cup [\frac{x+1}{3}, \frac{y+1}{3}] \cup [\frac{x+2}{3}, \frac{y+2}{3}]$ and

$\lambda(\tau^{-1}([x,y])) = \lambda([\frac{x}{3},\frac{y}{3}] \cup [\frac{x+1}{3},\frac{y+1}{3}] \cup [\frac{x+2}{3},\frac{y+2}{3}]) = y - x = \lambda([x,y])$. *By Theorem 2.2, the transformation τ is λ-invariant. Thus, $([0,1],\mathcal{B},\lambda,\tau)$ is a deterministic dynamical system.*

2.2.1 Ergodicity and Birkhoff individual ergodic theorem

Let $\tau : [0,1] \to [0,1]$ be a measure preserving transformation and $x_0 \in [0,1]$. The Birkhoff ergodic theorem allows us to study the statistical behavior of orbit $\{x_0, x_1 = \tau(x_0), \dots, x_n = \tau(x_{n-1}\}$. If τ is ergodic, then the Birkhoff ergodic theorem provides more specific information of the orbit. Let A be a measurable set of $[0,1]$ and χ_A be the characteristic function on A. For any $i \in \{0,1,\dots,n\}$, $x_i = \tau^i(x_0) \in A$ if and only if $\chi_A(\tau^i(x_0)) = 1$.

Definition 2.3 *A measure-preserving transformation $\tau : (X,\mathcal{B},\mu) \to (X,\mathcal{B},\mu)$ is* **ergodic** *if for any $B \in \mathcal{B}$ such that $\tau^{-1}B = B$, we have $\mu(B) = 0$ or $\mu(X \setminus B) = 0$.*

Ergodicity of a measure preserving transformation $\tau : [0,1] \to [0,1]$ is an indecomposability property such that if τ has this indecomposability property then the study of τ cannot be split into separate parts. The following Theorem (Theorem 3.2.1 in [2], see also [4]) is useful for checking whether a transformation is ergodic:

Theorem 2.4 *[2] Let $\tau : (I,\mathcal{B},\mu) \to (I,\mathcal{B},\mu)$ be a measure preserving transformation. Then the following statements are equivalent:*

1. *τ is ergodic.*
2. *If f is measurable and $(f \circ \tau)(x) = f(x)$ almost everywhere, then f is constant almost everywhere.*
3. *If $f \in L^2$ and $(f \circ \tau)(x) = f(x)$ almost everywhere, then f is constant almost everywhere.*

Theorem 2.5 ***Birkhoff's ergodic theorem for deterministic dynamical systems*** *[2, 9]: Let $\tau : (I,\mathcal{B},\mu) \to (I,\mathcal{B},\mu)$ be μ-invariant and $f \in L^1(I,\mathcal{B},\mu)$. Then there exists a function $f^* \in L^1(X,\mathcal{B},\mu)$ such that for $\mu-$almost all $x \in I$ the limit of the time averages $\frac{1}{n+1}\sum_{k=0}^{n} f(x_k)$ exists and*

$$\frac{1}{n+1}\sum_{k=0}^{n} f(x_k) \to f^*, \tag{2.2.1}$$

$\mu-$ almost everywhere. Moreover, if τ is ergodic and $\mu(X) = 1$, then f^* is constant μ a.e. and $f^* = \int_X f d\mu$.

Application of the Birkoff ergodic theorem: Let $A \in \mathcal{B}$. Then $\sum_{k=0}^{n} \chi_A(x_k)$ is the number of points of the orbit $\{x_0, x_1 = \tau(x_0), \dots, x_n = \tau(x_{n-1}\}$ in A and $\frac{1}{n+1}\sum_{k=0}^{n} \chi_A(x_k)$ is the relative frequency of the elements of $\{x_0, x_1 = \tau(x_0), \dots, x_n =$

$\tau(x_{n-1}\}$. If we replace $f \in L^1$ by the characteristic function χ_A on the measurable set $A \subset [0,1]$ and if τ is ergodic and $\mu(I) = 1$ then by the Birkoff ergodic theorem 2.5,

$$\frac{1}{n+1}\sum_{k=0}^{n-1} \chi_A(\tau^k(x)) \to \mu(A), \tag{2.2.2}$$

$\mu-$ almost everywhere and thus the orbit of almost every point of I occurs in the set A with asymptotic frequency $\mu(A)$.

Example 2.2 *Consider the transformation τ in Example 2.1. τ preserves the Lebesgue measure λ and τ is λ-ergodic. Consider the measurable sets E_i of $[0,1]$ where $E_i = [\frac{i}{5}, \frac{i+1}{5}], i = 0,1,2,3$. Let x_0 be any initial point in $[0,1]$. By the Birkoff ergodic theorem 2.5*

$$\frac{1}{n+1}\sum_{k=0}^{n-1} \chi_{A_i}(\tau^k(x_0)) \to \lambda(E_i) = \frac{1}{3}, \tag{2.2.3}$$

2.2.2 *Stationary (invariant) measures and the Frobenius–Perron operator for deterministic dynamical systems*

Consider the measure space $(I,\mathcal{B},\lambda)$ and let $\mathcal{M}(I) = \{m : m \text{ is a measure on I}\}$, that is, $\mathcal{M}(I)$ is the space of measures on $(I,\mathcal{B})$. Let $\tau : ([a,b],\mathcal{B},\lambda) \to (I,\mathcal{B},\lambda)$ be a piecewise monotonic non-singular transformation on the partition $\mathcal{P}$ of I where $\mathcal{P} = \{I_1, I_2, \ldots, I_N\}$ and $\tau_i = \tau|_{I_i}$. Let $\mu << \lambda$, that is, μ is absolutely continuous with respect to λ. The transformation τ induces an operator $\mathcal{O}$ on $\mathcal{M}(I)$ defined by

$$\mathcal{O}(\mu)(A) = \mu(\tau^{-1}(A)).$$

Non-singularity of τ implies that $\mathcal{O}(\mu) << \lambda$. Suppose that μ has a density $f \in \mathcal{D} = \{f \in L^1(I,\mathcal{B},\mu) : f \geq 0 \text{ and } \| \text{f} \|_1 = 1\}$ with respect to λ. Then by the Radon-Nikodyn Theorem, $\mu(A) = \int_A f d\lambda$ for any measurable set $A \in \mathcal{B}$. Since μ has a density f, the induced measure $\mathcal{O}(\mu)$ also has a density $P_\tau f$. Thus,

$$\mathcal{O}(\mu)(A) = \int_A P_\tau f d\lambda = \mu(\tau^{-1}(A)) = \int_{\tau^{-1}(A)} f d\lambda.$$

Clearly, $P_\tau : L^1(I,\mathcal{B},\lambda) \to L^1(I),\mathcal{B},\lambda)$ is a linear operator. The above operator P_τ defined by

$$\int_A P_\tau f d\lambda = \int_{\tau^{-1}(A)} f d\lambda \tag{2.2.4}$$

is known as the Frobenius-Perron operator. Let $A = [0,x]$. Then

$$\int_0^x P_\tau f d\lambda = \int_{\tau^{-1}([0,x])} f d\lambda.$$

Differentiating on both sides of (2.2.4) with respect to P_τ we get

$$P_\tau f d\lambda = \frac{d}{dx}\int_{\tau^{-1}([0,x))} f d\lambda. \tag{2.2.5}$$

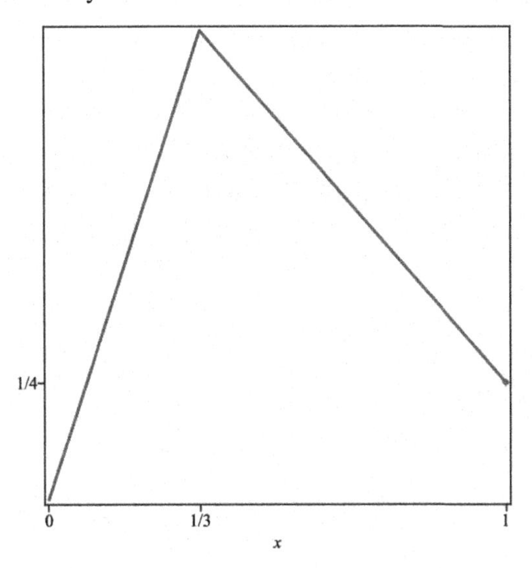

Figure 2.2 *The map τ.*

Example 2.3 *Let* $\tau : [0,1] \to [0,1]$ *be defined by*

$$\tau(x) = \begin{cases} 3x, & x \in [0, \frac{1}{3}] \\ -\frac{9}{8}x + \frac{11}{8}, & x \in [\frac{1}{3}, 1]. \end{cases}$$

See Figure 2.2. If $x >= \frac{1}{4}$, *then* $\tau^{-1}([0,x]) = [0, \frac{1}{3}x] \cup [\frac{11}{9} - \frac{8}{9}x, 1]$. *If* $0 \le x < \frac{1}{3}$, *then* $\tau^{-1}([0,x]) = [0, \frac{1}{3}x]$. *Therefore, Then* $\tau^{-1}([0,x]) = [0, \frac{1}{3}x] \cup \{[\frac{11}{9} - \frac{8}{9}x, 1] \cap A\}$, *where* $A = [\frac{1}{3}, 1]$. *For any* $f \in L^1(0,1)$,

$$\begin{aligned} P_\tau f d\lambda &= \frac{d}{dx} \int_{\tau^{-1}([0,x])} f d\lambda \\ &= \frac{d}{dx} \int_{[0,\frac{1}{3}x] \cup \{[\frac{11}{9} - \frac{8}{9}x, 1] \cap A\}} f(x) d\lambda \\ &= \frac{d}{dx} \left[\int_0^{\frac{x}{3}} f(x) d\lambda + \int_{\frac{11}{9} - \frac{8}{9}x}^{1} f(x) \chi_A(x) d\lambda \right] \\ &= f(\frac{x}{3}) + \frac{8}{9} f(\frac{11}{9} - \frac{8}{9}x) \chi_J(x), \end{aligned}$$

where $J = \tau(A) = [\frac{1}{4}, 1]$.

Properties of the Frobenius–Perron operator operator P_τ [2, 9]: It is not difficult to show that the Frobenius–Perron operator operator P_τ of a transformation τ has the following useful properties:

1. Linearity: the Frobenius-Perron operator operator is a linear operator, that is

$$P_\tau(\alpha f + \beta g) = \alpha P_\tau f + \beta P_\tau g,$$

where α, β are real numbers and $f, g \in L^1$.
2. Positivity: Let $f \in L^1$ and assume $f \geq 0$. Then $P_\tau f \geq 0$.
3. Contraction Property: $P_\tau : L^1 \to L^1$ is a contraction. It means that for any $f \in L^1$

$$\| P_\tau f \|_1 \leq \| f \|_1$$

4. Preservation of Integrals: P_τ preserves integrals, i.e., $\int_I f d\lambda = \int_I P_\tau f d\lambda$;
5. Composition Property: Let $\tau : I \to I$ and $\sigma : I \to$ I be non-singular, then

$$P_{\tau \cdot \sigma} f = P_\tau \cdot P_\sigma f$$

Moreover,

$$P_{\tau^n} f = P_\tau^n f$$

6. Adjoint Property: If $f \in L^1$ and $g \in L^\infty$, then

$$< P_\tau f, g > = < f, U_\tau g d\lambda >$$

For more details of the above properties see [2, 4, 9].

Definition 2.6 *A transformation $\tau : [0,1] \to [0,1]$ is piecewise monotonic if there exists a partition $0 = x_0 < x_1 < \cdots < x_n = 1$ and a constant $r \geq 1$ such that*

1. $|\tau'(x)| > 0$ *for* $x \in (x_{i-1}, x_i), i = 1, 2, \ldots, n$.
2. $\tau_{|(x_{i-1}, x_i)}$ *is a r times continuously differentiable function which can be extended to a r times continuously differentiable function on the closed interval* $[x_{i-1}, x_i], i = 1, 2, \ldots, n$.

A transformation $\tau : [0,1] \to [0,1]$ is piecewise expanding if τ is piecewise monotonic and $|\tau'(x)| > 1$ for $x \in (x_{i-1}, x_i), i = 1, 2, \ldots, n$.

Example 2.4 *The map $\tau : [0,1] \to [0,1]$ (see Figure 2.3) defined by*

$$\tau(x) = \begin{cases} \frac{x}{2}, & 0 \leq x < \frac{1}{2}, \\ 2x - 1, & \frac{1}{2} \leq x \leq 1, \end{cases}$$

is piecewise monotonic and the tent map $\tau : [0,1] \to [0,1]$ (see Figure 2.4) defined by

$$\tau(x) = \begin{cases} 2x, & 0 \leq x \leq \frac{1}{2}, \\ 2 - 2x, & \frac{1}{2} \leq x \leq 1, \end{cases}$$

is piecewise expanding.

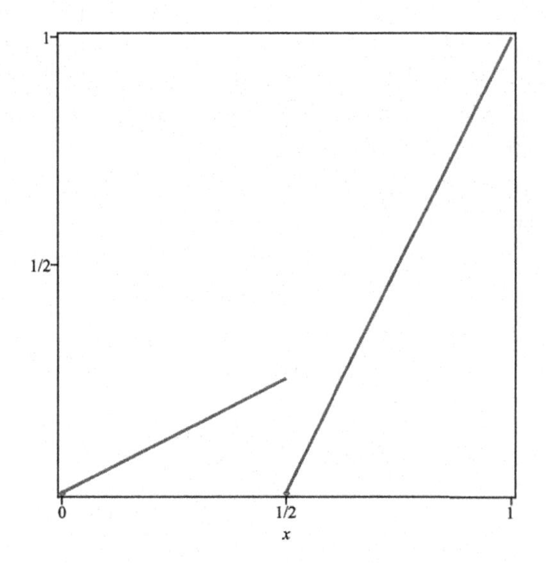

Figure 2.3 *The piecewise monotonic map τ.*

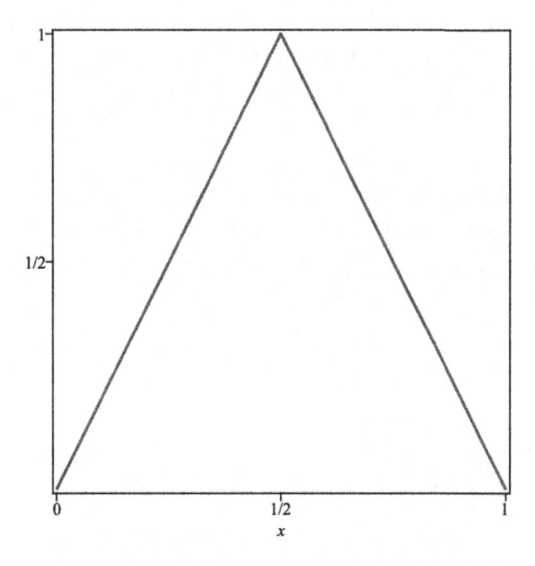

Figure 2.4 *The tent map τ.*

Representation of the Frobenius–Perron operator P_τ [2]: Let $\tau : [0,1] \to [0,1]$ be a piecewise monotonic transformation with respect to the partition $0 = x_0 < x_1 < \cdots < x_n = 1 = \{I_1, I_2, \ldots, I_n\}$. For each $1 \le i \le n$, let $D_i = \tau([x_{i-1}, i])$ and $g_i : D_i \to [x_{i-1}, x_i]$ is defined by $g_i(x) = \tau^{-1}_{|D_i}$. Piecewise monotonicity of τ implies that g_i exists for each $1 \le i \le n$. Let $A \in \mathcal{B}$. Then $\tau^{-1}(A) = \sum_{i=1}^{n} g_i(A \cap D_i)$ and $\{A \cap D_i\}_{1 \le i \le n}$ is a family of mutually disjoint sets. For $f \in L^1(I)$,

$$\begin{aligned}
\int_A (P_\tau(f))(x) d\lambda &= \int_{\tau^{-1}(A)} f d\lambda \\
&= \int_{\sum_{i=1}^{n} g_i(A \cap D_i)} f d\lambda \\
&= \sum_{i=1}^{n} \int_{g_i(A \cap D_i)} f d\lambda \\
&= \sum_{i=1}^{n} \int_{A \cap D_i} f(g_i(x)) |g_i'(x)| d\lambda \\
&= \sum_{i=1}^{n} \int_A f(g_i(x)) |g_i'(x)| \chi_{D_i}(x) d\lambda \\
&= \int_A \sum_{i=1}^{n} \frac{f(\tau_i^{-1}(x))}{\tau'(\tau_i^{-1}(x))} \chi_{\tau(x_{i-1}, x_i)}(x) d\lambda
\end{aligned}$$

$$(P_\tau(f))(x) = \sum_{i=1}^{n} \frac{f(\tau_i^{-1}(x))}{\tau'(\tau_i^{-1}(x))} \chi_{\tau(x_{i-1}, x_i)}(x), \tag{2.2.6}$$

for any measurable set A and $f \in L^1(I)$. Equation (2.2.6) is the representation of the Frobenius–Perron operator P_τ. Equation (2.2.6) can also be rewritten as

$$(P_\tau(f))(x) = \sum_{y \in \{\tau^{-1}(x)\}} \frac{f(y)}{\tau'(y)}, \tag{2.2.7}$$

Example 2.5 *Let $\tau : [0,1] \to [0,1]$ be the logistic map (see Figure 2.5) defined by $\tau(x) = 4x(1-x)$. It is not difficult to show that τ is piecewise monotonic with respect to the partition $\{x_0, x_1, x_2\} = \{0, \frac{1}{2}, 1\}$.*

$$\begin{aligned}
\tau_1^{-1}(x) &= \frac{1}{2} - \frac{1}{2}\sqrt{1-x}, \ \tau_2^{-1}(x) = \frac{1}{2} + \frac{1}{2}\sqrt{1-x}, \\
|\tau'(\tau_1^{-1}(x))| &= |\tau'(\tau_2^{-1}(x))| = 4\sqrt{1-x}.
\end{aligned}$$

Let $f \in L^1(I)$, then by (2.2.6)

$$\begin{aligned}
(P_\tau(f))(x) &= \sum_{i=1}^{2} \frac{f(\tau_i^{-1}(x))}{\tau'(\tau_i^{-1}(x))} \chi_{\tau(x_{i-1}, x_i)}(x) \\
&= \frac{f(\frac{1}{2} - \frac{1}{2}\sqrt{1-x})}{4\sqrt{1-x}} + \frac{f(\frac{1}{2} + \frac{1}{2}\sqrt{1-x})}{4\sqrt{1-x}} \\
&= \frac{1}{4\sqrt{1-x}} \left(f(\frac{1}{2} - \frac{1}{2}\sqrt{1-x}) + f(\frac{1}{2} + \frac{1}{2}\sqrt{1-x}) \right).
\end{aligned}$$

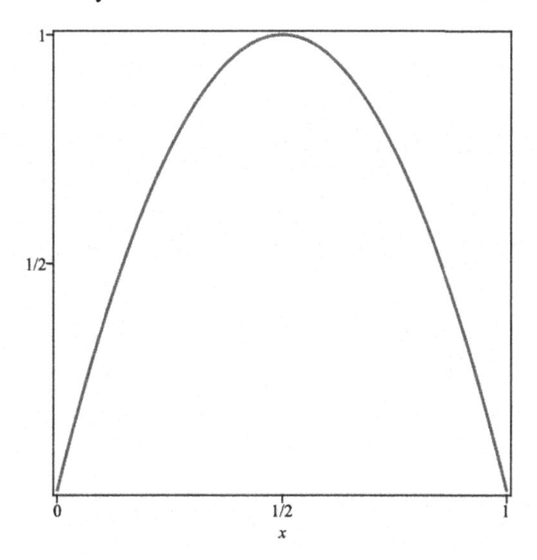

Figure 2.5 *The logistic map* $\tau = 4x(1-x)$.

Definition 2.7 *Let* $\mathcal{P} = \{I_1, I_2, \ldots, I_n\}, I_i = (x_{i-1}, x_i), i = 1, 2, \ldots, n$ *be a partition of* I, $\tau : I \to I$ *and* $\tau_i = \tau_{|_{I_i}}$. *For each* $i = 1, 2, \ldots, n$ *if* τ_i *is a homeomorphism from* I_i *to a connected union of intervals of* $\mathcal{P}$ *then* τ *is called a Markov transformation. For each* $i = 1, 2, \ldots, n$ *if* τ_i *is linear then* τ *is called a piecewise linear Markov transformation.*

Example 2.6 $\tau : [0,1] \to [0,1]$ *defined by*

$$\tau(x) = \begin{cases} \frac{1}{2} + x, & 0 \le x \le \frac{1}{2}, \\ 2 - 2x, & \frac{1}{2} \le x \le 1, \end{cases}$$

is a piecewise Markov transformation on the partition $\mathcal{P} = \{0, \frac{1}{2}, \frac{3}{4}, 1\}$.

The class of piecewise linear Markov transformations is a simple class of piecewise monotonic transformations and the matrix representation of the corresponding Frobenius–Perron operator can be calculated easily. In fact, it is a matrix which follows from the following theorem [2]:

Theorem 2.8 *(Theorem 9.2.1 in [2]) Let* $\tau : (I, \mathcal{B}, \lambda) \to (I, \mathcal{B}, \lambda)$ *be a piecewise linear Markov transformation with respect to the partition* $\{I_1, I_2, \ldots, I_n\} = \{x_0, x_1, \ldots, x_n\}$. *Then there exists a* $n \times n$ *matrix* M_τ *such that* $P_\tau f = f M_\tau^T$ *for every piecewise constant* $f = (f_1, f_2, \ldots, f_n)$. *The matrix* $M_\tau = (m_{ij})$ *is defined by*

$$m_{ij} = \frac{\lambda(I_i \cap \tau^{-1}(I_j))}{\lambda(I_i)}$$

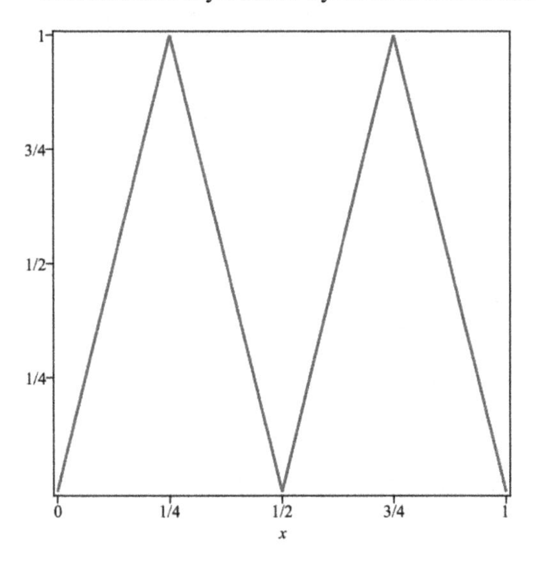

Figure 2.6 *The map τ^* which is the second iteration of the tent map in Figure 2.4.*

Example 2.7 *Let $\tau : [0,1] \to [0,1]$ be the tent map (see Figure 2.4)*

$$\tau(x) = \begin{cases} 2x, & 0 \le x \le \frac{1}{2}, \\ 2-2x, & \frac{1}{2} \le x \le 1, \end{cases}$$

and $\tau^* : [0,1] \to [0,1]$ (see Figure 2.6) is given by $\tau^*(x) = \tau^2(x)$. It can be easily checked that τ^* is a piecewise linear Markov on the partition $\{0, \frac{1}{4}, \frac{1}{2}, \frac{3}{4}, 1\}$. By the Theorem 2.8, the matrix representation of P_{τ^*} is M_τ^* where

$$M_{\tau^*} = \begin{bmatrix} \frac{1}{4} & \frac{1}{4} & \frac{1}{4} & \frac{1}{4} \\ \frac{1}{4} & \frac{1}{4} & \frac{1}{4} & \frac{1}{4} \\ \frac{1}{4} & \frac{1}{4} & \frac{1}{4} & \frac{1}{4} \\ \frac{1}{4} & \frac{1}{4} & \frac{1}{4} & \frac{1}{4} \end{bmatrix}.$$

Theorem 2.9 *[2] Let $\tau : I, \mathcal{B}, \lambda) \to (I, \mathcal{B}, \lambda)$ be a non-singular transformation. Then P_τ has a fixed point $f^* \in L^1, f^* \ge 0$ if and only if the measure $\mu = f^* \cdot \lambda$ defined by $\mu(A) = \int_A f^* d\lambda$ is $\tau-$invariant, that is, if and only if $\mu(\tau^{-1}(A) = \mu(A)$ for all measurable set A.*

Proof *Assume $\mu(\tau^{-1}(A)) = \mu(A)$ for any measurable set A. Then*

$$\int_{\tau^{-1}(A)} f^* d\lambda = \int_A f^* d\lambda$$

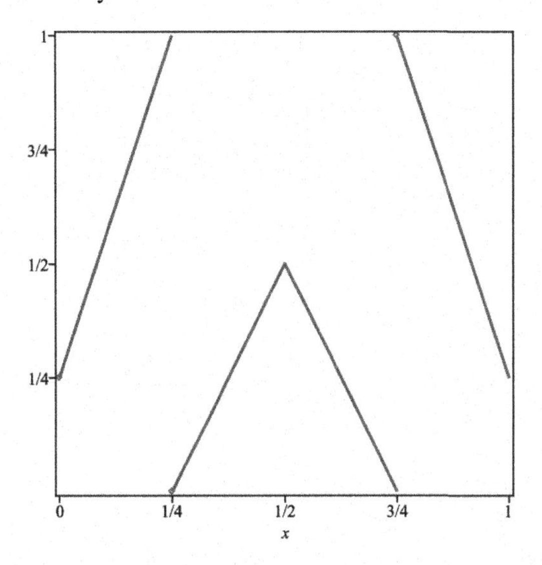

Figure 2.7 *The map τ in Example 2.8.*

and therefore

$$\int_A P_\tau f^* d\lambda = \int_A f^* d\lambda.$$

Since $A \in \mathcal{B}$ is arbitrary, $P_\tau f^* = f^*$ a.e.
Conversely, assume $P_\tau f^* = f^*$ a.e. Then

$$\int_A P_\tau f^* d\lambda = \int_A f^* d\lambda = \mu(A).$$

By definition,

$$\mu(A) = \int_A P_\tau f^* d\lambda = \int_{\tau^{-1}(A)} f^* d\lambda = \mu(\tau^{-1}(A)).$$

Example 2.8 *Let $\tau : [0,1] \to [0,1]$ be a piecewise linear Markov transformation on the partition $\{0, \frac{1}{4}, \frac{1}{2}, \frac{3}{4}, 1\}$ defined by*

$$\tau(x) = \begin{cases} 3x + \frac{1}{4}, & 0 \le x < \frac{1}{4}, \\ 2(x - \frac{1}{4}), & \frac{1}{4} \le x < \frac{1}{2}, \\ 2 - 2(x + \frac{1}{4}), & \frac{1}{2} \le x < \frac{3}{4}, \\ -3x + \frac{13}{4}, & \frac{3}{4} \le x \le 1. \end{cases}$$

It can be easily checked that τ is piecewise linear Markov on the partition

$\{0, \frac{1}{4}, \frac{1}{2}, \frac{3}{4}, 1\}$. *The matrix representation of* P_τ *is* M_τ *where*

$$M_\tau = \begin{bmatrix} 0 & \frac{1}{3} & \frac{1}{3} & \frac{1}{3} \\ \frac{1}{2} & \frac{1}{2} & 0 & 0 \\ \frac{1}{2} & \frac{1}{2} & 0 & 0 \\ 0 & \frac{1}{3} & \frac{1}{3} & \frac{1}{3} \end{bmatrix}$$

respectively. Let $f = [x_1, x_2, x_3, x_4]$, *where* $x_i = f|I_i, I_i = [\frac{i-1}{4}, \frac{i}{4}], i = 1,2,3,4$. *The normalized density of the map* τ *is the left eigenvector of* M_τ *with eigenvalue* 1. *The solution of* $fM = f$ *with* $x_1 + x_2 + x_3 + x_4 = 4$ *is*

$$f = \left[\frac{8}{7}, \frac{12}{7}, \frac{4}{7}, \frac{4}{7}\right].$$

Theorem 2.9 relates two problems: (i) existence of an invariant measure of a dynamical system τ and (ii) the fixed point problem of Frobenius–Perron operator P_τ. The measure μ in Theorem 2.9 is absolutely continuous with respect to λ. The Frobenius-Perron operator is one of the main tools for proving the existence of absolutely continuous invariant measures (acim). One of the advantages of studying P_τ is that P_τ is a contraction linear operator and we can apply the powerful tools of functional analysis. Lasota and Yorke [10] proved the following important result for the existence of an acim for a single transformation using bounded variation methods:

Theorem 2.10 *[2, 10] Let* $\tau : [0,1] \to [0,1]$ *be a piecewise* C^2 *transformation such that* $\inf|\tau'| > 1$. *Then for any* $f \in L^1[0,1]$ *the sequence* $\frac{1}{n}\sum_{k=1}^{n} \mathcal{P}_\tau^k f$ *is convergent in norm to* $f^* \in L^1[0,1]$. *The limit function has the following properties:*
(i) $f \geqq 0 \Rightarrow f^* \geqq 0$.
(ii) $\int_0^1 f^* d\lambda = \int_0^1 f d\lambda$.
(iii) $\mathcal{P}_\tau f^* = f^*$ *and consequently* $d\mu^* = f^* d\lambda$ *is invariant under* τ.
(iv) $f^* \in \mathrm{BV}[0,1]$. *Moreover there exists c independent to the choice of initial f such that*
$\bigvee_{[0,1]} f^* \leqq c \parallel f \parallel_1$.

Example 2.9 *The maps in Example 2.7 and Example 2.8 satisfy conditions of Theorem 2.10 and these maps have absolutely continuous invariant measures. In fact, in Example 2.8 we have shown that* $f = \left[\frac{8}{7}, \frac{12}{7}, \frac{4}{7}, \frac{4}{7}\right]$ *is a normalized invariant density of* τ *in Figure 2.7.*

2.3 Stochastic perturbations of deterministic dynamical systems

The Frobenius-Perron operator P_τ of a deterministic dynamical system τ on I defined in the previous section (see (2.2.6) and (2.2.7)) is an infinite dimensional linear operator from the space $L^1(I)$ into $L^1(I)$ and it is difficult to find a solution of the equation $P_\tau f = f$ except in some simple cases, for example, Markov cases. Approximation of invariant measures was suggested by Ulam [12]. In 1976, T–Y. Li [11] first proved convergence of Ulam's approximation. Since then, Ulam's method has been generalized in a number of directions [5, 6]. In this section, we first present stochastic perturbations of deterministic dynamical systems and then we present numerical techniques for approximations of invariant measures using Fourier approximations. We closely follow [8] and [7].

2.3.1 *Stochastic perturbations of deterministic systems and invariant measures*

Physical systems are usually subjected to small perturbations from external noise or round-off errors. There are well-known results [2,9] that study the stability of absolutely continuous invariant measures for measurable transformations. Consider the stochastically perturbed dynamical system $x \mapsto \tau(x) + \xi$ where ξ is an additive noise which is applied once per each iteration. Let $\mathcal{P}(x,y)$ be the transition density of a transition from point x to y induced by noise ξ. In [1] E. Bollt et al. proposed a numerical method based on basis Markov partitions to approximate density functions of stochastically perturbed dynamical system $x \mapsto \tau(x) + \xi$. In this section we consider Fourier approximation of ξ and obtain a finite approximation of the Frobenius–Perron operator associated with the perturbed system. We present a convergence analysis of our method.

Let $L^1 = L^1(I, \mathcal{B}, \lambda)$ and $\tau : [0,1] \to [0,1]$ be a piecewise monotonic mapping (see [2]) on a partition $\mathcal{P} = \{0 = b_0, b_1, \ldots, b_q = 1\}$ and $P_\tau : L^1 \to L^1$ be the Frobenius-Perron operator of τ defined in (2.2.4). For piecewise monotonic transformation τ the Frobenius-Perron operator P_τ has the following representation (see (2.2.6) and (2.2.7))

$$P_\tau f(x) = \sum_{z \in \{\tau^{-1}(x)\}} \frac{f(z)}{|\tau'(z)|}. \tag{2.3.1}$$

Let $\bigvee(\cdot)$ be the standard one dimensional variation of a function and $BV(I)$ be the space of functions of bounded variations on I equipped with the norm $\|\cdot\|_{BV} = \bigvee(\cdot) + \|\cdot\|_{L^1}$.

We consider Lasota-Yorke (see [10]) maps $\tau : [0,1] \to [0,1]$ such that $|\tau'| > 2$ and for every non-negative density function $f \in BV([0,1])$ there exist constants $\beta > 0$ and $0 < \alpha < 1$ such that

$$\bigvee P_\tau f \leq \alpha \bigvee f + \beta \, \| f \|_{L^1}. \tag{2.3.2}$$

It was proved in [10] that Lasota-Yorke map τ satisfying (2.3.2) has an invariant density $\hat{f}$ of bounded variation and thus, an absolutely continuous invariant measure $\hat{\mu} = \hat{f} \cdot \lambda$.

For small $r > 0$, let $w : \mathbb{R} \to \mathbb{R}^+$ be a bounded function satisfying the following conditions:

1. $w(t) = 0$ for $|t| > r$,
2. $w(-t) = w(t)$,
3. $\int_{-r}^{r} w(t) d\lambda(t) = 1$.

It is easy to see that w becomes Dirac's delta function as $r \to 0$. Let $q(x,y)$ be a kernel defined by

$$q(x,y) = \begin{cases} w(y-x) & , x \in [r, 1-r) \\ w(y-x) + w(\bar{y}-x) & , x \in I - [r, 1-r] \end{cases} , \tag{2.3.3}$$

where $\bar{y} = -y$ for $y \in [0,r)$ and $\bar{y} = 1 + (1-y)$ for $y \in (1-r, 1]$. The Markov process with transition density $p(x,\cdot) = q(\tau(x),\cdot)$ is called a stochastic perturbation of the map τ.

Let $Q : L^1 \to L^1$ be the operator induced by the kernel $q(x,y)$ defined by

$$(Qf)(y) = \int_0^1 q(x,y) f(x) d\lambda(x). \tag{2.3.4}$$

It is proved by Góra in [7] that for any positive $f \in L^1$

$$\bigvee(Qf) \leq 2 \bigvee f. \tag{2.3.5}$$

Treating $[0,1]$ as a circle and defining $q(x,y) = w(y-x) \pmod 1$, we show that the factor of 2 on the right-hand side of the above inequality does not occur.

Lemma 2.11 *For any $f \in L^1$ we have*

$$(Qf)(y) = (f * w)(y), y \in I,$$

where $g * h$ is the convolution of g and h defined by

$$g * h(x) = \int g(y) h(x-y) dy = \int g(x-y) h(y) dy .$$

Proof $(Qf)(y) = \int q(x,y) f(x) d\lambda(x) = \int w(x-y) f(x) d\lambda(x) = (f * w)(y)$.

Lemma 2.12 *For any positive $f \in L^1$ we have*

$$\bigvee(Qf) \leq \bigvee(f).$$

Proof *For a fixed integer $q \geq 1$ and a partition $0 = t_0 < t_1 < \ldots < t_q = 1$, we have*

$$\begin{aligned}\sum_{i=1}^{q}|(Qf)(t_i)-(Qf)(t_{i-1})| &= \sum_{i=1}^{q}|(f*w)(t_i)-(f*w)(t_{i-1})| \\ &= \sum_{i=1}^{q}|(w*f)(t_i)-(w*f)(t_{i-1})| \\ &= \sum_{i=1}^{q}|\int w(t)f(t_i-t)dt - \int w(t)f(t_{i-1}-t)dt| \\ &\leq \int\left(\sum_{i=1}^{q}|f(t_i-t)-f(t_{i-1}-t)|\right)w(t)dt \leq \int \bigvee(f)w(t)dt = \bigvee(f)\,.\end{aligned}$$

The time evolution under the densities of the stochastic perturbation $p(x,\cdot) = q(\tau(x),\cdot)$ of τ is given by

$$\begin{aligned}(P_{\text{pert}}f)(y) &= \int_I p(x,y)f(x)d\lambda(x) = \int_I q(\tau(x),y)f(x)d\lambda(x) \\ &= \int_I (P_\tau f)(x)q(x,y)d\lambda(x) = ((Q\circ P_\tau)f)(y)\,.\end{aligned}$$

Thus,

$$P_{\text{pert}} = Q \circ P_\tau. \tag{2.3.6}$$

and

$$\bigvee P_{\text{pert}}f = \bigvee Q\circ P_\tau f \leq \bigvee P_\tau f \leq \alpha \bigvee f + \beta \parallel f \parallel_{L^1}\,. \tag{2.3.7}$$

Lemma 2.13 *There is an $f^* \in L^1(0,1)$ of bounded variation such that $P_{\text{pert}}f^* = f^*$.*

Proof *From inequality (2.3.7), $\{\bigvee P^n_{\text{pert}}f\}_{n\geq 1}$ is uniformly bounded in BV. By Helly's Theorem (see Theorem 2.3.9 in [2]), $\{P^n_{\text{pert}}f\}$ is relatively compact, which implies by the Kakutani-Yoshida Theorem (see [2]), that*

$$\lim_{n\to\infty}\frac{1}{n}\sum_{i=0}^{n}P^i_{\text{pert}}f = f^*\,.$$

for some $f^* \in L^1(0,1)$. It is easy to see that f^* is a fixed point of P_{pert} and that it is of bounded variation.

Theorem 2.14 *Let $\tau : [0,1] \to [0,1]$ be a Lasota–Yorke (see [10]) map such that $|\tau'| > 2$ and for every non-negative density function $f \in L^1([0,1])$ there exist constants $\beta > 0$ and $0 < \alpha < 1$ such that $\bigvee_0^1 P_\tau f \leq \alpha \bigvee f + \beta \parallel f \parallel_{L^1}$. If the above kernel $q(x,y)$ satisfies (2.3.3), then the stochastic perturbation $p(x,.) = q(\tau(x),.)$ of the map τ has an invariant density f^*.*

Proof *The proof follows from Lemma 2.11, Lemma 2.12, and Lemma 2.13.*

In the following section we consider a family $q^N(\cdot,\cdot), N \geq 1$ of doubly stochastic kernels and corresponding stochastic perturbations $p^N(x,\cdot) = q^N(\tau(x),\cdot), N \geq 1$ of Lasota-Yorke map $\tau : [0,1] \to [0,1]$. They will be constructed in such a way that the corresponding operator P_{pert} is finite dimensional. We will prove the existence of invariant probability measures μ_N of the stochastic perturbations $p^N(x,\cdot) = q^N(\tau(x),\cdot)$ of the map τ. Our main objective is to show that the limit points (limit measures) μ of the set $\{\mu_N : N \geq 1\}$ are of the form $\mu = \hat{f} \cdot \lambda$, where $\hat{f}$ is the invariant density of τ.

2.3.2 *A family of stochastic perturbations and invariant measures*

Now, we define a family of probability densities $\bar{q}^N(x,y), N = 1,2,\ldots$ as follows: let $\{g_N\}_{N\geq 1}$ be a sequence of C^2 non-negative functions with support in $[-1/2,1/2]$ such that g_N is symmetric with respect to y axis, $g_N(-1/2) = g_N(1/2)$ for all $N \geq 1$ and which converges to Dirac's delta function as $N \to \infty$. Each g_N, which can be also seen as a 1−periodic on the whole real line, can be approximated by its partial Fourier sum arbitrary close in the supremum norm. Let

$$h_N(\xi) = c_S + a_{0,N} + 2\sum_{s=1}^{S} \left(a_{s,N}\cos(2s\pi\xi) + b_{s,N}\sin(2s\pi\xi)\right),$$

where S can be chosen independently of N, be an approximation obtained from Fourier approximation by shifting it up by a small constant c_S to ensure $h_N \geq 0$ on $[-1/2,1/2]$. We have $c_S \to 0$ as $S \to \infty$. We can also make h_N converge to Dirac's delta δ_0 as $N \to \infty$. Let $L = \int_{-1/2}^{1/2} h_N(t)dt$. Define a family of functions w^N:

$$w^N(t) = \frac{1}{L}h_N(t), \quad N = 1,2,3,\ldots, \tag{2.3.8}$$

Now we define a family of probability densities $q^N(x,y), N = 1,2,\ldots$ as follows:

$$q^N(x,y) = w^N(x-y), \quad N = 1,2,3,\ldots. \tag{2.3.9}$$

Thus,

$$\begin{aligned}
q^N(x,y) &= w^N(x-y)\\
&= \frac{1}{L}\left[c_S + a_{0,N} + 2\sum_{s=1}^{S}\left(a_{s,N}\cos(2s\pi(x-y)) + b_{s,N}\sin(2s\pi(x-y))\right)\right]\\
&= \frac{1}{L}\Bigg[c_S + a_{0,N} + 2\sum_{s=1}^{S}\big(a_{s,N}(\cos(2s\pi x)\cos(2s\pi y) + \sin(2s\pi x)\sin(2s\pi y))\\
&\qquad + b_{s,N}(\sin(2s\pi x)\cos(2s\pi y) - \cos(2s\pi x)\sin(2s\pi y))\big)\Bigg].
\end{aligned} \tag{2.3.10}$$

The family of transition densities $p^N(x,\cdot) = q^N(\tau(x),\cdot)$ induces a family of stochastic perturbation of the map τ. For $N = 1,2,\dots$ let $Q_N : L^1 \to L^1$ be the operator induced by the kernel $q^N(x,y)$ defined by

$$(Q_N f)(y) = \int_0^1 q^N(x,y) f(x) d\lambda(x) . \tag{2.3.11}$$

The time evolution of the densities of the stochastic perturbation $p^N(x,\cdot) = q^N(\tau(x),\cdot)$ of τ is given by

$$\begin{aligned}(P_N f)(y) &= \int_I p^N(x,y) f(x) d\lambda(x) = \int_I q^N(\tau(x),y) f(x) d\lambda(x) \\ &= \int_I (P_\tau f)(x) q^N(x,y) d\lambda(x) = ((Q_N \circ P_\tau) f)(y) .\end{aligned}$$

Thus,

$$P_N = Q_N \circ P_\tau . \tag{2.3.12}$$

From Section 2.3.1 we have

$$\bigvee_0^1 P_N f = \bigvee_0^1 Q_N \circ P_\tau f \le \bigvee_0^1 P_\tau f \le \alpha \bigvee f + \beta \parallel f \parallel_{L^1} . \tag{2.3.13}$$

Thus, by Theorem 2.14, for each $N \ge 1$, the operator P_N has a fixed point f_N^*.

2.3.3 *Matrix representation of P_N*

Let us define:

$$\begin{aligned}
u_0(x) &= 1; \\
u_{4s+1}(x) &= \cos(2(s+1)\pi x), \quad s = 0,1,2,,\dots S-1; \\
u_{4s+2}(x) &= \sin(2(s+1)\pi x), \quad s = 0,1,2,\dots S-1; \\
u_{4s+3}(x) &= \sin(2(s+1)\pi x), \quad s = 0,1,2,\dots S-1; \\
u_{4s+4}(x) &= \cos(2(s+1)\pi x), \quad s = 0,1,2,\dots S-1; \\
v_0(x) &= 1; \\
v_{4s+1}(x) &= \cos(2(s+1)\pi x), \quad s = 0,1,2,,\dots S-1; \\
v_{4s+2}(x) &= \sin(2(s+1)\pi x), \quad s = 0,1,2,\dots S-1; \\
v_{4s+3}(x) &= \cos(2(s+1)\pi x), \quad s = 0,1,2,\dots S-1; \\
v_{4s+4}(x) &= \sin(2(s+1)\pi x), \quad s = 0,1,2,\dots S-1;
\end{aligned} \tag{2.3.14}$$

Let $K = 4\ S$ and let the matrix $A = (A_{mn})_{0 \le m,n \le K}$ be the diagonal matrix with the diagonal

$$\frac{1}{L}(c_S + a_{0,N}, 2a_{1,N}, 2a_{1,N}, 2b_{1,N}, -2b_{1,N}, 2a_{2,N}, 2a_{2,N}, 2b_{2,N}, -2b_{2,N}, \\ \dots, 2a_{S,N}, 2a_{S,N}, 2b_{S,N}, -2b_{S,N}).$$

Thus, $q^N(x,y) = \sum_{m,n=0}^{K} A_{mn} u_n(x) v_m(y)$,

The kernel $q^N(\cdot,\cdot)$ defined above satisfies the following properties:

1. $q^N(x,y) \geq 0$.
2. $q^N(\cdot,\cdot)$ is measurable as functions of two variables,
3. For every $x \in I$ we have $\int_I q^N(x,y)dy = 1$,
4. For every $y \in I$ we have $\int_I q^N(x,y)dx = 1$,
5. $q^N(x,y) \equiv q^N(x \bmod 1, y \bmod 1)$,
6. $q^N(x,y) = \sum_{m,n=0}^{K} A_{mn} u_n(x) v_m(y)$,
7. Let $B(x,r) = \{y : |x-y| < r\}$ and $c_N(x,r) = \int_{I \setminus B(x,r)} q_n(x,y)dy$. Then for any $r > 0$,

$$c_N(r) = \sup_{x \in I} c_N(x,r) \to 0$$

as $N \to +\infty$.

We have

$$\begin{aligned}
[P_N f](y) &= \int_0^1 \sum_{m,n=0}^{K} A_{mn} u_n(\tau(x)) v_m(y) f(x) dx \\
&= \sum_{m,n=0}^{K} A_{mn} \left[\int_0^1 u_n(\tau(x)) f(x) dx \right] v_m(y) \\
&= \sum_{m,n=0}^{K} \left[\int_0^1 u_n(\tau(x)) f(x) \right] \bar{v}_m(y)
\end{aligned}$$

for $y \in I$, where,

$$\bar{v}_n(y) = \sum_{m=0}^{K} A_{mn} v_m(y), n = 0,1,2,\ldots K. \tag{2.3.15}$$

Thus, any initial density f is projected by the operator P_N into the vector space Δ_N spanned by the functions $\bar{v_n}, n = 0,\ldots,K$, that is,

$$(P_N f)(y) = \sum_{n=0}^{K} q_n' \bar{v}_n(x),$$

where

$$q_n' = \int_0^1 u_n(\tau(x)) f(x) dx.$$

We are interested in finding the matrix representation of the operator P_N.

Assuming that a given density $f(x)$ belongs to the space Δ_N , we can expand it in the basis,

$$f(x) = \sum_{m=0}^{K} q_m \bar{v}_m(x). \tag{2.3.16}$$

Let B denote a matrix of integrals,

$$B_{nm} = \int_0^1 u_n(\tau(x))v_m(x)dx, \tag{2.3.17}$$

where $n,m = 0,\ldots,K$. Observe that B depends directly on the system τ and on the noise via the basis functions u and v. Let us define

$$D = BA. \tag{2.3.18}$$

Lemma 2.15 *The matrix D in (2.3.18) is the representation of the operator P_N with respect to the basis $\{\bar{v}_l\}_{l=0}^K$.*

Proof *All we need to show is the following: $q'_n = \sum_{m=0}^K D_{nm}q_m, \quad n = 0,1,2,\ldots K$. Now,*

$$\begin{aligned}
\sum_{m=0}^K D_{nm}q_m &= D_{n0}q_0 + D_{n1}q_1 + \ldots D_{nK}q_K \\
&= \left(\sum_{l=0}^K B_{nl}A_{l0}\right)q_0 + \left(\sum_{l=0}^K B_{nl}A_{l1}\right)q_1 + \ldots + \left(\sum_{l=0}^K B_{nl}A_{lK}\right)q_K \\
&= \left(\sum_{l=0}^K \{\int_0^1 u_n(\tau(x))v_l(x)dx\}A_{l0}\right)q_0 \\
&\quad + \left(\sum_{l=0}^K \{\int_0^1 u_n(\tau(x))v_l(x)dx\}A_{l1}\right)q_1 \\
&\quad + \ldots + \left(\sum_{l=0}^K \{\int_0^1 u_n(\tau(x))v_l(x)dx\}A_{lK}\right)q_K.
\end{aligned}$$

On the other hand

$$\begin{aligned}
q'_n &= \int_0^1 u_n(\tau(x))f(x)dx = \int_0^1 u_n(\tau(x))\left(\sum_{m=0}^K q_m\bar{v}_m(x)\right)dx \\
&= \int_0^1 u_n(\tau(x))\left(\sum_{m=0}^K q_m\left(\sum_{l=0}^K A_{lm}v_l(x)\right)\right)dx \\
&= \int_0^1 u_n(\tau(x))[\left(\sum_{l=0}^K A_{l0}v_l(x)\right)q_0 + \left(\sum_{l=0}^K A_{l1}v_l(x)\right)q_1 \\
&\quad + \ldots + \left(\sum_{l=0}^K A_{lK}v_l(x)\right)q_K]dx
\end{aligned}$$

$$
\begin{aligned}
&= \left(\sum_{l=0}^{K}\{\int_0^1 u_n(\tau(x))v_l(x)dx\}A_{l0}\right)q_0 + \left(\sum_{l=0}^{K}\{\int_0^1 u_n(\tau(x))v_l(x)dx\}A_{l1}\right)q_1 + \dots \\
&\qquad \dots + \left(\sum_{l=0}^{K}\{\int_0^1 u_n(\tau(x))v_l(x)dx\}A_{lK}\right)q_K \\
&= \sum_{m=0}^{K} D_{nm}q_m \,.
\end{aligned}
$$

In this way we have arrived at a representation of the operator $P_N f$ by a matrix D of size $(K+1)\times(K+1)$ with respect to the basis $\{\bar{v_k}\}_{k=0}^{K}$, the elements of which read,

$$
D_{nm} = \int_0^1 u_n(\tau(x))\bar{v_m}(x)dx, \qquad n,m = 0,\dots,K. \tag{2.3.19}
$$

2.3.4 *Stability and convergence*

Recall from Section 2.3.2

$$
(Q_N f)(y) = \int_0^1 q^N(x,y)f(x)d\lambda(x).
$$

Lemma 2.16 *For any $f \in L^1$ we have $Q_N f \to f$ as $N \to \infty$ in the L^1 norm. The convergence is uniform on relatively compact subsets of L^1.*

Proof *It can be shown that for each $N \geq 1, \| Q_N \|_1 = 1$. Let $f \in L^1$ and $\varepsilon > 0$. Since continuous functions are dense in L^1, there exists a continuous function g in I such that $\| g - f \|_1 < \frac{\varepsilon}{3}$. Since g is continuous it is uniformly continuous in $[0,1]$. Thus,*

$$
\| Q_N f - f \|_1 \leq \| Q_N f - Q_N g \|_1 + \| Q_N g - g \|_1 + \| g - f \|_1 \,.
$$

Now,

$$
\begin{aligned}
\| Q_N g - g \|_1 &= \int |g(y) - (Q_N g)(y)|dy = \int |g(y) - \int g(x)q^N(x,y)dx|dy \\
&\leq \int \int |g(y) - g(x)|q^N(x,y)dxdy \leq \frac{\varepsilon}{3}\int \int q^N(x,y)dxdy = \frac{\varepsilon}{3} \,.
\end{aligned}
$$

This proves

$$
\| Q_N f - f \|_1 \leq \varepsilon \,.
$$

Lemma 2.17 *Let $f_N \in \Delta_N$ and $f_N = \sum_{j=0}^{N} c_j \bar{v}_j(x)$. Then $P_N f_N = f_N$ if and only if $Dc = c$ where c is the transpose of $(c_0, c_1, \dots, c_N)$.*

Proof *Let* $f_N = \sum_{j=0}^{N} c_j \bar{v}_j(x)$. *Then*

$$\begin{aligned}
P_N f_N &= P_N\left(\sum_{j=0}^{N} c_j \bar{v}_j(x)\right) = \sum_{j=0}^{N} c_j P_N \bar{v}_j(x) \\
&= c_0\left(\sum_{i=0}^{N} D_{i0}\bar{v}_i(x)\right) + c_1\left(\sum_{i=0}^{N} D_{i1}\bar{v}_i(x)\right) + \ldots + c_N\left(\sum_{i=0}^{N} D_{iN}\bar{v}_i(x)\right) \\
&= \left(\sum_{l=0}^{N} D_{0l}c_l\right)\bar{v}_0(x) + \left(\sum_{l=0}^{N} D_{1l}c_l\right)\bar{v}_1(x) + \ldots + \left(\sum_{l=0}^{N} D_{Nl}c_l\right)\bar{v}_N(x)\,.
\end{aligned}$$

Thus, $P_N f_N = f_N$ *if and only if*

$$\begin{aligned}
\sum_{l=0}^{N} D_{0l}c_l &= c_0 \\
\sum_{l=0}^{N} D_{1l}c_l &= c_1 \\
&\vdots \\
\sum_{l=0}^{N} D_{Kl}c_l &= c_N
\end{aligned}$$

That is,

$$Dc = c.$$

Now we prove the main theorem of this section.

Theorem 2.18 *Let* $\tau : [0,1] \to [0,1]$ *be a Lasota–Yorke (see [10]) map such that* $|\tau'| > 2$ *and for every non-negative density function* $f \in L^1([0,1])$ *there exist constants* $\beta > 0$ *and* $0 < \alpha < 1$ *such that* $\bigvee_0^1 P_\tau f \le \alpha \bigvee f + \beta \parallel f \parallel_{L^1}$. *Let* $f_N^* \in \Delta_N$ *be an invariant density of stochastic perturbation* $q^N(\tau(x),\cdot)$ *of* τ *such that* $P_N f_N^* = f_N^*$. *Then the set* $\{f_N^*\}_{N\ge1}$ *is relatively compact in* L^1 *and any limit point of* $\{f_N^*\}_{N\ge1}$ *is a* τ *invariant density* $\hat{f}$.

Proof *By inequality (2.3.7),*

$$\bigvee f_N^* = \bigvee(P_N f_N^*) = \bigvee((Q_N \circ P_\tau) f_N^* \le \bigvee(P_\tau f_N^*) f_N^*) \le \alpha \bigvee f + \beta \parallel f \parallel_{L^1}$$

Thus, the set $\{f_N^*\}_{N\ge1}$ *is uniformly bounded in variation. By Helly's Theorem,* $\{f_N^*\}_{N\ge1}$ *is relatively compact in* L^1. *Let* $f_{N_i^*}$ *be a subsequence of* f_N *and* $f_{N_i^*} \to f$ *in* L^1. *Then,*

$$\begin{aligned}
\parallel \hat{f} - P_\tau \hat{f} \parallel_1 \le\ & \parallel \hat{f} - f_{N_i}^* \parallel_1 + \parallel f_{N_i}^* - Q_{N_i} P_\tau f_{N_i}^* \parallel_1 \\
& + \parallel Q_{N_i} P_\tau f_{N_i}^* - Q_{N_i} P_\tau \hat{f} \parallel_1 + \parallel Q_{N_i} P_\tau \hat{f} - P_\tau \hat{f} \parallel_1\,.
\end{aligned}$$

Using Lemma 2.16 and the definition of P_N *it is easy to see that* $P_\tau \hat{f} = \hat{f}$.

2.3.5 Examples

Our approximation method uses as "building blocks" trigonometric functions which have the same values at 0 and at 1. Therefore, the method is not best suited to approximate densities which do not have this property. To go around this deficiency we use a symmetrization of the map.

Example 2.10 *For $0<\alpha<1, 0<p\le 1, q>0$, consider the deterministic dynamical system $\tau_1 : [0,1] \to [0,1]$ defined by*

$$\tau_1(x) = \begin{cases} \frac{\alpha x}{\alpha p+(\alpha-p)x} & , x \in [0,\alpha] \\ \frac{q(1-\alpha)(1-x)}{q-q\alpha-\alpha+(1-q+q\alpha)x} & , x \in (\alpha,1] \end{cases}$$

We set

$$\alpha = \frac{1}{2} \ , \ p = \frac{1}{3} \ , \ q = 6 \quad .$$

It can be shown that τ_1-invariant probability density is

$$f_1(x) = \frac{1+\beta}{\beta^2(x+\frac{1}{\beta})^2} \ ,$$

where for our values of constants $\beta = -\frac{1}{2}$. Let $\tau_2 : [0,1] \to [0,1]$ defined by $\tau_2(x) = 1-\tau_1(1-x)$. τ_2 is conjugated to τ_1 by homeomorphism $h(x) = 1-x$. It can be easily proved that τ_2-invariant density is $f_2(x) = f_1(1-x)$, where f_1 is τ_1-invariant. Let $\tau : [0,1] \to [0,1]$ be defined by

$$\tau(x) = \begin{cases} \frac{1}{2}\tau_1(2x) & , x \in [0,\frac{1}{2}] \\ \frac{1}{2}+\frac{1}{2}\tau_2(2(x-\frac{1}{2})) & , x \in (\frac{1}{2},1] \end{cases}$$

τ-invariant density is

$$f_\tau(x) = \begin{cases} f_1(2x) & , \text{ for } 0 \le x \le \frac{1}{2} \ ; \\ f_2(2(x-\frac{1}{2})) & , \text{ for } \frac{1}{2} < x \le 1 \ . \end{cases}$$

which is symmetric with respect $x = 1/2$ so $f_\tau(0) = f_\tau(1)$.

Consider the stochastic perturbation of the above deterministic dynamical systems τ by the noise $g_N(\xi) = Ng(N\xi)$, $g(\xi) = e^{-\xi^2}$ restricted to $[-1/2,1/2]$ and extended periodically to whole real line, $N \ge 1$. In particular, we consider the dynamical systems τ with $\alpha = \frac{1}{2}$, $p = \frac{1}{3}$, $q = 6$ and the noise g_N with $N = 15$. The Fourier approximation of g_{15}, with $S(15) = 10$ is

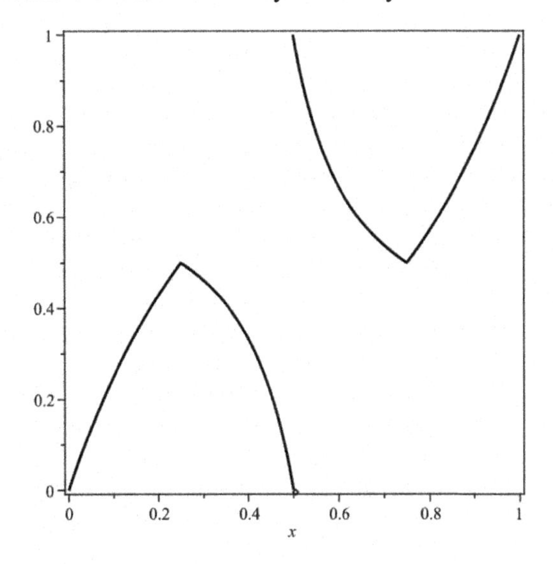

Figure 2.8 *The piecewise smooth map τ.*

$$
\begin{aligned}
&1.77245385090551602732 + 3.39277175796556683 60\cos(2\pi\xi)\\
\end{aligned}
$$

where we have chosen $C_{S(15)} = 0.0320895553170388570$ to ensure that the Fourier approximation is positive on $[-1/2, 1/2]$. After normalization we obtain

$$
\begin{aligned}
\mathcal{P}_{15}(\xi) \quad = \quad & 1.000000000 + 1.8801275415522674707\cos(2\pi\xi)\\
& +1.6483007197945023217\cos(4\pi\xi) + 1.3236861044793344190\cos(6\pi\xi)\\
& +.9737175462808609 6259\cos(8\pi\xi) + .6561156649946654 2801\cos(10\pi\xi)\\
& +.4049740937653220 3282\cos(12\pi\xi) + .2289673207467475 8962\cos(14\pi\xi)\\
& +.1185821136112685 0736\cos(16\pi\xi)\\
& +0.0562554102584195 19092\cos(18\pi\xi)\\
& +0.024446057568923656717\cos(20\pi\xi)
\end{aligned}
$$

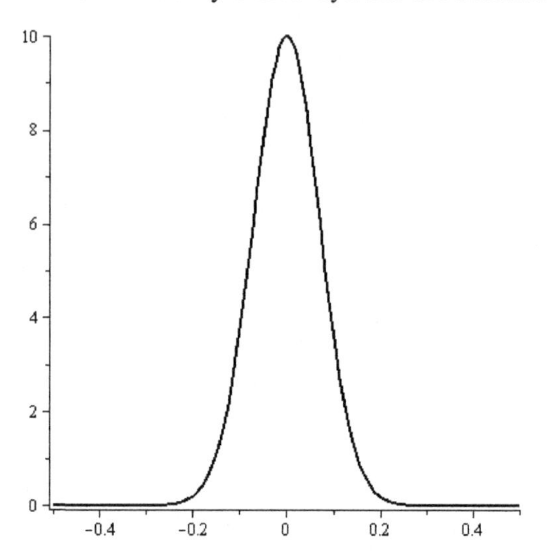

Figure 2.9 *The transition density P_{15}.*

$$
\begin{aligned}
\mathcal{P}_{15}(x-y) \quad = \quad & 1+1.8801275415522674707\cos(2\pi x)\cos(2\pi y)\\
& +1.8801275415522674707\sin(2\pi x)\sin(2\pi y)\\
& +1.6483007197945023217\cos(4\pi x)\cos(4\pi y)\\
& +1.6483007197945023217\sin(4\pi x)\sin(4\pi y)\\
& +1.3236861044793344190\cos(6\pi x)\cos(6\pi y)\\
& +1.3236861044793344190\sin(6\pi x)\sin(6\pi y)\\
& +.97371754628086096259\cos(8\pi x)\cos(8\pi y)\\
& +.97371754628086096259\sin(8\pi x)\sin(8\pi y)\\
& +.65611566499466542801\cos(10\pi x)\cos(10\pi y)\\
& +.65611566499466542801\sin(10\pi x)\sin(10\pi y)\\
& +.40497409376532203282\cos(12\pi x)\cos(12\pi y)\\
& +.40497409376532203282\sin(12\pi x)\sin(12\pi y)\\
& +.22896732074674758962\cos(14\pi x)\cos(14\pi y)\\
& +.22896732074674758962\sin(14\pi x)\sin(14\pi y)\\
& +.11858211361126850736\cos(16\pi x)\cos(16\pi y)\\
& +.11858211361126850736\sin(16\pi x)\sin(16\pi y)\\
& +0.056255410258419519092\cos(18\pi x)\cos(18\pi y)\\
& +0.056255410258419519092\sin(18\pi x)\sin(18\pi y)\\
& +0.024446057568923656717\cos(20\pi x)\cos(20\pi y)\\
& +0.024446057568923656717\sin(20\pi x)\sin(20\pi y)
\end{aligned}
$$

From Equation 2.3.14 we obtain u's and v's for $s = 0, 1, \ldots, 10$. *Then, the matrix* $A = (A_{mn})_{0 \le m,n \le 20}$, *is the diagonal matrix with diagonal*

$$[1, 1.8801275415522674707, 1.8801275415522674707, 1.6483007197945023217, 1.6483007197945023217, 1.3236861044793344190, 1.3236861044793344190, .9737175462808609625 9, .97371754628086096259, .65611566499466542801, .65611566499466542801, .40497409376532203282, -.40497409376532203282, .22896732074674758962, .22896732074674758962, .11858211361126850736, .11858211361126850736, .056255410258419519092, .056255410258419519092, 0.024446057568923656717, 0.024446057568923656717]$$

and we have

$$\bar{v}_m = A_{mm} v_m, \quad m = 0, 1, 2, \ldots 20.$$

For the above perturbed dynamical system we compute the matrix D in (2.3.18). The eigenvector of the matrix D for the eigenvalue 1 *is:*

$$\begin{aligned} w = & [1, -.29110520670549977218, 0.000001844000331524919 6190, \\ & 0.069565405956335630715, -0.0000013137353338792945233, \\ & -0.045604664886956196346, (8.0249929253956130018) \times 10^{-7}, \\ & 0.020642297111933035461, (-6.0489054621082258731) \times 10^{-7}, \\ & -0.017486330019405867864, (4.7949021303234341827) \times 10^{-7}, \\ & 0.0094180878541850472866, (-4.0168209587784764254) \times 10^{-7}, \\ & -4.0168209587784764254, (-4.0168209587784764254) \times 10^{-7}, \\ & -4.0168209587784764254, (-2.9853096972503892825) \times 10^{-7}, \\ & -0.0054895182849284534088, (2.6379713404167750709) \times 10^{-7}, \\ & 2.6379713404167750709, (-2.3751816076527746986) \times 10^{-7}] \end{aligned}$$

and it provides an approximation $f_{15}^* = \sum_0^{20} w_m \bar{v}_m$ *to the* $\tau-$*invariant density (Figure 2.10)* $\hat{f}$. *Much better approximations shown in Figure 2.10 and Figure 2.11 are obtained by taking* $N = 20$ *and* $S = 20$, *which results in matrix D of size* $2S + 1 = 41$.

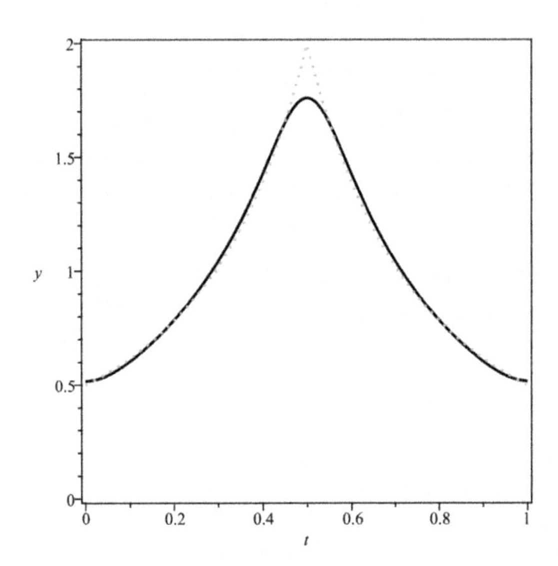

Figure 2.10 *An approximation f_{15}^* to the invariant density $\hat{f}$.*

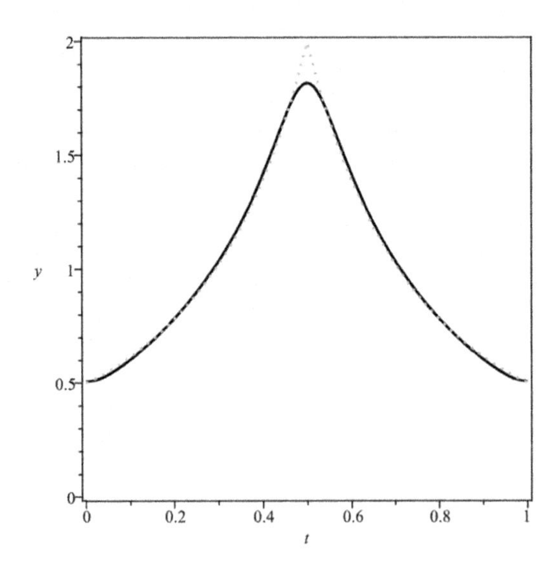

Figure 2.11 *An approximation f_{20}^* to the invariant density $\hat{f}$.*

Errors in L^1- norms are listed in the following table.

N	S	$\parallel f_N^* - \hat{f} \parallel$
15	*10*	*0.025044041879482*
20	*15*	*0.018413171411567*
30	*20*	*0.011280614132958*

References

[1] Bollt, E., Góra, P., Ostruszka, A., Zyczkowski, K., *Basis Markov partitions and transition matrices for stochastic systems, SIAM J. Appl. Dyn. Syst.* 7, no. 2, 2008, 341–360.

[2] Boyarsky, A. and Gora, P., *Laws of chaos*: invariant measures and dynamical systems in one dimension. Birkhauser, 1997.

[3] Ding, J., *The Projection Method for a class of Frobenius-Perron operators, Applied Mathematics and Computations*, 217, p. 3257–3262, 1999.

[4] Ding, J. and Zhou, A., *Statistical properties of deterministic systems*, Tsinghua University Texts. Springer-Verlag, Berlin; Tsinghua University Press, Beijing, 2009.

[5] Ding, J. and Zhou, A., *Finite approximations of Frobenius–Perron operators. A solution of Ulam's conjecture to multi-dimensional transformations*, *Phys. D* 92, 1996, no. 1-2, 61–68.

[6] Froyland, G., *Ulams method for random interval maps*, Nonlinearity, 12, 1029–1052, 1999.

[7] Góra, P., *On small stochastic perturbations of mappings of the unit intervals, Colloquium Mathematicum XLIX*, no. 3, 73–85, 1984.

[8] Islam, M. S. and Góra, P., *Invariant measures of stochastic perturbations of dynamical systems using Fourier approximations, Int. J. Bifur. Chaos*, vol. 21, no. 1, 2011, 113–123.

[9] Lasota, A. and Mackey, M. C., *Chaos, fractals, and noise. Stochastic aspects of dynamics, Applied Mathematical Sciences*, 97. Springer-Verlag, New York, 1994.

[10] Lasota, A. and Yorke, J. A., *On the existence of invariant measures for piecewise monotonic transformations, Trans. Amer. Math. Soc.* 186, 481–488.

[11] Li, T.-Y., *Finite approximation for the Frobenius–Perron operator: a solution to Ulam's conjecture, J. Approx. Theory*. 17, 177–186, 1976.

[12] Ulam, S. M., *A collection of mathematical problems: Interscience Tracts in pure and applied math.*, vol 8, *Interscience*, New York, 1960.

Chapter 3

Random Dynamical Systems and Random Maps

3.1 Chapter overview

In Chapter 2, we presented one dimensional single deterministic dynamical systems. In particular, we have presented invariant measures of piecewise expanding maps on $[0,1]$. In this chapter, we present random dynamical systems with some applications in finance. We present special classes of random dynamical systems which are known as random maps or iterated function systems. We closely follow [1, 8, 10, 19, 20, 25, 26, 32, 34] and the references therein. We focus our special attention on the existence and properties of invariant measures. We also present some applications in finance using random maps with constant probabilities.

3.2 Random dynamical systems

In this section, we present a general definition of a random dynamical system. We closely follow Arnold's definition (see [1]) of random dynamical systems.

Definition 3.1 (**Dynamical system**) *[1]: Let X be a nonempty set and $\mathbb{T}$ be the set of two sided or one sided discrete or continuous times ($\mathbb{T} = \mathbb{R}, \mathbb{R}^+, \mathbb{R}^-, \mathbb{Z}, \mathbb{Z}^+, \mathbb{Z}^-$). A dynamical system on X is a mapping $\phi : \mathbb{T} \times X \to X, (t,x) \mapsto (\phi(t))(x)$ for which the family of mappings $\phi(t) : X \to X, x \mapsto (\phi(t))(x)$ forms a flow, that is, (i) $\phi(0) = \mathrm{id}_X$, (if $0 \in X$) (ii) $\phi(t+s) = \phi(t) \circ \phi(s)$ for all $t,s \in \mathbb{T}$. In other words, a dynamical system is a family of mappings $\{\phi(t) : t \in \mathbb{T}\}$ on X into X satisfying (i) and (ii) above.*

Definition 3.2 (**Measure preserving or metric dynamical systems**) *[1]: In the above definition $\phi : \mathbb{T} \times X \to X, (t,x) \mapsto (\phi(t))(x)$ is called a metric dynamical system if $X = (X, \mathcal{B}, \mu)$ is a probability space and ϕ satisfies (a) $\phi(\cdot,\cdot)$ is $\mathcal{F} \times \mathcal{B}, \mathcal{B}$ measurable where $\mathcal{F}$ is the Borel σ-algebra of $\mathbb{T}$; (b) $\mu(B) = \mu((\phi(t))^{-1}(B)) = \mu(\{x : (\phi(t))(x) \in B\})$ for every $t \in \mathbb{T}$ and $B \in \mathcal{B}$. A measure preserving dynamical system is denoted by $(X, \mathcal{B}, \mu, \{\phi(t), t \in \mathbb{T}\})$. For simplicity, we write $(X, \mathcal{B}, \mu, \phi)$ for $(X, \mathcal{B}, \mu, \{\phi(t), t \in \mathbb{T}\})$.*

Example 3.1 *Let $\mathbb{T}$ be discrete. Then $(X, \mathcal{B}, \mu, \{\phi(n) : n \in \mathbb{T}\}) = (X, \mathbb{B}, \mu, \phi^n =$*

$\phi^{n-1} \circ \phi$), *where* $\phi = \phi(1)$ *is the time one mapping. In particular,* $([0,1],\mathcal{B},\lambda,\tau)$ *is a metric dynamical system where* $\tau : [0,1] \to [0,1]$ *is defined by* $\tau(x) = 2x(\text{mod } 1)$ *and* λ *is the Lebesgue measure on* $[0,1]$.

A random dynamical system is a system which is perturbed by noise [1]. The system is modeled by a difference or differential equation and noise is modeled by a metric or measurable dynamical system.

Definition 3.3 (**Random dynamical systems**) *[1]: Let* $(X,\mathcal{B},\mu,\{\phi(t),t \in \mathbb{T}\})$ *be a metric dynamical system and* $(Y,\mathcal{F})$ *be a measurable space. A random dynamical system on* $(Y,\mathcal{F})$ *over* $(X,\mathcal{B},\mu,\{\phi(t),t \in \mathbb{T}\})$ *is a mapping* $\theta := \mathbb{T} \times X \times Y \to Y, (t,x,y) \mapsto \theta(t,x,y)$ *satisfying the following condition: the mappings* $\theta(t,x) : \theta(t,x,\cdot) : Y \to Y$ *form a cocycle, that is, they satisfy*
(i) $\theta(0,x) = \mathrm{id}_Y$ *for all* $x \in X$ *(if* $0 \in X$*);*
(ii) $\theta(t+s,x) = \theta(t,\phi(s)x \circ \theta(s,x)$ *for all* $s,t \in \mathbb{R}, x \in X$.

Definition 3.4 (**Measurable, continuous, and smooth random dynamical systems**)*[1]*

1. *A random dynamical system on* $(Y,\mathcal{F})$ *over* $(X,\mathcal{B},\mu,\{\phi(t),t \in \mathcal{T}\})$ *is a* **measurable random dynamical system** *if* $\mathcal{B}(\mathbb{T}) \times \mathcal{B} \times \mathcal{F},\mathcal{F}$ *measurable.*

2. *A measurable random dynamical system on* $(Y,\mathcal{F})$ *over* $(X,\mathcal{B},\mu,\{\phi(t),t \in \mathcal{T}\})$ *is* **a continuous or topological random dynamical system** *if* Y *is a topological space and* $\theta(\cdot,x,\cdot) : \mathbb{T} \times Y \to Y$ *is continuous for every* $x \in X$.

3. *A continuous or topological random dynamical system on* $(Y,\mathcal{F})$ *over* $(X,\mathcal{B},\mu,\{\phi(t),t \in \mathcal{T}\})$ *is* **a smooth random dynamical system** *of class* C^k *if* Y *is a manifold and* $\theta(t,x) = \theta(t,x,\cdot) : Y \to Y$ *is* $C^k, 1 \leq k \leq \infty$ *for every* $(t,x) \in \mathbb{T} \times X$.

3.3 Skew products

In this section, we present skew products. Skew products are important tools for the study of random dynamical systems.

Definition 3.5 *Let* θ *be a random dynamical system on a measurable space* $(Y,\mathcal{F})$ *over the metric dynamical systems* $(X,\mathcal{B},\mu,\{\phi(t),t \in \mathbb{T}\})$. *A skew product of* $(X,\mathcal{B},\mu,\{\phi(t),t \in \mathbb{T}\})$ *and* θ *is a transformation* $S : X \times Y \to X \times Y$ *defined by*

$$S(x,y) = (\phi(t)x, \theta(t,x)y),$$

$$x \in X, y \in Y.$$

It can be shown that S is a measurable dynamical system on the product space $(X \times Y, \mathcal{B} \times \mathcal{F})$. Moreover, every measurable skew product S defines a measurable random dynamical system θ.

Example 3.2 *Let* $(\Omega, \mathcal{B}, \sigma, \mu)$ *be a deterministic dynamical system and* $(Y, \mathcal{F}, \tau_\omega, \nu_\omega)_{\omega \in \Omega}$ *be a family of deterministic dynamical systems. Then* $S : \Omega \times Y \to \Omega \times Y$ *defined by*

$$S(\omega, y) = (\sigma(\omega), \tau_\omega(y))), \ \omega \in \Omega, y \in Y$$

is a skew product.

The following lemma is proved in [8].

Lemma 3.6 *If* μ *is* σ *invariant and* ν_ω *is* τ_ω *invariant, then the measure on* $\mathcal{B} \times \mathcal{F}$ *defined by*

$$\nu(A \times B) = \int_A \nu_\omega(B) d\mu(\omega) \tag{3.3.1}$$

is $S-$ *invariant. If* ν *is a* $T-$*invariant measure and* $\mathcal{F}$ *is countably generated, then there is a* $\mu-$ *invariant measure* μ *on* $\mathcal{B}$ *and a family of measures* $\{\nu_\omega\}$ *on* $\mathcal{F}$ *such that* ν_ω *is* $\tau_\omega-$ *invariant and the representation (3.3.1) holds.*

In the following section we consider a special class of random dynamical systems. These random dynamical systems are known as random maps. Random maps can be viewed as a skew product.

3.4 Random maps: Special structures of random dynamical systems

A random dynamical system of special interest is a random map consisting of a number of dynamical systems from a set into itself where the process switches from one dynamical system to another according to fixed probabilities [34] or, more generally, position dependent probabilities [16]. Random maps have application in the study of fractals [7], in modeling interference effects in quantum mechanics [9], in computing metric entropy [38], and in forecasting the financial markets [3]. Invariant measures [4–7, 14, 32, 34, 42] of random dynamical systems and random maps reflect their long-time behavior and play an important role in understanding their chaotic nature. In 1984, Pelikan [34] proved sufficient conditions for existence of acim for random maps with constant probabilities. Morita [32] proved a spectral decomposition theorem. Góra and Boyarsky [16] proved sufficient conditions for the existence of acim for random maps with position dependent probabilities. Bahsoun and Góra [5, 6] proved sufficient average expanding conditions for the existence of acim for position dependent random maps in one and higher dimensions, weakly convex and concave position dependent random maps. Recently, Froyland [14] and Bahsoun, Góra and Boyarsky [3] proved the sufficient condition for the existence of Markov switching random map with constant switching matrix, and position dependent switching matrix, respectively.

3.4.1 Random maps with constant probabilities

Random maps with constant probabilities are an important special case of skew products. Let $(X,\mathcal{B},\lambda)$ be a measure space and $\Omega=\{1,2,3,\ldots,K\}^{\{0,1,2,\ldots\}}=\{\omega=\{\omega_i\}_{i=0}^{\infty}:\omega_i\in\{1,2,3,\ldots,K\}\}$ be the set of set of all one sided infinite sequences. Let $\tau_k:X\to X, k=1,2,\ldots,K$ be non-singular piecewise one-to-one transformations and $p_1,p_2,\ldots,p_K$ be constant probabilities such that $\sum_{i=1}^{K}p_i=1$. The topology on Ω is the product of the discrete topology on $\{1,2,3,\ldots,n\}$ and the Borel probability measure μ_p on Ω is defined as $\mu_p(\{\omega:\omega_0=i_0,\omega_1=i_1,\ldots,\omega_n=i_n\})=p_{i_0}p_{i_1}\cdots p_{i_n}$. Let $\sigma:\Omega\to\Omega$ be the left shift. Now consider the skew product $S:\Omega\times X\to\Omega\times X$ defined by

$$S(\omega,x)=\left(\sigma(\omega),\tau_{\omega_0}(x)\right),\omega\in\Omega,x\in X.$$

Now,

$$S^2(\omega,x)=\left(\sigma^2(\omega),\tau_{\omega_1}\circ\tau_{\omega_0}(x)\right)$$

and for any integer $N\geq 1$,

$$S^N(\omega,x)=\left(\sigma^N(\omega),\tau_{\omega_{N-1}}\circ\tau_{\omega_{N-2}}\circ\ldots\circ\tau_{\omega_1}\circ\tau_{\omega_0}(x)\right)$$

A random map

$$T=\{\tau_1,\tau_2,\ldots,\tau_K;p_1,p_2,\ldots,p_K\},$$

with constant probabilities $p_1,p_2,\ldots,p_K$ is defined as follows: for any $x\in X, T(x)=\tau_k(x)$ with probability p_k and for any non-negative integer N, $T^N(x)=\tau_{k_N}\circ\tau_{k_{N-1}}\circ\ldots\circ\tau_{k_1}(x)$ with probability $\Pi_{j=1}^{N}p_{k_j}$. $T^N(x)$ can be viewed as the second component of the S^N of the skew product S. It can be easily shown that a measure μ is $T-$invariant if and only if the measure $\mu_p\times\mu$ is $S-$invariant. Pelikan [34] defined a $T-$invariant measure μ as follows:

Definition 3.7 *Let T be a random map on X and μ be a measure on X. The measure μ is invariant under the random map T if*

$$\mu(E)=\sum_{k=1}^{K}p_k\mu(\tau_k^{-1}(E)), \tag{3.4.1}$$

for any measurable set $E\in\mathcal{B}$.

Lemma 3.8 *Let μ be a measure on X. Let μ_p be the Borel probability measure on $\Omega=\{1,2,3,\ldots,K\}^{\{0,1,2,\ldots\}}$. Then μ is $T-$invariant if and only if the measure $\mu_p\times\mu$ on $\mathcal{B}(\Omega)\times\mathcal{B}$ is S invariant.*

Proof *By definition of S and μ_p,*

$$\begin{aligned}(\mu_p\times\mu)(S^{-1}(A\times B)) &= \sum_{k=1}^{K}p_k\mu_p(A)\mu(\tau_k^{-1}(B))\\ &= \mu_p(A)\sum_{k=1}^{K}p_k\mu(\tau_k^{-1}(B))\end{aligned}$$

If μ is T invariant, then

$$(\mu_p \times \mu)(S^{-1}(A \times B)) = \mu_p(A)\mu(B).$$

Thus, $\mu_p \times \mu$ is S invariant. The proof of the converse is easy.

3.4.2 The Frobenius–Perron operator for random maps with constant probabilities

Let f be the density of μ. Then $d\mu = f \cdot d\lambda$. Let $A \times B$ be a measurable subset of $\Omega \times X$. Then

$$\begin{aligned}(\mu_p \times \mu)(S^{-1}(A \times B)) &= \sum_{k=1}^{K} p_k \mu_p(A)\mu(\tau_k^{-1}(B)) \\ &= \sum_{k=1}^{K} p_k \mu_p(A) \int_B P_{\tau_k} f d\lambda \\ &= \mu_p(A) \sum_k p_k \int_B P_{\tau_k} f d\lambda.\end{aligned}$$

Thus, the density on the second component is $\sum_i p_i P_{\tau_i} f$. Hence the Frobenius–Perron operator P_T for the random map T is given by

$$P_T f = \sum_{k=1}^{K} p_k P_{\tau_k} f, \tag{3.4.2}$$

where P_{τ_k} is the Frobenius–Perron operator(see Chapter 2, also [8]) of the transformation τ_k.

3.4.3 Properties of the Frobenius–Perron operator

The Frobenius–Perron of a random map has properties resembling those of the traditional Frobenius–Perron operator.

(a) **Linearity** : Let α, β be constants. Then if $f, g \in L^{(}[0,1])$,

$$\begin{aligned}(P_T(\alpha f + \beta g))(x) &= \sum_{k=1}^{K} p_k (P_{\tau_k}(\alpha f + \beta g)(x) \\ &= \alpha \sum_{k=1}^{K} p_k (P_{\tau_k} f)(x) + \beta \sum_{k=1}^{K} p_k (P_{\tau_k} g)(x) \\ &= \alpha (P_T f)(x) + \beta (P_T f)(x).\end{aligned}$$

That is, $P_T : L^1 \to L^1$ is a linear operator.

(b) **(Positivity)** Let $f \in L^1([0,1])$ and $f \geq 0$. Then, for each $1 \leq k \leq K, P_{\tau_k} f \geq 0$.

Thus, $P_T f = \sum_{k=1}^{K} p_k P_{\tau_k} f \geq 0$, that is, if$f \geq 0$, then $P_T f \geq 0$.

(c) **(Preservation of integrals)**:

$$\begin{aligned} \int_{[0,1]} P_T f d\lambda &= \int_{[0,1]} \left(\sum_{k=1}^{K} p_k P_{\tau_k} f \right) d\lambda \\ &= \sum_{k=1}^{K} \left(p_k \int_{[0,1]} f d\lambda \right) \\ &= \left(\sum_{k=1}^{K} p_k \right) \int_{[0,1]} f d\lambda \\ &= \int_{[0,1]} f d\lambda . \end{aligned}$$

That is, $\int_{[0,1]} P_T f d\lambda = \int_{[0,1]} f d\lambda$

(d) **(Contraction property)** $P_T : L^1([0,1]) \to L^1([0,1])$ is a contraction: let $f \in L^1([0,1])$. Then $f = f^+ - f^-$ and $|f| = f^+ + f^-$, where $f^+ = \max(f,0)$ and $f^- = -\min(f,0)$.

$$\begin{aligned} \| P_T f \|_1 &= \int_{[0,1]} |P_T f| = \int_{[0,1]} |P_T (f^+ - f^-)| d\lambda \\ &= \int_{[0,1]} |(P_T f^+ - P_T f^-)| d\lambda \\ &\leq \int_{[0,1]} |P_T f^+| d\lambda + \int_{[0,1]} |P_T f^-| d\lambda \\ &= \int_{[0,1]} (P_T f^+ d\lambda + \int_{[0,1]} P_T f^- d\lambda \\ &= \int_{[0,1]} P_T f d\lambda \\ &= \int_{[0,1]} f d\lambda \\ &= \| fd \|_1 . \end{aligned}$$

That is, $\| P_T f \|_1 \leq \| fd \|_1$.

(e) **(Composition property)** P_T satisfies the composition property, i.e., if T and R are two random maps on $[0,1]$, then $P_{T \circ R} = P_T \circ P_R$. In particular, for any $n \geq 1, P_T^n = P_{T^n}$.

Lemma 3.9 *$P_T f^* = f^*$ if and only if $\mu = f^* \lambda$ is T invariant.*

Proof *Assume that $\mu(A) = \sum_{k=1}^{K} p_k \mu(\tau_k^{-1}(A))$, for any $A \in \mathcal{B}$. Then*

$$\begin{aligned}
\int_A f^* d\lambda &= \sum_{k=1}^{K} p_k \int_{\tau_k^{-1}(A)} f^* d\lambda \\
&= \sum_{k=1}^{K} p_k \int_A P_{\tau_k} f^* d\lambda \\
&= \int_A \sum_{k=1}^{K} p_k P_{\tau_k} f^* d\lambda \\
&= \int_A P_T f^* d\lambda.
\end{aligned}$$

Therefore, $P_T f^* = f^*$.

Conversely, assume that $P_T f^* = f^*$ almost everywhere. Then

$$\begin{aligned}
\mu(A) = \int_A f^* d\lambda &= \int_A P_T f^* d\lambda \\
&= \int_A \sum_{k=1}^{K} p_k P_{\tau_k} f^* d\lambda \\
&= \sum_{k=1}^{K} p_k \int_A P_{\tau_k} f^* d\lambda \\
&= \sum_{k=1}^{K} p_k \int_{\tau_k^{-1}(A)} f^* d\lambda \\
&= \sum_{k=1}^{K} p_k \mu(\tau_k^{-1}(A))
\end{aligned}$$

3.4.4 Representation of the Frobenius–Perron operator

Let $T = \{\tau_1, \tau_2, \ldots, \tau_K; p_1, p_2, \ldots, p_K\}$ be a random map on the measure space $([0,1], \mathcal{B}, \lambda)$, where $\tau_k : [0,1] \to [0,1]$, $k = 1, 2, \ldots, K$ is a piecewise monotonic transformation on a common partition $\mathcal{J} = \{J_i = [x_{i-1}, x_i], i = 1, 2, \ldots, q\}$ of $[0,1]$. From Chapter 2, (see 2.2.6), for any $f \in L^1([0,1])$,

$$(P_{\tau_k}(f))(x) = \sum_{i=1}^{n} \frac{f(\tau_{k,i}^{-1}(x))}{\tau'(\tau_{k,i}^{-1}(x))} \chi_{\tau(x_{i-1}, x_i)}(x), \tag{3.4.3}$$

Therefore,

$$(P_T(f))(x) = \sum_{k=1}^{K} p_k (P_{\tau_k}(f))(x) = \sum_{k=1}^{K} p_k \sum_{i=1}^{n} \frac{f(\tau_{k,i}^{-1}(x))}{\tau'(\tau_{k,i}^{-1}(x))} \chi_{\tau(x_{i-1}, x_i)}(x), \tag{3.4.4}$$

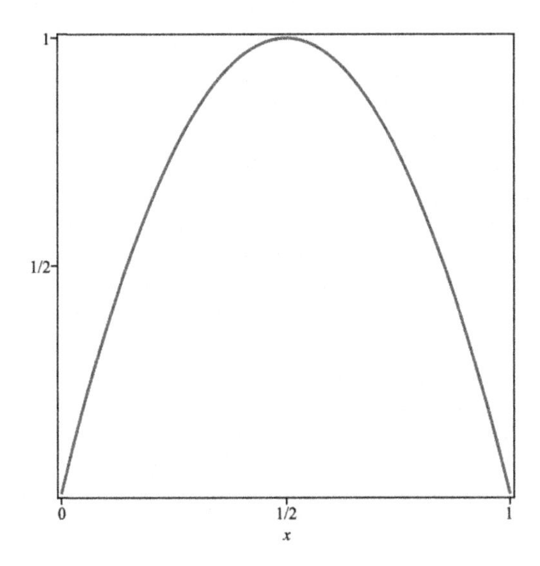

Figure 3.1 *The map τ_1 for the random maps $T = \{\tau_1, \tau_2; \frac{1}{4}, \frac{3}{4}\}$.*

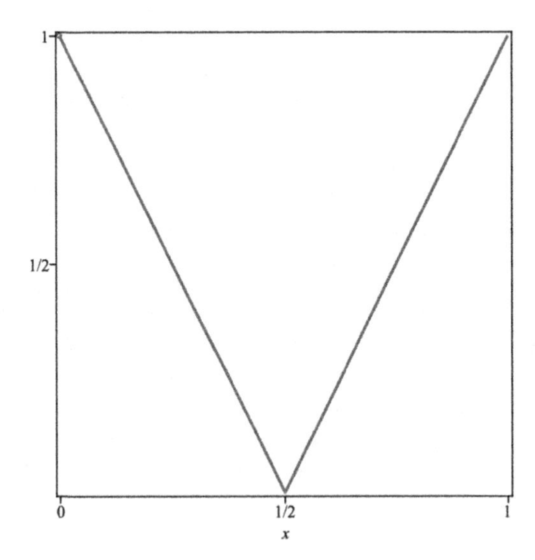

Figure 3.2 *The map τ_2 for the random maps $T = \{\tau_1, \tau_2; \frac{1}{4}, \frac{3}{4}\}$.*

Example 3.3 *Consider the random map $T = \{\tau_1, \tau_2; \frac{1}{4}, \frac{3}{4}\}$ where $\tau_1 : [0,1] \to [0,1]$ is the logistic map defined by $\tau_1(x) = 4x(1-x)$ and $\tau_2 : [0,1] \to [0,1]$ defined by*

$$\tau_2(x) = \begin{cases} -2x+1, & 0 \le x \le \frac{1}{2}, \\ 2x-1, & \frac{1}{2} \le x \le 1, \end{cases}$$

See Figure 3.1 for τ_1 and Figure 3.2 for τ_2. Let $f \in L^1(I)$. In Chapter 2 (see Example logistic), it was shown that

$$(P_{\tau_1}(f))(x) = \frac{1}{4\sqrt{1-x}}\left(f(\frac{1}{2} - \frac{1}{2}\sqrt{1-x}) + f(\frac{1}{2} + \frac{1}{2}\sqrt{1-x})\right).$$

It can be shown that

$$(P_{\tau_2}(f))(x) = \frac{1}{2}\left(f(\frac{1-x}{2} + \frac{1+x}{2}\right).$$

Therefore,

$$\begin{aligned} (P_T(f))(x) &= \frac{1}{4} \cdot \frac{1}{4\sqrt{1-x}}\left(f(\frac{1}{2} - \frac{1}{2}\sqrt{1-x}) + f(\frac{1}{2} + \frac{1}{2}\sqrt{1-x})\right) \\ &\quad + \frac{3}{4} \cdot \frac{1}{2}\left(f(\frac{1-x}{2} + \frac{1+x}{2}\right). \end{aligned}$$

3.4.5 *Existence of invariant measures for random maps with constant probabilities*

Let $\mathcal{T}_0(I)$ denote the class of transformations $\tau : I = [0,1] \to I$ that satisfy the following conditions:

(i) τ is piecewise monotonic, i.e., there exists a partition $\mathcal{J} = \{J_i = [x_{i-1}, x_i], i = 1, 2, \dots, q\}$ of I such that $\tau_i = \tau | J_i$ is C^1, and

$$|\tau_i'(x)| \ge \alpha > 0, \tag{3.4.5}$$

for any i and for all $x \in (x_{i-1}, x_i)$;
(ii) $g(x) = \frac{1}{|\tau_i'(x)|}$ is a function of bounded variation, where $\tau_i'(x)$ is the appropriate one-sided derivative at the end points of $\mathcal{J}$.

We say that $\tau \in \mathcal{T}_1(I)$ if $\tau \in \mathcal{T}_0(I)$ and $\alpha > 1$ in condition (3.4.5), i.e., τ is piecewise expanding.

Lemma 3.10 *[34] Let $T = \{\tau_1, \tau_2, \dots, \tau_K; p_1, p_2, \dots, p_K\}$ be a random map, where $\tau_k \in \mathcal{T}_0(I)$, with the common partition $\mathcal{J} = \{J_1, J_2, \dots, J_q\}$, $k = 1, 2, \dots, K$. If, for all $x \in [0,1]$, the following Pelikan's condition*

$$\sum_{k=1}^{K} \frac{p_k}{|\tau_k'(x)|} \le \gamma < 1, \tag{3.4.6}$$

is satisfied, then, for any $f \in BV(I)$,

$$V_I P_T f \leq AV_I f + B \parallel f \parallel_1, \text{where } 0 < \text{A} < 1, \text{ and B} > 0 \tag{3.4.7}$$

Theorem 3.11 *[34] Let $T = \{\tau_1, \tau_2, \ldots, \tau_K; p_1, p_2, \ldots, p_K\}$ be a random map, where $\tau_k \in \mathcal{T}_0(I)$, with the common partition $\mathcal{J} = \{J_1, J_2, \ldots, J_q\}$, $k = 1, 2, \ldots, K$. If, for all $x \in [0,1]$, the following Pelikan's condition*

$$\sum_{k=1}^{K} \frac{p_k}{|\tau_k'(x)|} \leq \gamma < 1, \tag{3.4.8}$$

is satisfied, then for all $f \in L^1 = L^1([0,1], \lambda)$:
(i) The limit

$$\lim_{n\to\infty} \frac{1}{n} \sum_{i=1}^{n-1} P_T^i(f) = f^* \text{ exists in } \mathrm{L}^1;$$

(ii) $P_T(f^*) = f^*$;
(iii) $V_{[0,1]}(f^*) \leq C \cdot \|f\|_1$, for some constant $C > 0$, which is independent of $f \in L^1$.

Example 3.4 *Let $T = \{\tau_1, \tau_2, \ldots, \tau_K; p_1, p_2, \ldots, p_K\}$ be a random map on the measure space $([0,1], \mathcal{B}, \lambda)$, where $\tau_k : [0,1] \to [0,1]$, $k = 1, 2, \ldots, K$ are piecewise expanding. It can be easily shown that T satisfies Pelikan's average expanding condition(3.4.8) and hence by Theorem 3.11 the Frobenius–Perron operator P_T has a fixed point $f^* \in L^1$ and the measure $\nu = f^* \cdot \lambda$ is T invariant.*

Example 3.5 *Consider the random map $T = \{\tau_1(x), \tau_2(x); \frac{1}{2}, \frac{1}{2}\}$ where $\tau_1, \tau_2 : [0,1] \to [0,1]$ are defined by*

$$\tau_1(x) = \begin{cases} 2x, & 0 \leq x \leq \frac{1}{2}, \\ x, & \frac{1}{2} < x \leq 1. \end{cases}$$

$$\tau_2(x) = \begin{cases} x + \frac{1}{2}, & 0 \leq x \leq \frac{1}{2}, \\ 2x - 1, & \frac{1}{2} < x \leq 1. \end{cases}$$

See Figure 3.3 for tau_1 and Figure 3.4 for tau_2. The maps τ_1 and τ_2 are piecewise linear and Markov. The random map T satisfies Pelikan's average expanding condition (3.4.8) and thus T has an invariant density f^* and the measure $\nu = f^* \cdot \lambda$ is T invariant.

3.4.6 Random maps of piecewise linear Markov transformations and the Frobenius–Perron operator

Let $T = \{\tau_1, \tau_2, \ldots, \tau_K; p_1, p_2, \ldots, p_K\}$ be a random map on the measure space $([0,1], \mathcal{B}, \lambda)$, where $\tau_k : [0,1] \to [0,1]$, $k = 1, 2, \ldots, K$ is a piecewise linear and Markov on a common partition $\mathcal{J} = \{J_i = [x_{i-1}, x_i], i = 1, 2, \ldots, q\}$ of $[0,1]$. Then,

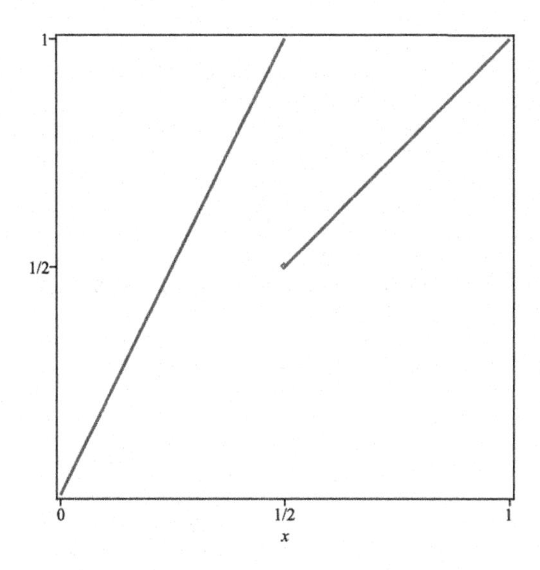

Figure 3.3 τ_1 *in Example 3.5.*

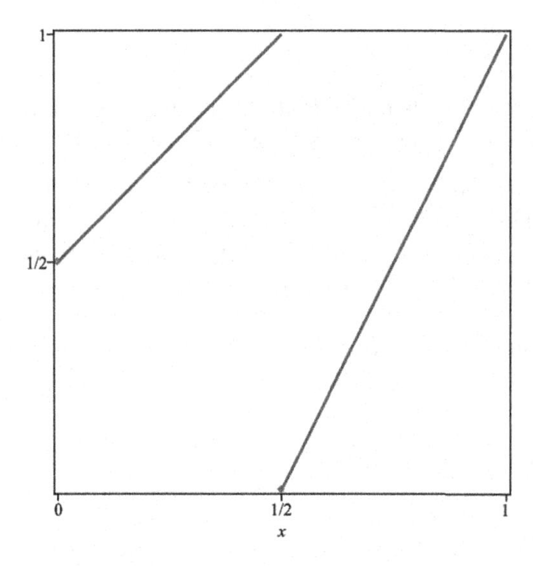

Figure 3.4 τ_2 *in Example 3.5.*

there exists $q \times q$ matrices $M_{\tau_k} = (m_{k,ij})_{1 \le i,j \le q}$ such that $P_{\tau_k} f = M^c_{\tau_k} \pi^f$, where π^f is the column vector obtained from f, c denotes the transpose, and

$$m_{k,ij} = \frac{\lambda(J_i \cap \tau_k^{-1}(J_j))}{\lambda(J_i)}.$$

Therefore,

$$P_T f = \sum_{k=1}^{K} p_k M^c_{\tau_k} \pi^f.$$

Example 3.6 *Consider the random map in Example 3.5. Then*

$$M_{\tau_1} = \begin{bmatrix} \frac{1}{2} & \frac{1}{2} \\ 0 & 1 \end{bmatrix} \tag{3.4.9}$$

and

$$M_{\tau_2} = \begin{bmatrix} 0 & 1 \\ \frac{1}{2} & \frac{1}{2} \end{bmatrix}. \tag{3.4.10}$$

Therefore,

$$P_T f = \frac{1}{2}\left(M^c_{\tau_1} + M^c_{\tau_2}\right)\pi^f = M_T \pi^f \begin{bmatrix} \frac{1}{4} & \frac{1}{4} \\ \frac{3}{4} & \frac{3}{4} \end{bmatrix} \pi^f.$$

It can be easily shown that the solution of the matrix equation $M_T \pi^f = \pi^f$ is $\pi^f = \left[\frac{1}{2}, \frac{3}{2}\right]$ which is an invariant density of T.

3.5 Necessary and sufficient conditions for the existence of invariant measures for a general class of random maps with constant probabilities

In this section we consider random maps with constant probabilities and present necessary and sufficient conditions for the existence of invariant measures. Our approach is different than Pelikan's approach [34]. Our results are a generalization of the results of Straube [39]. We closely follow [25].

We recall some definitions and results from [39, 40] which will be used to prove our main results in the next section.

Definition 3.12 *A set function $\phi : \mathcal{B} \to R$ is a finitely additive measure if*
(i) $-\infty < \phi(E) < \infty$, for all $E \in \mathcal{B}$;
(ii) $\phi(\emptyset) = 0$;
(iii) $\sup_{E \in \mathcal{B}} |\phi(E)| < \infty$;
(iv) $\phi(E_1 \cup E_2) = \phi(E_1) + \phi(E_2)$, for all $E_1, E_2 \in \mathcal{B}$ such that $E_1 \cap E_2 = \emptyset$.

Definition 3.13 *A finitely additive positive measure μ is a purely additive measure if every countably additive measure such that $\nu \ge 0$, $\nu \le \mu$ is identically zero.*

The following theorems (Theorem 3.14 and Theorem 3.16) and lemma (Lemma 3.15) are proved in [40] and will be useful for this section.

Theorem 3.14 *Let ϕ be a finitely additive (positive) measure. Then ϕ has a unique representation $\phi = \phi_c + \phi_p$, where ϕ_c is countably additive ($\phi_c \geq 0$) and ϕ_p is purely additive ($\phi_p \geq 0$).*

Lemma 3.15 *If μ is a finitely additive positive measure on $\mathcal{B}$, then μ_c is the greatest measure among countably additive measures ν with $0 \leq \nu \leq \mu$.*

Theorem 3.16 *Let ϕ be a finitely additive positive measure on a $\sigma-$algebra $\mathcal{B}$ and ν be a countably additive positive measure on $\mathcal{B}$. Then there exists a decreasing sequence $\{E_n\}_{n\geq 1}$ of elements of $\mathcal{B}$ such that $\lim_{n\to\infty} \nu(E_n) = 0$ and $\phi(E_n) = \phi(X)$.*

The following theorem, Theorem 3.17, was proved in [39] and we will use this theorem in our main result of this section.

Theorem 3.17 *Let (X, B, λ) be a measure space with normalized measure λ, $f : X \to X$ be a non-singular transformation. Then the following conditions are equivalent:*

(i) there exists an $f-$invariant normalized measure μ which is absolutely continuous with respect to λ;
(ii) there exists $\delta > 0$, and $\alpha, 0 < \alpha < 1$ such that

$$\lambda(E) < \delta \Rightarrow \sup_{k\in N} \lambda(f^{-k}(E)) < \alpha, E \in \mathcal{B}. \tag{3.5.1}$$

Now we present necessary and sufficient conditions for existence of an absolutely continuous invariant measure for random maps. For notational convenience, we consider $K = 2$; that is, we consider only two transformations τ_1, τ_2. The proof for a larger number of maps is analogous. Our presentation is based on [25].

Theorem 3.18 *Let $(X, \mathcal{B}, \lambda)$ be a measure space with normalized measure λ and $\tau_i : X \to X$, $i = 1, 2$ be non-singular transformations. Consider the random map $T = \{\tau_1, \tau_2; p_1, p_2\}$ with constant probabilities p_1, p_2. Then, there exists a normalized absolutely continuous (w.r.t. λ) $T-$ invariant measure μ if and only if there exists $\delta > 0$ and $0 < \alpha < 1$ such that for any measurable set E and any positive integer k, $\lambda(E) < \delta$ implies*

$$\begin{aligned}
p_1\lambda\left(\tau_1^{-1}(E)\right) &+ p_2\lambda\left(\tau_2^{-1}(E)\right) < \alpha; \\
p_1^2\lambda\left(\tau_1^{-2}(E)\right) &+ p_1p_2\lambda\left(\tau_2^{-1}\tau_1^{-1}(E)\right) + p_1p_2\lambda\left(\tau_1^{-1}\tau_2^{-1}(E)\right) \\
&+ p_2^2\lambda\left(\tau_2^{-2}(E)\right) < \alpha; \\
&\vdots \\
&\sum_{(i_1,i_2,i_3,\ldots,i_k)} p_{i_1}p_{i_2}\cdots p_{i_k}\lambda\left(\tau_{i_1}^{-1}\tau_{i_2}^{-1}\ldots\tau_{i_k}^{-1}(E)\right) < \alpha.
\end{aligned} \tag{3.5.2}$$

The following two lemmas are useful lemmas for proving the above theorem.

Lemma 3.19 *Let $(X, \mathcal{B}, \lambda)$ be a probability measure space and μ be absolutely continuous with respect to λ, $\mu = f \cdot \lambda$, for f an $L^1(X, \mathcal{B}, \lambda)$ function. Then there exists a constant $M \geq 0$ and a measurable set A_0 such that $\mu(A_0) \leq \frac{1}{10}$ and $f \leq M$ on $X \setminus A_0$.*

Proof *Consider the following sets :*

$$B_n = \{x \in X : n \leq f(x) < n+1\},\ n = 0, 1, \ldots. \tag{3.5.3}$$

Clearly, $\{B_n\}$ are disjoint measurable sets and $X = \cup_{n=0}^{\infty} B_n$ and $1 = \mu(X) = \sum_{n=0}^{\infty} \mu(B_n)$. Thus, there exists an $M \geq 0$ such that $\sum_{n=M}^{\infty} \mu(B_n) < \frac{1}{10}$. Let $A_0 = \cup_{n=M}^{\infty} B_n$. Then on $X \setminus \cup_{n=M}^{\infty} B_n$, $f(x) \leq M$.

For any measure ϕ, any integer k, and any measurable set E, define

$$\Lambda_k^{\phi}(E) := \sum_{(i_1, i_2, i_3, \ldots, i_k)} p_{i_1} p_{i_2} \cdots p_{i_k} \phi\left(\tau_{i_1}^{-1} \tau_{i_2}^{-1} \ldots \tau_{i_k}^{-1}(E)\right). \tag{3.5.4}$$

It can be easily shown that Λ_k^{λ} and Λ_k^{μ} are normalized measures and Λ_k^{μ} are measures absolutely continuous with respect to Λ_k^{λ}.

Lemma 3.20 *Let M be the constant from the previous lemma and δ be such that $M\delta + \frac{1}{10} < \frac{1}{4}$. Then, for any $n \geq 1$, and any measurable set A, we have $\Lambda_n^{\lambda}(A) < \delta \Rightarrow \Lambda_n^{\mu}(A) < \frac{1}{4}$.*

Proof *Let M and A_0 be as in the previous lemma. We have*

$$\begin{aligned}
\Lambda_n^{\mu}(A) &= \sum_{(i_1, i_2, i_3, \ldots, i_n)} p_{i_1} p_{i_2} \cdots p_{i_n} \mu\left(\tau_{i_1}^{-1} \tau_{i_2}^{-1} \ldots \tau_{i_n}^{-1}(A)\right) \\
&= \sum_{(i_1, i_2, i_3, \ldots, i_n)} p_{i_1} p_{i_2} \cdots p_{i_n} \mu\left(\tau_{i_1}^{-1} \tau_{i_2}^{-1} \ldots \tau_{i_n}^{-1}(A) \cap A_0\right) \\
&\quad + \sum_{(i_1, i_2, i_3, \ldots, i_n)} p_{i_1} p_{i_2} \cdots p_{i_n} \mu\left(\tau_{i_1}^{-1} \tau_{i_2}^{-1} \ldots \tau_{i_n}^{-1}(A) \cap (X \setminus A_0)\right) \\
&\leq \sum_{(i_1, i_2, i_3, \ldots, i_n)} p_{i_1} p_{i_2} \cdots p_{i_n} \frac{1}{10} \\
&\quad + \sum_{(i_1, i_2, i_3, \ldots, i_n)} p_{i_1} p_{i_2} \cdots p_{i_n} M\lambda\left(\tau_{i_1}^{-1} \tau_{i_2}^{-1} \ldots \tau_{i_n}^{-1}(A)\right) \\
&\leq \frac{1}{10} + M\Lambda_n^{\lambda}(A) < \frac{1}{10} + M\delta < \frac{1}{4}.
\end{aligned}$$

Proof of Theorem 3.18:
Suppose

$$\mu(E) = \sum_{i=1}^{2} p_i \mu(\tau_i^{-1}(E)),\ E \in B,\ \mu(X) = 1,\ \mu << \lambda.$$

We want to prove that there exist $\delta > 0$, $0 < \alpha < 1$, such that for any $E \in \mathcal{B}$ and for any positive integer k

$$\lambda(E) < \delta \Rightarrow \Lambda_k^{\lambda}(E) < \alpha. \tag{3.5.5}$$

Suppose not. Then, for any α, $0 < \alpha < 1$, there exists $E \in \mathcal{B}$ and there exists a positive integer n_0 such that

$$\lambda(E) < \delta \Rightarrow \Lambda_{n_0}^{\lambda}(E) > \alpha, \tag{3.5.6}$$

where $E \in \mathcal{B}$.

Choose $\delta > 0$ such that $M\delta + \frac{1}{10} < \frac{1}{4}$ where M is the constant of Lemma 3.19. Let n_0 be the index corresponding to δ in (3.5.5). Then by Lemma 3.19, we have for $A \in \mathcal{B}$

$$\begin{aligned} \lambda(A) < \delta &\Rightarrow \mu(A) < \frac{1}{4}; \\ \Lambda_{n_0}^{\lambda}(A) < \delta &\Rightarrow \Lambda_{n_0}^{\mu}(A) < \frac{1}{4}. \end{aligned} \tag{3.5.7}$$

Let $\alpha = 1 - \frac{\delta}{2}$. Then,

$$\Lambda_{n_0}^{\lambda}(X \setminus E) = 1 - \Lambda_{n_0}^{\lambda}(E) < 1 - 1 + \delta = \delta.$$

By our choice of δ, we get

$$\Lambda_{n_0}^{\mu}(X \setminus E) < \frac{1}{4}.$$

Since μ is invariant, we have

$$\mu(X \setminus E) = \Lambda_{n_0}^{\mu}(X \setminus E) < \frac{1}{4}.$$

Thus,

$$1 = \mu(X) = \mu(E) + \mu(X \setminus E) < \frac{1}{4} + \frac{1}{4},$$

a contradiction.

Conversely, suppose that there exists $\delta > 0$ and $0 < \alpha < 1$ such that for any measurable set E and any positive integer k, $\lambda(E) < \delta$ implies

$$\begin{aligned} p_1\lambda\left(\tau_1^{-1}(E)\right) &+ p_2\lambda\left(\tau_2^{-1}(E)\right) < \alpha; \\ p_1^2\lambda\left(\tau_1^{-2}(E)\right) &+ p_1p_2\lambda\left(\tau_2^{-1}\tau_1^{-1}(E)\right) + p_1p_2\lambda\left(\tau_1^{-1}\tau_2^{-1}(E)\right) \\ &+ p_2^2\lambda\left(\tau_2^{-2}(E)\right) < \alpha; \\ &\vdots \\ &\sum_{(i_1,i_2,i_3,\ldots,i_k)} p_{i_1}p_{i_2}\cdots p_{i_k}\lambda\left(\tau_{i_1}^{-1}\tau_{i_2}^{-1}\ldots\tau_{i_k}^{-1}(E)\right) < \alpha. \end{aligned}$$

We want to show that there exists a measure μ such that $\mu(E) = \sum_{i=1}^{2} p_i\mu\left(\tau_i^{-1}(E)\right), E \in \mathcal{B}, \mu(X) = 1$ and $\mu << \lambda$.

Consider the measures λ_n defined by

$$\lambda_n(E) := \frac{1}{n}\sum_{k=0}^{n-1} \Lambda_k^{\lambda}(E),\ E \in \mathcal{B}. \tag{3.5.8}$$

It can be shown that, for all n, λ_n are normalized measures. Moreover, if $\lambda(E) = 0$, then

$$\begin{aligned}
\lambda_n(E) &= \lambda(E) + p_1\lambda\left(\tau_1^{-1}(E)\right) + p_2\lambda\left(\tau_2^{-1}(E)\right) \\
&\quad + p_1^2\lambda\left(\tau_1^{-2}(E)\right) + p_1p_2\lambda\left(\tau_2^{-1}\tau_1^{-1}(E)\right) \\
&\quad + p_1p_2\lambda\left(\tau_1^{-1}\tau_2^{-1}(E)\right) + p_2^2\lambda\left(\tau_2^{-2}(E)\right) \\
&\quad + \ldots + \sum_{(i_1,i_2,i_3,\ldots,i_n)} p_{i_1}p_{i_2}\ldots p_{i_n}\lambda\left(\tau_{i_1}^{-1}\tau_{i_2}^{-1}\ldots\tau_{i_n}^{-1}(E)\right) \\
&= 0,
\end{aligned}$$

by non-singularity of τ_1 and τ_2. Hence $\lambda_n << \lambda$. We imbed λ_n in the dual space $L_\infty(\lambda)^*$ of $L_\infty(\lambda)$ in the following way:

$$g_n(f) = \int_X f d\lambda_n, f \in L_\infty(\lambda).$$

For every n,

$$|g_n(f)| = |\int_X f d\lambda_n| \leq ||f||_\infty \int_X d\lambda_n = ||f||_\infty.$$

Hence, for each n, $||g_n|| \leq 1$. Thus, the λ_n can be thought of as elements of the unit ball of $L_\infty(\lambda)^*$. This unit ball is weak*– compact by Alaoglu's Theorem [13, 40]. Let ν be a cluster point in the weak*– topology of $L_\infty(\lambda)^*$ of the sequence $\{\lambda_n\}_{n\geq 1}$.

Define a set function μ on $\mathcal{B}$ by

$$\mu(E) = \nu(\chi_E). \tag{3.5.9}$$

We claim that μ is finitely additive, bounded, and that it vanishes on sets of $\lambda-$ measure zero: $\mu(\emptyset) = \nu(\chi_\emptyset) = \nu(0) = 0$, since ν is a linear functional. For any $E \in \mathcal{B}$,

$$\begin{aligned}
\mu(E) &= \nu(\chi_E) = \lim_{s\to\infty} g_{n_s}(\chi_E) = \lim_{s\to\infty}\int_E d\lambda_{n_s} = \lim_{s\to\infty}\lambda_{n_s}(E) \\
&= \lim_{s\to\infty}\frac{1}{n_s}\sum_{k=0}^{n_s-1}\Lambda_k^{\lambda}(E) \geq 0,
\end{aligned}$$

since Λ_k^{λ} is a measure. Thus,

$$0 \leq \mu(E) \leq \mu(X) = \lim_{s\to\infty}\lambda_{n_s}(X) = 1.$$

Now,

$$\begin{aligned}\mu(\cup_{i=1}^m E_i) &= \lim_{s\to\infty} \lambda_{n_s}(\cup_{i=1}^m E_i) = \lim_{s\to\infty} \sum_{i=1}^m \lambda_{n_s}(E_i) \\ &= \sum_{i=1}^m \lim_{s\to\infty} \lambda_{n_s}(E_i) = \sum_{i=1}^m \mu(E_i).\end{aligned}$$

Let $\lambda(E)=0$. Then $\mu(E)=\lim_{s\to\infty}\lambda_{n_s}(E)=0$, because $\lambda_{n_s} << \lambda$. Hence, μ is finitely additive, bounded, and it vanishes on sets of $\lambda-$ measure zero.

μ is $T-$invariant:

$$\begin{aligned}\mu(E) &= \lim_{s\to\infty} \lambda_{n_s}(E) = \lim_{s\to\infty} \frac{1}{n_s}\sum_{k=0}^{n_s-1} \Lambda_k^\lambda(E) \\ &= \lim_{s\to\infty} \frac{1}{n_s}[\Lambda_0^\lambda(E)+\Lambda_1^\lambda(E)\ldots+\Lambda_{n_s-1}^\lambda(E)] \\ &= \lim_{s\to\infty} \frac{1}{n_s}[\lambda(E)+p_1\lambda\left(\tau_1^{-1}(E)\right)+p_2\lambda\left(\tau_2^{-1}(E)\right)] \\ &+\ldots+ \sum_{(i_1,i_2,i_3,\ldots,i_{n_s-1})} p_{i_1}p_{i_2}\cdots p_{i_{n_s-1}}\lambda\left(\tau_{i_1}^{-1}\tau_{i_2}^{-1}\ldots\tau_{i_{n_s-1}}^{-1}(E)\right)].\end{aligned}$$

On the other hand,

$$\sum_{i=1}^2 p_i\mu\left(\tau_i^{-1}(E)\right) = p_1\mu\left(\tau_1^{-1}(E)\right)+p_2\mu\left(\tau_2^{-1}(E)\right)$$

Using definition of Λ_k^λ we get,

$$= p_1 \lim_{s\to\infty}\frac{1}{n_s}\sum_{k=0}^{n_s-1}\Lambda_k^\lambda\left(\tau_1^{-1}(E)\right)+p_2 \lim_{s\to\infty}\frac{1}{n_s}\sum_{k=0}^{n_s-1}\Lambda_k^\lambda\left(\tau_2^{-1}(E)\right)$$

Splitting the sum we get,

$$= \lim_{s\to\infty}\frac{1}{n_s}[p_1\{\lambda\left(\tau_1^{-1}(E)\right)+p_1\lambda\left(\tau_1^{-2}(E)\right)+p_2\lambda\left(\tau_2^{-1}\tau_1^{-1}(E)\right)+\ldots$$

By rearranging we get

$$\begin{aligned}&+ \sum_{(i_1,i_2,i_3,\ldots,i_{n_s-1})} p_{i_1}p_{i_2}\cdots p_{i_{n_s-1}}\lambda\left(\tau_{i_1}^{-1}\tau_{i_2}^{-1}\ldots\tau_{i_{n_s-1}}^{-1}(\tau_1^{-1}(E))\right)\} \\ &+p_2\{\lambda(\tau_2^{-1}(E))+p_1\lambda\left(\tau_1^{-1}\tau_2^{-1}(E)\right)+p_2\lambda\left(\tau_2^{-2}(E)\right)+\ldots \\ &+ \sum_{(i_1,i_2,i_3,\ldots,i_{n_s-1})} p_{i_1}p_{i_2}\cdots p_{i_{n_s-1}}\lambda\left(\tau_{i_1}^{-1}\tau_{i_2}^{-1}\ldots\tau_{i_{n_s-1}}^{-1}(\tau_2^{-1}(E))\right)\}]\end{aligned}$$

$$
\begin{aligned}
= & \lim_{s\to\infty} \frac{1}{n_s}[p_1\lambda(\tau_1^{-1}(E)) + p_1^2\lambda\left(\tau_1^{-2}(E)\right) + p_1p_2\lambda\left(\tau_2^{-1}\tau_1^{-1}(E)\right) + \dots \\
& + p_1 \sum_{(i_1,i_2,i_3,\dots,i_{n_s-1})} p_{i_1}p_{i_2}\cdots p_{i_{n_s-1}}\lambda\left(\tau_{i_1}^{-1}\tau_{i_2}^{-1}\dots\tau_{i_{n_s-1}}^{-1}(\tau_1^{-1}(E))\right) \\
& + p_2\lambda\left(\tau_2^{-1}(E)\right) + p_2p_1\lambda\left(\tau_1^{-1}\tau_2^{-1}(E)\right) + p_2^2\lambda\left(\tau_2^{-2}(E)\right) + \dots \\
& + p_2 \sum_{(i_1,i_2,i_3,\dots,i_{n_s-1})} p_{i_1}p_{i_2}\cdots p_{i_{n_s-1}}\lambda\left(\tau_{i_1}^{-1}\tau_{i_2}^{-1}\dots\tau_{i_{n_s-1}}^{-1}(\tau_2^{-1}(E))\right)].
\end{aligned}
$$

Clearly,

$$\mu(E) = \sum_{i=1}^{2} p_i\mu\left(\tau_i^{-1}(E)\right).$$

Thus, we have shown that μ is a finitely additive $T-$ invariant measure. By Theorem 3.16, μ has a unique representation

$$\mu = \mu_c + \mu_p,$$

where μ_c is countably additive and $\mu_c \geq 0$ and μ_p is purely additive and $\mu_p \geq 0$. We claim that $\mu_c \neq 0$. Suppose $\mu_c = 0$. Then by Theorem 3.17, there exists a decreasing sequence $\{E_n\}_{n\geq 1}$ of elements of $\mathcal{B}$ such that $\lim_{n\to\infty}\lambda(E_n) = 0$ and $\mu(E_n) = \mu(X) = 1$. Thus, there exists an integer n_0 such that for all $n \geq n_0$, $\lambda(E_n) < \delta$ and, as a consequence of our hypothesis, we have for all k,

$$\Lambda_k^{\lambda}(E_n) < \alpha.$$

Hence,

$$\lambda_k(E_n) < \alpha, k = 1,2,3,\dots.$$

Thus, $\mu(E_n) = \lim_{s\to\infty} g_{n_s}(E_n) < \alpha < 1$, a contradiction. Now,

$$
\begin{aligned}
\mu(E) &= p_1\mu\left(\tau_1^{-1}(E)\right) + p_2\mu\left(\tau_2^{-1}(E)\right) \\
&= p_1\{\mu_c\left(\tau_1^{-1}(E)\right) + \mu_p\left(\tau_1^{-1}(E)\right)\} + p_2\{\mu_c\left(\tau_2^{-1}(E)\right) + \mu_p\left(\tau_2^{-1}(E)\right)\} \\
&= \{p_1\mu_c\left(\tau_1^{-1}(E)\right) + p_2\mu_c\left(\tau_2^{-1}(E)\right)\} \\
&\quad + \{p_1\mu_p\left(\tau_1^{-1}(E)\right) + p_2\mu_p\left(\tau_2^{-1}(E)\right)\}.
\end{aligned}
$$

Clearly $m : \mathcal{B} \to R$, defined by

$$m(E) = p_1\mu_c(\tau_1^{-1}(E)) + p_2\mu_c(\tau_2^{-1}(E)),$$

is a countably additive measure, and $m \leq \mu$. Thus, by Lemma 3.15, we have $m \leq \mu_c$ and hence

$$E \mapsto \mu_c(E) - m(E) = \mu_c(E) - \{p_1\mu_c\left(\tau_1^{-1}(E)\right) + p_2\mu_c\left(\tau_2^{-1}(E)\right)\}$$

is a positive measure. But this measure has total mass zero. Hence, it is a zero measure. Thus, μ_c is $T-$ invariant. Because μ vanishes on sets of $\lambda-$ measure zero and $0 \le \mu_c \le \mu$, we have $\mu_c << \lambda$. Finally, $\gamma(E) = \frac{\mu_c(E)}{\mu_c(X)}$ is a normalized, $T-$ invariant and absolutely continuous with respect to λ.

3.6 Support of invariant densities for random maps

Invariant measures for random maps with constant probabilities reflect their long time behavior and play an important role in understanding their chaotic nature. It is, therefore, important to establish properties of their absolutely continuous invariant measures. In this section we present an important property of absolutely continuous invariant measures for random maps. In fact, we generalize to random maps results of Keller [28] and Kowalski [29], which state that the density of an acim of a non-singular map is strictly positive on its support. Our main results are proven under the assumption that the individual maps used to construct the random map are piecewise expanding. We also give an example satisfying Pelikan's condition (3.4.8), showing that the assumption of expanding cannot be removed. We follow closely [19].

Definition 3.21 *Let $\tau : (X, \mathcal{B}, \lambda) \to (X, \mathcal{B}, \lambda)$ be a non-singular transformation and μ an acim with respect to Lebesgue measure λ possessing density function f. We define the support of μ as follows:*

$$\text{supp}(\mu) = \text{supp}(f) = \{x \in X : f(x) > 0\}$$

Definition 3.22 *A function $f : \mathbb{R} \to \mathbb{R}$ is said to be a lower semicontinuous function if and only if $f(y) \le \liminf_{x \to y} f(x)$ for any $y \in \mathbb{R}$.*

Theorem 3.23 *[8] If f is lower semicontinuous on $I = [a,b] \subset \mathbb{R}$, then it is bounded below and assumes its minimum value. For any $c \in \mathbb{R}$, the set $\{x : f(x) > c\}$ is open.*

Lemma 3.24 *[8] If f is of bounded variation on I, then it can be redefined on a countable set to become a lower semicontinuous function.*

Recall the classes $\mathcal{T}_0(I)$ and $\tau \in \mathcal{T}_1(I)$ which are defined in Section 3.4.5.

Theorem 3.25 *[28, 29] Let $\tau \in \mathcal{T}_1(I)$ and f be a $\tau-$invariant density which can be assumed to be lower semicontinuous. Then there exists a constant $\beta > 0$ such that $f|_{\text{supp}(f)} \ge \beta$.*

In the following we prove that the invariant density of an acim of the random map $T = \{\tau_1, \tau_2, \ldots, \tau_K; p_1, p_2, \ldots, p_K\}$, $\tau_1, \tau_2, \ldots, \tau_K \in \mathcal{T}_1$, is strictly positive on its support. For notational convenience, we consider $K = 2$; that is, we consider only two transformations τ_1, τ_2. The proofs for a larger number of maps are analogous. We consider random maps with constant probabilities.

Let $\mathcal{Q}$ denote the set of end points of intervals of partition $\mathcal{P}$ except the point 0 and 1. The main result of this note applies to random maps, where each component map is in $\mathcal{T}_1(I)$, but the first two lemmas are proved under more the general assumptions of Theorem 3.11:

Lemma 3.26 *Let the random map $T = \{\tau_1, \tau_2; p_1, p_2\}$ satisfy the assumptions of Theorem 3.11. In particular,*

$$\frac{p_1}{|\tau_1'(x)|} + \frac{p_2}{|\tau_2'(x)|} \leq \gamma < 1, \tag{3.6.1}$$

for all $x \in I \setminus \mathcal{Q}$. Then, for any interval J disjoint with $\mathcal{Q}$, at least one of the images $\tau_1(J)$, $\tau_2(J)$ is longer than J.

Proof *First, let us note that if ν is the normalized Lebesgue measure on J, then*

$$\begin{aligned} 1 = \left(\int_J 1 d\nu\right)^2 &= \left(\int_J \sqrt{|\tau'(x)|} \cdot \frac{1}{\sqrt{|\tau'(x)|}} d\nu(x)\right)^2 \\ &\leq \int_J |\tau'(x)| d\nu(x) \cdot \int_J \frac{1}{|\tau'(x)|} d\nu(x), \end{aligned} \tag{3.6.2}$$

or

$$\frac{1}{\int_J |\tau'(x)| d\nu(x)} \leq \int_J \frac{1}{|\tau'(x)|} d\nu(x), \tag{3.6.3}$$

or

$$\frac{1}{\frac{1}{\lambda(J)} \int_J |\tau'(x)| dx} \leq \frac{1}{\lambda(J)} \int_J \frac{1}{|\tau'(x)|} dx. \tag{3.6.4}$$

Integrating (3.6.1) over J, we obtain

$$p_1 \int_J \frac{1}{|\tau_1'(x)|} dx + p_2 \int_J \frac{1}{|\tau_2'(x)|} dx \leq \gamma \cdot \lambda(J), \tag{3.6.5}$$

and, using (3.6.4), we obtain

$$\frac{p_1 \lambda(J)}{\lambda(\tau_1(J))} + \frac{p_2 \lambda(J)}{\lambda(\tau_2(J))} \leq \gamma.$$

Thus, at least one of the numbers $\lambda(\tau_1(J))$, $\lambda(\tau_2(J))$ is larger than $\lambda(J)$.

Corollary 3.27 *Under assumptions of Lemma 3.26, if J is an interval disjoint with end points of the partition $\mathcal{P}^{(n)} = \mathcal{P} \vee \bigvee \tau_{j_{n-1}}^{-1} \tau_{j_{n-2}}^{-1} \cdots \tau_{j_1}^{-1} \mathcal{P}$ for an $n > 1$, then*

$$\sum_{(j_n, j_{n-1}, \ldots, j_1)} p_{j_n} p_{j_{n-1}} \cdots p_{j_1} \frac{\lambda(J)}{\lambda(\tau_{j_n} \tau_{j_{n-1}} \cdots \tau_{j_1}(J))} \leq \gamma^n,$$

where $j_n, j_{n-1}, \ldots, j_1 \in \{1, 2\}$. In particular, we have

$$\max_{J \in \mathcal{P}^{(n)}} \operatorname{diam}(J) \leq \gamma^n, \quad n \geq 1.$$

Proof *It follows from the fact that Pelikan's condition (3.4.8) implies*

$$\sup_x \sum_{(j_n,j_{n-1},\ldots,j_1)} \frac{p_{j_n}p_{j_{n-1}}\cdots p_{j_1}}{|(\tau_{j_n}\circ\tau_{j_{n-1}}\circ\cdots\circ\tau_{j_1})'(x)|} \le \gamma^n\,,$$

$n \ge 1,\ j_n, j_{n-1}, \ldots, j_1 \in \{1,2\}$.

Remark 3.28 *Instead of Pelikan's condition (3.4.8) we can use throughout the paper a weaker condition*

$$p_1 \ln|\tau_1'(x)| + p_2 \ln|\tau_2'(x)| \ge \beta > 0\,, \quad x \in [0,1], \tag{3.6.6}$$

which one could call Morita's condition [32]. Condition (3.6.6) implies the existence of absolutely continuous invariant measure but its density may not be of bounded variation. If we assume additionally that this density is of bounded variation or at least is a lower semicontinuous function, then we can reprove the results of this paper under assumption of Condition (3.6.6).

Condition (3.6.6) implies

$$\sum_{(j_n,j_{n-1},\ldots,j_1)} p_{j_n}p_{j_{n-1}}\cdots p_{j_1} \ln|(\tau_{j_n}\circ\tau_{j_{n-1}}\circ\cdots\circ\tau_{j_1})'(x)| \ge n\beta > 0\,, \quad x\in[0,1]\,,$$

$n \ge 1,\ j_n, j_{n-1}, \ldots, j_1 \in \{1,2\}$. *Using Condition (3.6.6) we can prove an analogue of Lemma 3.26:*

For any interval J contained in an element of partition $\mathcal{P}$ we have

$$p_1 \ln\frac{\lambda(\tau_1(J))}{\lambda(J)} + p_2 \ln\frac{\lambda(\tau_2(J))}{\lambda(J)} \ge \beta > 0\,,$$

which implies that at least one of the images is longer than interval J.

Analogously, for any interval J contained in an element of partition $\mathcal{P}^{(n)}$ we have

$$\sum_{(j_n,j_{n-1},\ldots,j_1)} p_{j_n}p_{j_{n-1}}\cdots p_{j_1} \ln\frac{\lambda(\tau_{j_n}\circ\tau_{j_{n-1}}\circ\cdots\circ\tau_{j_1}(J))}{\lambda(J)} \ge n\beta > 0\,,$$

$n \ge 1,\ j_n, j_{n-1}, \ldots, j_1 \in \{1,2\}$. *This, in particular, implies that*

$$\max_{J\in\mathcal{P}^{(n)}} \operatorname{diam}(J) \le e^{-n\beta}\,, \quad n \ge 1.$$

Lemma 3.29 *Let $T = \{\tau_1, \tau_2; p_1, p_2\}$ be a random map on $[0,1]$ satisfying the conditions of Theorem 3.11. Then, the support of the invariant density of T contains an interval which is not disjoint with $\mathcal{Q}$.*

Proof *Let* $\operatorname{supp}(f) = \{x \in [0,1] : f(x) > 0\}$. *The density function f of the acim μ is a function of bounded variation by Theorem 3.11 and thus, by Lemma 3.24, f can be redefined on a countable set to become a lower semicontinuous function $\bar{f}$*

and $\bar{f} = f$ *a.e. Thus* $\text{supp}(f) = \text{supp}(\bar{f}) = \{x : \bar{f} > 0\}$ *is an open set by Theorem 3.23. Thus,* $\text{supp}(f) = \cup_{i=1}^{\infty} I_i$*, where* I_i*'s are open disjoint intervals. Without loss of generality, let us assume that* $\lambda(I_i) \geq \lambda(I_{i+1})$ *for* $i = 1, 2, \ldots$ *We will prove that* $\mathcal{Q} \cap I_1 \neq \emptyset$*. Suppose* $\mathcal{Q} \cap I_1 = \emptyset$*. Then* I_1 *is contained in one of the subintervals,* J_**, of the partition* $\mathcal{P}$ *and* $\tau_1(I_1)$ *and* $\tau_2(I_1)$ *are both open intervals. Since* f *is an invariant density of the random map* T*, we have,*

$$f(x) = p_1 \cdot \sum_{i=1}^{q} \frac{f(\tau_{1,i}^{-1}(x))}{|\tau_1'(\tau_{1,i}^{-1}(x))|} \chi_{\tau_1(J_i)}(x) + p_2 \cdot \sum_{i=1}^{q} \frac{f(\tau_{2,i}^{-1}(x))}{|\tau_2'(\tau_{2,i}^{-1}(x))|} \chi_{\tau_2(J_i)}(x).$$

Let $x \in \tau_1(I_1)$*. It is clear that at least one element of* $\{\tau_{1,*}^{-1}(x)\}$ *is in* I_1 *and since* $I_1 \subset \text{supp}(f)$ *we have* $f(x) > 0$*. Thus,* $\tau_1(I_1)$ *is a subset of* $\text{supp}(f)$*. Similarly,* $\tau_2(I_1)$ *is a subset of* $\text{supp}(f)$*. By Lemma 3.26, at least one of the intervals* $\tau_1(I_1)$*,* $\tau_2(I_1)$ *has larger length than the length of* I_1*. This is a contradiction because* $\text{supp}(f)$ *does not contain an interval of length greater than* $\lambda(I_1)$*. This proves that* $\mathcal{Q} \cap I_1 \neq \emptyset$*.*

Corollary 3.30 *The number of different ergodic acim for the random map* T *satisfying the assumptions of Theorem 3.11 is at most equal to the cardinality of the partition* $\mathcal{P}$ *minus one.*

For Theorem 3.31 we assume that $\tau_1, \tau_2 \in \mathcal{T}_1$. After the theorem we give an example showing that it fails if we assume only $\tau_1, \tau_2 \in \mathcal{T}_0$.

Theorem 3.31 *Let* $T = \{\tau_1, \tau_2; p_1, p_2\}$ *be a random map on* $[0,1]$*, where* $\tau_1, \tau_2 \in \mathcal{T}_1$ *and have a common partition* $\mathcal{P} = \{J_1, J_2, \ldots, J_q\}$*. Then the support of the invariant density* f *of* T*,* $\text{supp}(f)$*, is a finite union of open intervals almost everywhere.*

Proof *Again we can assume that* $\text{supp}(f) = \cup_{i=1}^{\infty} I_i$*, where* I_i*'s are open disjoint intervals. Let* $\mathcal{D} = \{j \geq 1 : I_j$ *contains a discontinuity of* τ_1 *or* τ_2 *or both*$\}$*. By Lemma 3.29,* $\mathcal{D}$ *is not empty. Also,* $\mathcal{D}$ *is a finite set. If* $j \in \mathcal{D}$*, then* $\tau_i(I_j), i = 1, 2$*, is a finite union of intervals. Let* J *be the shortest interval of the family*

$$\{I_j\}_{j \in \mathcal{D}} \cup \{I : I \text{ is a connected component of } \tau_i(\mathrm{I_j}), \mathrm{i} = 1, 2, \mathrm{j} \in \mathcal{D}\}.$$

Let $\mathcal{F} = \{i \geq 1 : \lambda(I_i) \geq \lambda(J)\}$*, where* i *is not necessarily in* $\mathcal{D}$ *and let*

$$S = \cup_{i \in \mathcal{F}} I_i \subset \cup_i I_i = \text{supp}(f).$$

S *is a finite union of open disjoint intervals since it is a family of disjoint intervals with length* $\geq \lambda(J) > 0$*. For any* $j \in \mathcal{D}, I_j \subseteq S$*.*

We will prove that $\tau_i(S) \subseteq S$*,* $i = 1, 2$*. Let* $I_k \subset S$*. If* $k \notin \mathcal{D}$*, then* $\tau_i(I_k)$ *is contained in an interval* I_{k_i}*,* $i = 1, 2$ *and*

$$\lambda(I_{k_i}) \geq \lambda(\tau_i(I_k)) = \int_{I_k} |\tau_i'(x)| d\lambda > \inf_{x \in [0,1]} |\tau_i'(x)| \cdot \lambda(I_k), \; i = 1, 2.$$

Since $\inf_{x \in [0,1]} |\tau_i'(x)| > 1$ *for* $i = 1, 2$*, we have*

$$\lambda(I_{k_i}) \geq \lambda(I_k) > \lambda(J).$$

Thus, by the definition of S, we get $I_{k_i} \subseteq S$, $i=1,2$ and hence $\tau_i(I_k) \subset I_{k_i} \subseteq S$, $i=1,2$. If $k \in \mathcal{D}$, then by the definition of S, $\tau_i(I_k) \subset S$, $i=1,2$. Thus, $\tau_i(S) \subseteq S$, $i=1,2$.

Now we will prove that $\mathrm{supp}(f) \subseteq S$*. Suppose not. Let I_s be the largest interval of* $\mathrm{supp}(f) \setminus S$*. Thus, $s \notin \mathcal{D}$ and*

$$\lambda(\tau_i(I_s)) = \int_{I_s} |\tau_i'(x)| d\lambda > \inf_{x\in[0,1]} |\tau_i'(x)| \cdot \lambda(I_s) > \lambda(I_s),\ i=1,2.$$

Then $\tau_i(I_s) \subset S$, $i=1,2$. Thus, $I_s \subset \tau_i^{-1}(S)$, $i=1,2$. But $I_s \not\subset S$, so $I_s \subset \tau_i^{-1}(S) \setminus S$, $i=1,2$. Let $\mu = f\cdot\lambda$ be the T-invariant absolutely continuous measure. We will show that $\mu(\tau_i^{-1}(S)\setminus S) = 0$, $i=1,2$. Since $\tau_i(S) \subseteq S$, $i=1,2$, we have

$$S \subseteq \tau_i^{-1}(\tau_i(S)) \subseteq \tau_i^{-1}(S),\ i=1,2.$$

Thus,

$$\begin{aligned} 0 &= \mu(S) - \mu(S) = [p_1\mu(\tau_1^{-1}S) + p_2\mu(\tau_2^{-1}S)] - [p_1\mu(S) + p_2\mu(S)] \\ &= p_1[\mu(\tau_1^{-1}S) - \mu(S)] + p_2[\mu(\tau_2^{-1}S) - \mu(S)] \\ &= p_1\mu(\tau_1^{-1}S\setminus S) + p_2\mu(\tau_2^{-1}S\setminus S). \end{aligned} \tag{3.6.7}$$

Thus, both $\mu(\tau_1^{-1}S\setminus S) = 0$ and $\mu(\tau_2^{-1}S\setminus S) = 0$ if $p_1, p_2 > 0$. Thus, $\mu(I_s) = 0$, which contradicts the fact that $I_s \subset$ $\mathrm{supp}(f)$*.*

Example 3.7 *We present a random map satisfying the assumptions of Theorem 3.11 where the support of absolutely continuous T- invariant measure is an infinite countable union of disjoint intervals. In this case the invariant density cannot be bounded away from 0 on its support. Let us define the maps:*

$$\tau_1(x) = (8x - E(8x))/4\ ;$$

$$\tau_2(x) = \begin{cases} 3/4 + x/2 & \textit{for}\ \ 0 \le x \le 1/2\ ; \\ x/2 & \textit{for}\ \ 1/2 < x \le 1\ ; \end{cases}$$

where $E(t)$ denotes the integral part of t, as shown in Figure 1.

Let us consider the random map $T = \{\tau_1, \tau_2; 3/4, 1/4\}$. It satisfies all the assumptions of Pelikan's Theorem 3.11. In particular, T admits an absolutely continuous invariant measure μ with density f of bounded variation. Let S denote the support of f. Lebesgue measure restricted to the interval $L_0 = [0,1/4]$ is invariant for τ_1. Since $\tau_1([0,1]) = L_0$ and τ_1 is exact on L_0, we have $L_0 \subset S$. Since the interval $K_0 = (1/2, 3/4)$ is not in the image of τ_1 or τ_2 it is disjoint from S. Every image of L_0 is in S. Since $\tau_1([0,1]) = L_0$ we consider only images by τ_2. We have

$$\begin{aligned} \tau_2(L_0) &= [1-1/4, 1-1/8] = L_1^{(1)}\ ; \\ \tau_2(L_1^{(1)}) &= [1/2-1/8, 1/2-1/16] = L_1^{(2)}\ ; \\ \tau_2(L_1^{(2)}) &= [1-1/16, 1-1/32] = L_2^{(1)}\ ; \\ \tau_2(L_2^{(1)}) &= [1/2-1/32, 1/2-1/64] = L_2^{(2)}\ ; \end{aligned}$$

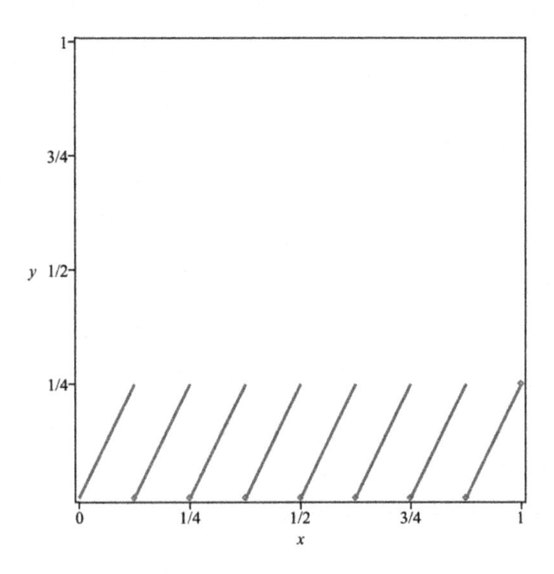

Figure 3.5 *The first map* τ_1 *for the random maps* $T = \{\tau_1, \tau_2; 3/4, 1/4\}$.

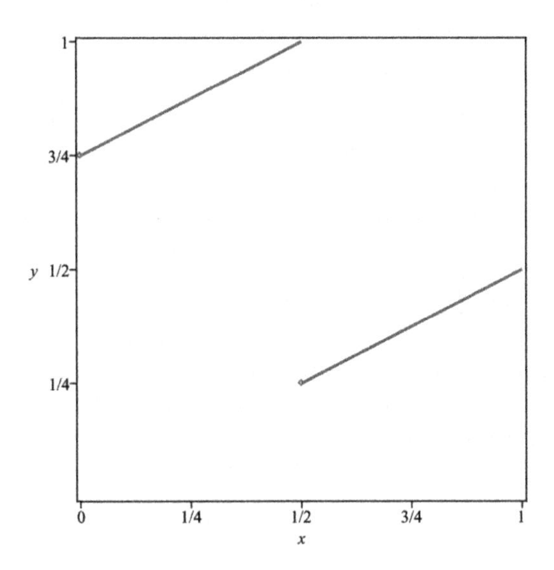

Figure 3.6 *The second map* τ_2 *for the random maps* $T = \{\tau_1, \tau_2; 3/4, 1/4\}$.

and in general, for $L_n^{(1)} = [1-1/4^n, 1-1/(2\cdot 4^n)]$ *and* $L_n^{(2)} = [1/2-1/(2\cdot 4^n), 1/2-1/(4\cdot 4^n)]$,

$$\tau_2(L_n^{(1)}) = L_n^{(2)} ;$$
$$\tau_2(L_n^{(2)}) = L_{n+1}^{(1)} .$$

Let $K_n^{(1)}$ *be the open interval between* $L_n^{(1)}$ *and* $L_{n+1}^{(1)}$, *and let* $K_n^{(2)}$ *be the open interval between* $L_n^{(2)}$ *and* $L_{n+1}^{(2)}$, $n \geq 1$. *Then, obviously, we have for all* $n \geq 1$

$$\tau_2(K_n^{(1)}) = K_n^{(2)} ;$$
$$\tau_2(K_n^{(2)}) = K_{n+1}^{(1)} .$$

Since τ_2 *is an injective map we also have*

$$\tau_2^{-1}(K_n^{(2)}) = K_n^{(1)} ;$$
$$\tau_2^{-1}(K_{n+1}^{(1)}) = K_n^{(2)} ; \quad n \geq 1 .$$

We have $\tau_2^{-2}(K_1^{(1)}) = K_0$ *and* $\tau_1^{-1}(K_n^{(i)}) = \emptyset$ *for* $i = 1,2$, $n \geq 1$. *Thus,*

$$(K_n^{(i)}) \cap S = \emptyset , \quad i = 1,2 ; \quad n \geq 1 .$$

We have proved that

$$S = L_0 \cup \bigcup_{i=1,2} \bigcup_{n\geq 1} L_n^{(i)} ,$$

i.e., that the support of the absolutely continuous T*- invariant measure is an infinite countable union of disjoint intervals. Since the invariant density* f *is of bounded variation we have*

$$\sup_{x \in L_n^{(i)}} f(x) \to 0, \quad n \to +\infty ,$$

for $i = 1,2$, *so the density* f *is not bounded away from 0.*

In Lemma 3.32 and Theorem 3.33 we return to the assumption $\tau_1, \tau_2 \in \mathcal{T}_0$, but we assume additionally that the support of the invariant density is a finite union of disjoint intervals. This was proved in Theorem 3.31 using the assumption $\tau_1, \tau_2 \in \mathcal{T}_1$.

Lemma 3.32 *Let* $T = \{\tau_1, \tau_2; p_1, p_2\}$ *be a random map on* $[0,1]$, *where* $\tau_1, \tau_2 \in \mathcal{T}_0$ *and have a common partition* $\mathcal{P} = \{J_1, J_2, \ldots, J_q\}$. *Let* f *be the invariant density of an acim* μ *of the random map* T *and* $S = \text{supp}(f) = \{x : f(x) > 0\}$. *We assume that* S *is a finite union of disjoint intervals. Then*
(i) $\tau_i(S \setminus \{a_0, a_1, \ldots, a_q\}) \subseteq S$, $i = 1,2$;
(ii) $\lambda(S \setminus \tau_i(S \setminus \{a_0, a_1, \ldots, a_q\})) = 0$, $i = 1,2$;
where $\{a_0, a_1, \ldots, a_q\}$ *are the end points of the intervals in the partition* $\mathcal{P}$.

Proof *We assume that f is lower semicontinuous. If it is not, we modify it on at most a countable set. By assumption, $S = \cup_{i=1}^{r} I_i$. Let $x \in S \setminus \{a_0, a_1, \dots, a_q\}$. Then $x \in \mathrm{Int}\ \mathrm{I_k}$, for some $k \in \{1, 2, \dots, r\}$ and there exists an $\varepsilon > 0$ such that $B(x, \varepsilon) \subset I_k$ and $f(y) > \frac{1}{2} f(x) > 0$ for all $y \in B(x, \varepsilon)$ since f is lower semicontinuous. We may assume that $\tau_i | I_k, i = 1, 2$ is increasing and that $f(\tau_i(x)) = \lim_{y \to \tau_i(x)^+} f(y)$, $i = 1, 2$. Now, for any $\delta > 0$, $\tau_i([x, x+\delta)) = [\tau_i(x), \tau_i(x) + \delta')$, $i = 1, 2$ and $\delta' \to 0$ as $\delta \to 0$. Then, for $i = 1, 2$, we have*

$$
\begin{aligned}
\int_{[\tau_i(x), \tau_i(x)+\delta')} f d\lambda = \mu(\tau_i([x, x+\delta))) &\geq \mu([x, x+\delta)) \\
&= \int_{[x, x+\delta)} f d\lambda \geq \frac{1}{2} f(x) \lambda([x, x+\delta)) \\
&\geq \frac{1}{2} f(x) \frac{1}{\max|\tau_i'|} \lambda([\tau_i(x), \tau_i(x) + \delta')).
\end{aligned}
$$

Since f is lower semicontinuous,

$$
\begin{aligned}
f(\tau_i(x)) &= \lim_{\delta' \to 0} \frac{1}{\lambda([\tau_i(x), \tau_i(x) + \delta'))} \int_{[\tau_i(x), \tau_i(x)+\delta')} f d\lambda \\
&\geq \frac{1}{2} f(x) \frac{1}{\max|\tau_i'|} > 0, i = 1, 2.
\end{aligned}
$$

Hence, $\tau_i(x) \in S$, $i = 1, 2$, and part (i) is proved.

Part (ii) is proved using reasoning similar to the end of Theorem 3.31 (equation (3.6.7)), which does not depend on the expanding property of τ_1, τ_2.

Theorem 3.33 *Let $T = \{\tau_1, \tau_2; p_1, p_2\}$ be a random map on $[0, 1]$, where $\tau_1, \tau_2 \in \mathcal{T}_1$ and have a common partition $\mathcal{P} = \{J_1, J_2, \dots, J_q\}$. Let f be the invariant density of an acim μ of the random map T and let $S = \mathrm{supp} f = \{x : f(x) > 0\}$. We assume that S is a finite union of disjoint intervals. Then there exists a constant $a > 0$ such that $f_{|_S} \geq a$.*

Proof *Since $S = \{x : f(x) > 0\}$ is a finite union of open intervals, $S = \cup_{i=1}^{r} I_i$, we can assume they are separated by intervals of positive measure. Then $\widehat{S} = S \setminus \{a_0, a_1, \dots, a_q\}$ is also a finite union of intervals: $\widehat{S} = \cup_{j=1}^{s} J_i$. Let $\mathcal{F} = \{I_i\}_{i=1}^{r}$, and $\mathcal{C} = \{J_j\}_{j=1}^{s}$. For any $J_k \in \mathcal{C}$, $\tau_j|_{J_k}$, $j = 1, 2$ is of class C^1. Therefore, there exist $I_{i_j} \in \mathcal{F}$, $j = 1, 2$ such that $\tau_j(J_k) \subseteq I_{i_j}$,*
$j = 1, 2$.

Let (c, d) be any interval in $\mathcal{F}$ or $\mathcal{C}$. We associate with its end points two classes of standard intervals:

$$\eta_c = \{(c, c+\varepsilon) : \varepsilon > 0\} \text{ and } \eta_d = \{(d - \varepsilon, d) : \varepsilon > 0\}.$$

The points c and d are referred to as the end points of the classes η_c and η_d, respectively. Let

$$\mathcal{K} = \{\eta_c, \eta_d : c, d \text{ are end points of intervals of } \mathcal{F} \text{ and } \mathcal{C}\}.$$

We now define a relation $\mapsto$ *between elements of* $\mathcal{K}$. *For* $\eta, \eta' \in \mathcal{K} : \eta \mapsto \eta'$ *if and only if* $\tau_j(U) \in \eta'$ *for at least one of* $j = 1,2$ *and for sufficiently small* $U \in \eta$. *The relation has the following two properties:*

(1) *If* η' *is associated with an end point of* $I_i \in \mathcal{F}$, *then there exists at least one* η *such that* $\eta \mapsto \eta'$. *To prove this, let us fix an* I_i *and* η' *associated with one of its end points. We claim that for each* $J_k \in \mathcal{C}$, *either* $\tau_j(J_k) \subseteq I_i$, $j = 1,2$ *or* $\tau_j(J_k) \cap I_i = \emptyset$, $j = 1,2$. *To show this, let us note that since* $\tau_j(J_k)$, $j = 1,2$ *is contained in* $S = \cup_{i=1}^r I_i$ *and* $\{I_i\}$ *are separated,* $\tau_j(J_k)$, $j = 1,2$ *is contained in one of* I_{i_j}, $j = 1,2$. *Now, since*

$$\lambda(I_{i_j} \setminus \tau_j(\widehat{S})) \leq \lambda(S \setminus \tau_j(\widehat{S})) = 0, \; j = 1,2,$$

there must exist a J_k *with* $\tau_j(J_k) \in \eta'$, $j = 1,2$ *and this implies (1).*

(2) *If* c' *is an end point of* I_i *such that* $\lim_{x \to c'} f(x) = 0, \eta'$ *is associated with* $c', \eta \mapsto \eta'$, *and* c *is an end point of* η, *then for any* $U \in \eta$,

$$\lim_{\substack{x \to c \\ x \in U}} f(x) = 0.$$

To prove this let us suppose that $\lim_{\substack{x \to c \\ x \in U}} f(x) = a > 0$, *for some* $U \in \eta$. *By the definition of the relation* $\mapsto$, *for at least one of* $j = 1,2$, *say* $j = 1$, *if* $\eta = \{(c, c+\varepsilon)\}$, *then we have* $\tau_1(c, c+\varepsilon) = (c', c+\varepsilon')$, *for* ε *small enough. Then,*

$$\begin{aligned}
\lim_{\substack{x \to c' \\ x \in U'}} f(x) &= \lim_{\varepsilon \to 0} \frac{1}{\lambda((c', c'+\varepsilon'))} \int_{(c', c'+\varepsilon')} f(t)dt = \lim_{\varepsilon \to 0} \frac{\mu((c', c'+\varepsilon'))}{\lambda((c', c'+\varepsilon'))} \\
&\geq \lim_{\varepsilon \to 0} p_1 \cdot \frac{\mu((c, c+\varepsilon))}{\max|\tau_1'| \cdot \lambda((c, c+\varepsilon))} = \frac{p_1}{\max|\tau_1'|} a > 0,
\end{aligned} \tag{3.6.8}$$

which is a contradiction.

We make the following observations:

(3) *In the setting of (2) above,* $c \notin S$. *Therefore,* c *is an end point of an interval* $I_i \in \mathcal{F}$.

Now we define

$\mathcal{K}_0 = \{\eta : \eta$ *is associated with an end point* c *of an* $I_i \in \mathcal{F}$ *and* $\lim_{\substack{x \to c \\ x \in I_i}} f(x) = 0\}$.

From (2) and (3) we obtain:

(4) *If* $\eta' \in \mathcal{K}_0$ *and* $\eta \mapsto \eta'$, *then* $\eta \in \mathcal{K}_0$.

We note that by (1) and (4), for each $\eta' \in \mathcal{K}_0$ *there exists at least one* $\eta \in \mathcal{K}_0$ *such that* $\eta \mapsto \eta'$.

Now let $\eta \in \mathcal{K}_0$. For any $n \geq 1$, any η' such that $\eta' \mapsto \eta$ in n-steps also belongs to $\mathcal{K}_0$. Choose $U \in \eta$ to be sufficiently small, i.e., completely contained in an element of partition $\mathcal{P}$. Then, all the preimages $\tau_{j_n}^{-1} \circ \tau_{j_{n-1}}^{-1} \circ \ldots \circ \tau_{j_2}^{-1} \circ \tau_{j_1}^{-1}(U)$, $j_i \in \{1,2\}$, touch an end point of some η' in $\mathcal{K}_0$. Let M be the number of elements in $\mathcal{K}_0$ and $\gamma < 1$ be the constant from Pelikan's condition 3.4.8. Then,

$$\begin{aligned} \mu(U) &= \sum_{(j_n, j_{n-1}, \ldots, j_1)} p_{j_n} p_{j_{n-1}} \cdots p_{j_1} \mu(\tau_{j_n}^{-1} \circ \tau_{j_{n-1}}^{-1} \circ \ldots \circ \tau_{j_1}^{-1}(U)) \\ &\leq \sup f \cdot \sum_{(j_n, j_{n-1}, \ldots, j_1)} p_{j_n} p_{j_{n-1}} \cdots p_{j_1} \lambda(\tau_{j_n}^{-1} \circ \tau_{j_{n-1}}^{-1} \circ \ldots \circ \tau_{j_1}^{-1}(U)). \end{aligned}$$

Each $\tau_{j_n}^{-1} \circ \tau_{j_{n-1}}^{-1} \circ \ldots \circ \tau_{j_1}^{-1}(U)$ consists of intervals contained in elements of partition $\mathcal{P}^{(n)}$ and touching one of M end points of classes in $\mathcal{K}_0$. Thus, using Corollary 3.27, we have

$$\lambda(\tau_{j_n}^{-1} \circ \tau_{j_{n-1}}^{-1} \circ \ldots \circ \tau_{j_2}^{-1} \circ \tau_{j_1}^{-1}(U)) \leq M \cdot \gamma^n$$

and

$$\mu(U) \leq \sup f \cdot \sum_{(j_n, j_{n-1}, \ldots, j_1)} p_{j_n} p_{j_{n-1}} \cdots p_{j_1} M \gamma^n = \sup f \cdot M \gamma^n.$$

Thus, $\mu(U) = 0$ which implies that $\lambda(U) = 0$ since $U \subseteq S$. This contradicts the fact that $U \in \eta$ is an open, nonempty interval. Hence, $\mathcal{K}_0 = \emptyset$ and $\lim_{\substack{x \to c \\ x \in U}} f(x) > 0$ for each of finitely many end points of intervals $I_i \in \mathcal{F}$. On the other hand, since f is lower semicontinuous, it assumes its infimum on any closed interval. Hence, there exists $a > 0$ such that $f(x) \geq a$ for all $x \in S$.

3.7 Smoothness of density functions for random maps

In this section we present another important property of invariant measure for random maps with constant probabilities. In fact, we present the smoothness property of random maps with constant probabilities. Halfant [22] proved that the density of an acim of a non-singular map of an interval inherits the smoothness properties of the map itself. A random map is a far more complicated system than an individual deterministic chaotic map. Although the methods of exploring are both similar, an extra complication of a random choice of the acting map is involved at every step. Our main results are proven under the assumption that the individual maps used to construct the random map are piecewise onto and piecewise expanding.

Recall the classes $\mathcal{T}_0(I)$ and $\tau \in \mathcal{T}_1(I)$ which are defined in Section 3.4.5. We also recall the following result of Lasota and Yorke [32]:

Theorem 3.34 *Let τ be piecewise monotonic, piecewise C^2 map of an interval $I = [0,1]$ into itself satisfying $\inf_{x \in I} |\tau'(x)| > 1$. Then,*

1. *If $f \in L^1([0,1])$ is of bounded variation, $P_\tau f$ is also of bounded variation and $V_0^1 P_\tau f \leq \alpha \parallel f \parallel + \beta V_0^1 f$ with $\alpha > 0$ and $\beta = \frac{2}{M}$ where $M = \inf|\tau'(x)|$.*

2. *τ has an acim whose density f is of bounded variation and satisfies $P_\tau f = f$.*

Thus, the fixed points of the operator P_τ are the density functions of the acims for the map τ.

We present a generalization of the following theorem established in [22].

Theorem 3.35 *Let $\tau \in \mathcal{T}_1(I)$ be piecewise onto and piecewise $C^r, r \geq 2$. Then $\tau-$invariant density f^* is of class C^{r-2} and, for any $s \leq r-2$, $(P_\tau^n \mathbf{1})^{(s)} \to f^{*(s)}$ uniformly as $n \to +\infty$*

We consider random maps $T = \{\tau_1, \tau_2, \ldots, \tau_K; p_1, p_2, \ldots, p_K\}$ satisfying conditions A and B below:

CONDITION A:

1. τ_k, $k = 1, 2, \ldots, K$, have a common defining partition $\mathcal{P} = \{I_i = (a_{i-1}, a_i), i = 1, 2, \ldots, q\}$ of I;
2. For each $i = 1, 2, \ldots, q$, $\tau_{k,i} = \tau_k|_{I_i}$, $k = 1, 2, \ldots, K$ has a C^2-extension to the closure $\overline{I_i}$ of I_i;
3. For each $i = 1, 2, \ldots, q$, $\tau_{k,i}$, $k = 1, 2, \ldots, K$ is strictly monotone on $\overline{I_i}$ and therefore determines a $1-1$ mapping of $\overline{I_i}$ onto some closed subinterval $\tau_k(\overline{I_i})$ of I.
4. For each $J \in \mathcal{P}$ and for each k, $k = 1, 2, \ldots, K$,

$$\tau_k(J) = I,$$

i.e., each τ_k is piecewise onto.

For each $n \geq 1$ we define

$$\Omega_n = \{\omega_n = (k_1, k_2, \ldots, k_n) : k_j \in \{1, 2, \ldots, K\}, j = 1, 2, \ldots, n\}.$$

For $\omega_{n-1} \in \Omega_{n-1}$, $\omega_{n-1} = (k_1, k_2, \ldots, k_{n-1})$ let $\mathcal{P}^{(n)}_{\omega_{n-1}}$ be the common refinement of

$$\mathcal{P}, \tau_{k_1}^{-1}(\mathcal{P}), (\tau_{k_2} \circ \tau_{k_1})^{-1}(\mathcal{P}), \ldots, (\tau_{k_{n-1}} \circ \tau_{k_{n-2}} \circ \ldots \circ \tau_{k_2} \circ \tau_{k_1})^{-1}(\mathcal{P}),$$

$n \geq 1$, where $\sigma^{-1}\mathcal{P} = \{\sigma^{-1}(J) : J \in \mathcal{P}\}$. Then, $\mathcal{P}^{(n)}$ is the union of all $\mathcal{P}^{(n)}_{\omega_{n-1}}$, $\omega_{n-1} \in \Omega_{n-1}$. Let $I^{(n)}$ denote a generic element of $\mathcal{P}^{(n)}$. We have $I^{(n)} \in \mathcal{P}^{(n)}_{\omega_{n-1}}$, for some $\omega_{n-1} = (k_1, k_2, \ldots, k_{n-1})$ or $I^{(n)} = (\tau_{k_{n-1}} \circ \tau_{k_{n-2}} \circ \ldots \circ \tau_{k_2} \circ \tau_{k_1})^{-1} I_s$, for some $I_s \in \mathcal{P}$. We will write $I^{(n)} = I^{(n)}_{\omega_{n-1}}$. Let $T^{i-1}_{\omega_{i-1}} = \tau_{k_{i-1}} \circ \tau_{k_{i-2}} \circ \ldots \circ \tau_{k_2} \circ \tau_{k_1}$. Then, $T^{n-1}_{\omega_{n-1}}$ is well defined on $I^{(n)}_{\omega_{n-1}}$ and $T^{n-1}_{\omega_{n-1}}(I^{(n)}_{\omega_{n-1}}) = I_s$. Moreover $T^{i-1}_{\omega_{i-1}}(I^{(n)}_{\omega_{n-1}}) \in \mathcal{P}^{(n-i)}_{\omega_{n-i-1}}$, where $\omega_{n-i-1} = (k_{i+1}, k_{i+2}, \ldots, k_{n-1})$.

Two points x, y are in the same $I^{(n)}_{\omega_{n-1}}$ if and only if $T^i_{\omega_i}(x), T^i_{\omega_i}(y)$ lie in the same element of $\mathcal{P}$ for $0 \le i \le n-1$. $T^n_{\omega_n}$ has a C^2−extension to $\overline{I}^{(n+1)}_{\omega_n}$, also denoted by $T^n_{\omega_n}$, which maps $\overline{I}^{(n+1)}_{\omega_n}$ monotonically onto some $\overline{I_s} \in \mathcal{P}$. Let

$$M(\overline{I}^{(n+1)}_{\omega_n}) = \sup_{x,y \in I^{(n+1)}} |(T^n_{\omega_n})'(x)/(T^n_{\omega_n})'(y)|$$

and

$$M_n = \sup_{\omega_n \in \Omega_n} \sup_{I^{(n+1)}_{\omega_n} \in \mathcal{P}^{(n+1)}_{\omega_n}} M(\overline{I}^{(n+1)}_{\omega_n}). \tag{3.7.1}$$

CONDITION B:

There exists an $\varepsilon > 0$ and a positive integer $p \geq 1$ such that for all $\omega_p \in \Omega_p$, all $I^{(p+1)}_{\omega_p}$ and all $x \in I^{(p+1)}_{\omega_p}$ we have

$$|(T^p_{\omega_p})'(x)| > 1 + \varepsilon.$$

For such random maps we conclude from the chain rule that

$$\alpha = \inf |(T^j_{\omega_j})'(x)| > 0, \tag{3.7.2}$$

where the infimum is taken over all $\omega_j \in \Omega_j$, all $I^{(j+1)}_{\omega_j}$ and all $x \in I^{(j+1)}_{\omega_j}$ for $0 \le j \le p$.

Furthermore,

$$\inf_{x \in I^{(n+1)}_{\omega_n},\ I^{(n+1)}_{\omega_n} \in \mathcal{P}^{(n+1)}} |(T^n)'(x)| \geq \alpha(1+\varepsilon)^{[n/p]} \geq \alpha(1+\varepsilon)^{(n/p)-1}, \quad n \geq 1. \tag{3.7.3}$$

For each $I^{(n+1)}$ there is an I_s such that $T^n_{\omega_n}(\overline{I}^{(n+1)}) = \overline{I_s}$. Thus by virtue of the mean value theorem, we know that there exists an $x \in \overline{I}^{(n+1)}$ such that $\lambda(I^{(n+1)}) = \lambda(I_s)/|(T^n_{\omega_n})'(x)|$. It follows from Condition B and (3.7.3) that

$$\lambda(I^{(n+1)}) \le B\theta^n, \tag{3.7.4}$$

where $\theta = (1+\varepsilon)^{-1/p} < 1$ and $B = \frac{(1+\varepsilon)}{\alpha} \max_{I_s \in \mathcal{P}} \lambda(I_s)$.

Let $T = \{\tau_1, \tau_2, \ldots, \tau_K; p_1, p_2, \ldots, p_K\}$ be a random map such that $\tau_1, \tau_2, \ldots, \tau_K \in \mathcal{T}_1(I)$ are piecewise $C^r, r \geq 2$, piecewise onto and T satisfies the above equations: Condition A and B. Thus, by Pelikan's Theorem (Theorem 3.11) T has an absolutely continuous invariant probability measure μ with respect to Lebesgue measure. In the following we generalize Theorem 3.35 for a single transformation to a theorem on random map.

The proof of the main result (Theorem 3.40) proceeds by following lemmas.

Lemma 3.36 *There exists $c>0$ such that for all $N\geq 0$, all $\omega_N\in\Omega_N$ and all $I^{(N+1)}_{\omega_N}\in P^{(N+1)}$*

$$|(\frac{d}{dx}|(T^N)'(x)|)^{-1}|<c, \tag{3.7.5}$$

where

$$T^N(x)=T^N_{\omega_N}(x)=\tau_{k_N}\circ\tau_{k_{N-1}}\circ\ldots\circ\tau_{k_1}(x),\;\; x\in I^{(N+1)}_{\omega_N}.$$

Proof: By (3.7.2), $(T^N)'(x)$ does not vanish on $\bar{I}^{(N+1)}$ and so $|(T^N)'(x)|$ is C^1 on $\bar{I}^{(N+1)}$. Using chain rule,

$$(T^m)'(x)=(\tau_{k_m})'(\tau_{k_{m-1}}\circ\tau_{k_{m-2}}\circ\cdots\circ\tau_{k_1}(x))\cdots(\tau_{k_2})'(\tau_{k_1}(x))\cdot\tau'_{k_1}(x)$$

and

$$(T^N)'(x)=(T^{N-m})'(T^m(x))\cdot(T^m)'(x).$$

Logarithmic differentiation gives

$$\begin{aligned}
&\left|\left(\frac{d}{dx}|(T^N)'(x)|\right)^{-1}\right| = |(T^N)'(x)|^{-1}\left|\frac{d}{dx}\log|(T^N)'(x)|\right| \\
=\;& |(T^N)'(x)|^{-1}\left|\frac{d}{dx}\log|(\tau_{k_N})'(\tau_{k_{N-1}}\circ\tau_{k_{N-2}}\circ\ldots\circ\tau_{k_1}(x))|\right. \\
&\quad +\frac{d}{dx}\log|(\tau_{k_{N-1}})'(\tau_{k_{N-2}}\circ\tau_{k_{N-3}}\circ\ldots\circ\tau_{k_1}(x))| \\
&\quad \left.+\ldots+\frac{d}{dx}\log|(\tau_{k_2})'(\tau_{k_1}(x))|+\frac{d}{dx}\log|(\tau_{k_1})'(x)|\right| \\
=\;& |(T^N)'(x)|^{-1}\left|\sum_{m=0}^{N-1}\frac{d}{dx}\log(\tau_{k_{m+1}})'(T^m(x))\right| \\
=\;& |(T^N)'(x)|^{-1} \\
&\left|\frac{|(\tau_{k_N})''(\tau_{k_{N-1}}\circ\tau_{k_{N-2}}\circ\ldots\circ\tau_{k_1}(x))|(\tau_{k_{N-1}}\circ\tau_{k_{N-2}}\circ\ldots\circ\tau_{k_1})'(x)}{|(\tau_{k_N})'(\tau_{k_{N-1}}\circ\tau_{k_{N-2}}\circ\ldots\circ\tau_{k_1}(x))|}\right. \\
&+\frac{|(\tau_{k_{N-1}})''(\tau_{k_{N-2}}\circ\tau_{k_{N-3}}\circ\ldots\circ\tau_{k_1}(x))|(\tau_{k_{N-2}}\circ\tau_{k_{N-3}}\circ\ldots\circ\tau_{k_1})'(x)}{|(\tau_{k_{N-1}})'(\tau_{k_{N-2}}\circ\tau_{k_{N-3}}\circ\ldots\circ\tau_{k_1}(x))|} \\
&\left.+\ldots+\frac{|(\tau_{k_2})''(\tau_{k_1}(x))|(\tau_{k_1})'(x))}{|(\tau_{k_2})'(\tau_{k_1}(x))|}+\frac{|(\tau_{k_1})''(x)|}{|(\tau_{k_1})'(x)|}\right| \\
=\;& \sum_{m=0}^{N-1}\left|\frac{(\tau_{k_{m+1}})''(T^m(x))}{(\tau_{k_{m+1}})'(T^m(x))}\frac{(T^m)'(x)}{(T^N)'(x)}\right|
\end{aligned}$$

$$\leq \max\{\sup\left|\frac{\tau_1''(x)}{\tau_1'(x)}\right|, \sup\left|\frac{\tau_2''(x)}{\tau_2'(x)}\right|, \ldots, \sup\left|\frac{\tau_K''(x)}{\tau_K'(x)}\right|\}\sum_{m=0}^{N-1}\left|\frac{(T^m)'(x)}{(T^N)'(x)}\right|$$

$$\leq \max\{\sup\left|\frac{\tau_1''(x)}{\tau_1'(x)}\right|, \sup\left|\frac{\tau_2''(x)}{\tau_2'(x)}\right|, \ldots, \sup\left|\frac{\tau_K''(x)}{\tau_K'(x)}\right|\}$$

$$\sum_{m=0}^{N-1}|(T^{(N-m)})'(T^m(x))|^{-1}.$$

Using (3.7.4), we obtain

$$\left|\left(\frac{d}{dx}|(T^N)'(x)|\right)^{-1}\right|$$

$$\leq \max\{\sup\left|\frac{\tau_1''(x)}{\tau_1'(x)}\right|, \sup\left|\frac{\tau_2''(x)}{\tau_2'(x)}\right|, \ldots, \sup\left|\frac{\tau_K''(x)}{\tau_K'(x)}\right|\}\sum_{m=0}^{N-1}\frac{1+\varepsilon}{\alpha}\theta^{N-m}$$

$$\leq \max\{\sup\left|\frac{\tau_1''(x)}{\tau_1'(x)}\right|, \sup\left|\frac{\tau_2''(x)}{\tau_2'(x)}\right|, \ldots, \sup\left|\frac{\tau_K''(x)}{\tau_K'(x)}\right|\}$$

$$\frac{1+\varepsilon}{\alpha}\sum_{m=0}^{N-1}\theta^{N-m} < \infty.$$

□

Lemma 3.37 *There exists $M > 0$ such that for all $N \geq 0$*

$$M_N \leq M,$$

where M_N is as in (3.7.1).

Proof: Let us fix N, $\omega_N \in \Omega_N$ and $I_{\omega_N}^{(N+1)}$. We will skip the ω_N subscript. For $x, y \in I^{(N+1)}$, we get by monotonicity of T^N on $I^{(N+1)}$ and Lemma 3.36

$$\begin{aligned}
\log\left|\frac{(T^N)'(x)}{(T^N)'(y)}\right| &= \log\frac{(T^N)'(x)}{(T^N)'(y)} = \int_y^x \frac{(T^N)''(t)}{(T^N)'(t)}dt \\
&= -\int_y^x (T^N)'(t)\frac{d}{dt}\frac{1}{(T^N)'(t)}dt \\
&\leq c\cdot\left|\int_y^x (T^N)'(t)dt\right| \leq c\cdot\lambda(T^N(\bar{I}^{(N+1)})) \leq c.
\end{aligned}$$

Setting $M = e^c$ completes the proof.
□

Now, we prove the main result (Theorem 3.40) of this chapter. We split the proof into a number of lemmas.

Lemma 3.38 *Let $T = \{\tau_1, \tau_2, \ldots, \tau_K; p_1, p_2, \ldots, p_K\}$ be a random map such that T satisfies conditions of Theorem 3.40 with $r = 2$. Then the $T-$invariant density f^* is a uniform limit of $\{P_T^n\mathbf{1}\}_{n\geq 0}$ and continuous.*

Proof: Consider the sequence $f_n = P_T^n \mathbf{1}, n = 1,2,\ldots$. We have

$$\begin{aligned} f_0(x) &= \mathbf{1}(x) \\ f_1(x) &= P_T\mathbf{1}(x) = \sum_{k=1}^{K} p_k P_{\tau_k}\mathbf{1}(x) \\ &= \sum_{k=1}^{K} p_k \sum_{i=1}^{q} \frac{1}{|\tau_k'(\tau_{k,i}^{-1}(x))|}. \end{aligned}$$

Since $\tau_k, k = 1,2,\ldots,K$ is piecewise onto and

$$\tau_{k,i}' = \tau_k'|_{I_i}, k = 1,2,\ldots,K, i = 1,2,\ldots,q$$

are continuous, $f_1(x)$ is continuous.

$$\begin{aligned} f_2(x) &= P_T^2\mathbf{1}(x) = P_T\left(P_T\mathbf{1}(x)\right) = P_T\left(\sum_{k=1}^{K} p_k P_{\tau_k}\mathbf{1}(x)\right) \\ &= \sum_{k=1}^{K} p_k P_T\left(P_{\tau_k}\mathbf{1}(x)\right) \\ &= \sum_{k=1}^{K} p_k \sum_{l=1}^{K} p_l P_{\tau_l}\left(P_{\tau_k}\mathbf{1}(x)\right) \\ &= \sum_{k=1}^{K} p_k \sum_{l=1}^{K} p_l \sum_{j=1}^{q}\sum_{i=1}^{q} \frac{1}{|\tau_l'(\tau_{l,j}^{-1}(\tau_{k,i}^{-1})(x))||\tau_k'(\tau_{k,i}^{-1}(x))|}, \end{aligned}$$

where $\tau_{k,i}' = \tau_k'|_{I_i}, \tau_{l,j}' = \tau_l'|_{I_j}, k,l = 1,2,\ldots K,$ and i,j $= 1,2,\ldots,$q. Thus, we have

$$f_2(x) = \sum_{1\le k_1,k_2\le K}\sum_{1\le i_1,i_2\le q} \frac{p_{k_1}p_{k_2}}{|\tau_{k_2}'(\tau_{k_2,i_2}^{-1}(\tau_{k_1,i_1}^{-1})(x))||\tau_{k_1}'(\tau_{k_1,i_1}^{-1}(x))|},$$

and again f_2 is continuous. In general, we have

$$\begin{aligned} f_n(x) = &\sum_{1\le k_1,k_2,\ldots,k_n\le K}\sum_{1\le i_1,i_2,\ldots,i_n\le q} \\ &\left[\frac{p_{k_1}p_{k_2}\cdots p_{k_n}}{|\tau_{k_n}'(\tau_{k_n,i_n}^{-1}(\tau_{k_{n-1},i_{n-1}}^{-1}(\ldots(\tau_{k_2,i_2}^{-1}(\tau_{k_1,i_1}^{-1}(x)))\ldots))))|}\right. \\ &\left.\frac{1}{|\tau_{k_{n-1}}'(\tau_{k_{n-1},i_{n-1}}^{-1}(\ldots(\tau_{k_2,i_2}^{-1}(\tau_{k_1,i_1}^{-1}(x)))\ldots))|\ldots|\tau_{k_1}'(\tau_{k_1,i_1}^{-1}(x))|}\right] \\ = &\sum_{1\le k_1,k_2,\ldots,k_n\le K}\sum_{1\le i_1,i_2,\ldots,i_n\le q} \\ &\left[\frac{p_{k_1}p_{k_2}\cdots p_{k_n}}{|(\tau_{k_n,i_n}\circ\tau_{k_{n-1},i_{n-1}}\circ\ldots\circ\tau_{k_2,i_2}\circ\tau_{k_1,i_1})'(\phi_{n,\mathbf{k}_n,\mathbf{j}_n}(x))|}\right] \end{aligned}$$

where f_n is continuous and

$$\phi_{n,\mathbf{k}_n,\mathbf{j}_n} = (\tau_{k_n,i_n} \circ \tau_{k_{n-1},i_{n-1}} \circ \ldots \circ \tau_{k_2,i_2} \circ \tau_{k_1,i_1})^{-1},$$

with

$$\mathbf{j}_n = (i_n, i_{n-1}, \ldots, i_2, i_1) \in \{1, 2\ldots, q\}^n, \mathbf{k}_n = (k_n, k_{n-1}, \ldots, k_2, k_1) \in \{1, 2, \ldots, K\}^n.$$

We want to show that f_n's are uniformly bounded and equicontinuous. We have,

$$f_n(x) = \sum_{\mathbf{k}_n} \sum_{\mathbf{j}_n} p_{k_1} \cdot p_{k_2} \cdots p_{k_n} |\phi'_{n,\mathbf{k}_n,\mathbf{j}_n}(x)|. \tag{3.7.6}$$

In Lemma 3.37 we have proved that there exists a constant $M > 1$ such that for any $n \geq 1$, any $\mathbf{k}$, any $\mathbf{j}$ and any $x \in I$,

$$\frac{1}{M} < \frac{\sup|\phi'_{n,\mathbf{k},\mathbf{j}}(x)|}{\inf|\phi'_{n,\mathbf{k},\mathbf{j}}(x)|} < M. \tag{3.7.7}$$

We can apply inequality (3.7.7) to equation (3.7.6) to obtain

$$\frac{1}{M} < \frac{\sup f_n(x)}{\inf f_n(x)} < M. \tag{3.7.8}$$

Since f_n is a density function of normalized measure, we get

$$\frac{1}{M} < f_n(x) < M. \tag{3.7.9}$$

Next, we show that f_n is equicontinuous. It is easy to see that for $l \geq 0$,

$$f_{n+l}(x) = \sum_{\mathbf{k}_n} \sum_{\mathbf{j}_n} p_{k_1} p_{k_2} \cdots p_{k_n} f_l(\phi_{n,\mathbf{k}_n,\mathbf{j}_n}(x)) |\phi'_{n,\mathbf{k}_n,\mathbf{j}_n}(x)|. \tag{3.7.10}$$

Differentiating both sides we obtain,

$$|f'_{n+l}(x)| \leq \sum_{\mathbf{k}_n} \sum_{\mathbf{j}_n} p_{k_1} p_{k_2} \cdots p_{k_n} \left[f'_l(\phi_{n,\mathbf{k}_n,\mathbf{j}_n}(x)) |(\phi'_{n,\mathbf{k}_n,\mathbf{j}_n}(x))^2| \right.$$
$$\left. + f_l(\phi_{n,\mathbf{k}_n,\mathbf{j}_n}(x)) |\phi''_{n,\mathbf{k}_n,\mathbf{j}_n}(x)| \right] \tag{3.7.11}$$

Now, using (3.7.3) and (3.7.9), we have

$$\sum_{\mathbf{k}_n} \sum_{\mathbf{j}_n} p_{k_1} p_{k_2} \cdots p_{k_n} |(\phi'_{n,\mathbf{k}_n,\mathbf{j}_n}(x))^2|$$
$$\leq \sup_{\mathbf{k}_n} \sup_{\mathbf{j}_n} |\phi'_{n,\mathbf{k}_n,\mathbf{j}_n}(x)| \sum_{\mathbf{k}} \sum_{\mathbf{j}} p_{k_1} p_{k_2} \cdots p_{k_n} |\phi'_{n,\mathbf{k}_n,\mathbf{j}_n}(x)| \leq \frac{(1+\varepsilon)M}{\alpha(1+\varepsilon)^{n/p}}.$$

Thus, for n_0 big enough, we have

$$\sup_{x \in I} \sum_{\mathbf{k}_{n_0}} \sum_{\mathbf{j}_{n_0}} p_{k_1} p_{k_2} \cdots p_{k_{n_0}} |(\phi'_{n_0,\mathbf{k}_{n_0},\mathbf{j}_{n_0}}(x))^2| \leq \theta < 1. \tag{3.7.12}$$

Let

$$d = \sup_{x\in I}\sum_{\mathbf{k}_{n_0}}\sum_{\mathbf{j}_{n_0}} p_{k_1}p_{k_2}\cdots p_{k_{n_0}}|\phi''_{n_0,\mathbf{k}_{n_0},\mathbf{j}_{n_0}}(x)| < +\infty$$

$$B_n = \sup_{x\in I}|f'_n(x)|, n=0,1,2,\ldots.$$

Then (3.7.11) with $n=n_0$ implies

$$B_{l+n_0} \le B_l\theta + Md, l=0,1,2,\ldots.$$

Thus,

$$B_{l+2n_0} \le B_{l+n_0}\theta + Md \le (B_l\theta+Md)\theta + Md,$$

and

$$B_{l+mn_0} \le B_l\theta^{m-1} + Md(1+\theta+\ldots\theta^{(m-1)}) \le B_l + Md\frac{1}{1-\theta} = \bar{B}_l,$$

for $m=1,2,\ldots$. Thus the sequence $\{B_n\}_{n=0}^{\infty}$ is bounded by $\max\{\bar{B}_0,\bar{B}_1,\ldots,\bar{B}_{n_0-1}\}$. We have proved that the sequence $\{f_n\}$ is uniformly bounded and equicontinuous. By the Ascoli-Arzela theorem it contains a subsequence $\{f_{n_l}\}_{l\ge 0}$ uniformly convergent to a continuous function g. By our assumption, the random map has a unique invariant density f^*. Thus $\{f_{n_l}\}_{l\ge 0}$ converges to f^* in L^1. Thus, $g=f^*$ and $f_{n_l}\to f^*$ uniformly, as $l\to+\infty$. Since this argument applies to any subsequence of $\{f_n\}_{n\ge 0}$, the entire sequence converges uniformly to f^*, which is continuous. □

The following lemma can be proved by induction:

Lemma 3.39 *Let $F(x) = f(\phi(x))\phi'(x), x\in I$. Then*

$$F^{(s+1)} = f^{(s+1)}(\phi)(\phi')^{s+2} + \sum_{i\le s} f^{(i)}(\phi)\left[P_{s,i}(\phi^{(1)},\phi^{(2)},\ldots,\phi^{(s+2)})\right],$$

where $P_{s,i}$ is a polynomial of order $i+1$, $i=0,1,\ldots,s$.

Theorem 3.40 *Let $T=\{\tau_1,\tau_2,\ldots,\tau_K;p_1,p_2,\ldots,p_K\}$ be a random map such that $\tau_1,\ldots,\tau_K\in\mathcal{T}_1(I)$ are piecewise $C^r, r\ge 2$, piecewise onto and T satisfy Conditions A and B above (Section 3.7). Then, $T-$invariant density f^* is of class C^{r-2} and, for any $s\le r-2, (P_T^n\mathbf{1})^{(s)}\to f^{*(s)}$ uniformly as $n\to+\infty$.*

Proof *We proceed by induction. The first step, i.e., uniform boundedness of $(P_T^n\mathbf{1})^{(1)}$ and uniform convergence $P_T^n\mathbf{1}\to f^*$, has been proved in Lemma 3.38. Let us assume that $r\ge 3$ and that $\{(P_T^n\mathbf{1})^{(j)}\}_{n\ge 0}$ is uniformly bounded for $j=0,1,\ldots,s\le r-2, (P_T^n\mathbf{1})^{(j)}\to f^{*(j)}$ uniformly as $n\to+\infty$ for $j=0,1,\ldots,s-1$. We will show that* *the same is true for* $s+1$. *Let* $f_n = P_T^n \mathbf{1}, n = 0,1,\ldots$. *Using Lemma 3.39 and formula (3.7.10), we can write*

$$f_{n+l}^{(s+1)}(x) =$$

$$\sum_{\mathbf{k}_n}\sum_{\mathbf{j}_n} p_{k_1}p_{k_2}\cdots p_{k_n}\Bigg[f_l^{(s+1)}(\phi_{n,\mathbf{k}_n,\mathbf{j}_n}(x))(|\phi'_{n,\mathbf{k}_n,\mathbf{j}_n}(x)|)^{s+1}$$

$$+\sum_{i=0}^{s} f_l^{(i)}(\phi_{n,\mathbf{k},\mathbf{j}}(x))\left[P_{s,i}(\phi_{n,\mathbf{k}_n,\mathbf{j}_n}^{(1)}(x),\phi_{n,\mathbf{k}_n,\mathbf{j}_n}^{(2)}(x),\ldots,\phi_{n,\mathbf{k}_n,\mathbf{j}_n}^{(s+2)}(x))\right]\Bigg]$$

For n_0 *of formula (3.7.12), we have*

$$\sum_{\mathbf{k}_{n_0}}\sum_{\mathbf{j}_{n_0}} p_{k_1}p_{k_2}\cdots p_{k_{n_0}}(|\phi'_{n_0,\mathbf{k}_{n_0},\mathbf{j}_{n_0}}(x)|)^{(s+1)} \tag{3.7.13}$$

$$\leq \left(\frac{(1+\varepsilon)}{\alpha(1+\varepsilon)^{n_0/p}}\right)^s M = \theta_{s+1} < 1.$$

By the inductive assumption, $f_l^{(i)}$ *are uniformly bounded for* $i = 0,1,\ldots,s$. *Also* $\phi_{n_0,\mathbf{k}_{n_0},\mathbf{j}_{n_0}}^{(i)}, i = 1,2,\ldots s+2$ *are bounded. Thus, we can find a constant* $A_{s+1} > 0$ *such that*

$$\Bigg|\sum_{\mathbf{k}_{n_0}}\sum_{\mathbf{j}_{n_0}} p_{k_1}p_{k_2}\cdots p_{k_{n_0}}\sum_{i=0}^{s} f_l^{(i)}(\phi_{n_0,\mathbf{k}_{n_0},\mathbf{j}_{n_0}}(x)) \tag{3.7.14}$$

$$\left[P_{s,i}(\phi_{n_0,\mathbf{k}_{n_0},\mathbf{j}_{n_0}}^{(1)}(x),\ldots,\phi_{n_0,\mathbf{k}_{n_0},\mathbf{j}_{n_0}}^{(s+2)}(x))\right]\Bigg| \leq A_{s+1}$$

for all $l \geq 0$ *and* $x \in I$. *If we denote* $D_n^{(s+1)} = \sup_{x\in I}|f_n^{(s+1)}(x)|$, *we obtain*

$$D_{n_0+l}^{(s+1)} \leq D_l^{(s+1)}\theta_{s+1} + A_{s+1}$$

for all $l = 0,1,\ldots$.

As in Lemma 3.38 this implies that the sequence $\{f_n^{(s+1)}\}$ *is uniformly bounded. Thus, the sequence* $\{f_n^{(s)}\}$ *is uniformly bounded and equicontinuous. By the Ascoli-Arzela theorem it contains a subsequence* $\{f_{n_l}\}_{l\geq 0}$ *convergent uniformly to a continuous function g. Since* $f_{n_l}^{(s-1)} \to (f^*)^{(s-1)}$ *uniformly as* $l \to \infty, g = (f^*)^{(s)}$. *Since this argument applies to any subsequence of* $\{f_n^{(s)}\}_{n\geq 0}$, *the entire sequence* $\{f_n^{(s)}\}_{n\geq 0}$, *converges uniformly to* $(f^*)^{(s)}$ *which is continuous. This completes the proof of the theorem.*

Remark: If we omit the assumption **A** (4), then the same reasoning proves the smoothness of the density function on the subintervals of the set

$$I \setminus \bigcup_{n\geq 1} T^n(\{a_0,a_1,\ldots,a_q\}),$$

where $T^n(\{a_0,a_1,\ldots,a_q\}) = \bigcup_{\omega_n\in\Omega_n} T^n_{\omega_n}(\{a_0,a_1,\ldots,a_q\})$.

Example 3.8 *In [31] Lasota and Rusek created a model of an oil drilling operation using eventually piecewise expanding maps of an interval (see [8, Chapter 13] for a detailed description). The map τ models the process of the drill jumping up and falling back down. The more uniform is the invariant density of τ the more uniform is the tear of the drill, so knowing the invariant density is of practical importance. Since the parameters of drill movement are measured only with a certain accuracy and they vary during the operation it is more realistic to model the system with a random map. Instead of considering just one map τ we will consider a family of approximations τ_i to τ applied at random on each step of the iteration.*

For Froude number $\Lambda = 3$ we approximate τ by three eventually piecewise expanding, piecewise onto maps τ_1, τ_2 and τ_3. We define τ_i by formula (3.7.15) using constants: $a_1 = -1.40$, $a_2 = -1.41$, $a_3 = -1.39$, $b_1 = 6.5888$, $b_2 = -8.7850$, $b_3 = 11.7134$, $e_1 = 10$, $e_2 = 11$, $e_3 = 12$. The graph of τ_1 is shown in Figure 3.7a. The graphs of the others are indistinguishable from the one at the precision we can show. We define random map $T = \{\tau_1,\tau_2,\tau_3;p_1,p_2,p_3\}$, with $p_1 = 0.5$, $p_2 = 0.25$, $p_3 = 0.25$. We calculated densities $f_n = P_T^n\mathbf{1}$ numerically using Maple 11. f_5 is shown in Figure 3.7b. The Maple 11 program is available on request.

$$\begin{aligned}
\tau_i(x) =& \qquad\qquad\qquad\qquad\qquad\qquad\qquad\qquad\qquad\qquad\qquad (3.7.15)\\
& a_i(x-0.25)+(30.6667a_i+66.3382)(x-0.25)^2\\
& +(240a_i+562.5680)(x-0.25)^3+(533.3333a_i+1444.8607)(x-0.25)^4\,,\\
& \qquad\qquad\qquad\qquad\qquad\qquad\qquad\qquad \textit{for } 0\le x\le 0.25\,;\\
& (1+b_i(x-1)^{e_i})\left(0.9(x-1)-0.17(x-1)^2+\frac{3}{2}(1-\sqrt{1-\frac{4}{3}(1-x)}\right)\,,\\
& \qquad\qquad\qquad\qquad\qquad\qquad\qquad\qquad \textit{for } 0.25< x\le 1\,.
\end{aligned}$$

3.8 Applications in finance

In this section, we present some applications of the above results to the field of finance. First, we present a classical binomial model for the evolution of stock prices with which we evaluate one period option prices. Then, we present a binomial model for the evolution of interest rates with which we evaluate bond prices. We provide an example that shows that random maps with constant probabilities are useful alternative models for classical binomial models. Random map models are also useful for the evolution of any binomial type asset prices provided the random maps possess invariant measures. Given predetermined stationary densities of asset prices, one can construct binomial trees induced by random maps and analyze these predetermined densities using the theory of random maps (see [4] in Chapter 4). The density functions of invariant measures for random maps allow us to find the statistical behavior

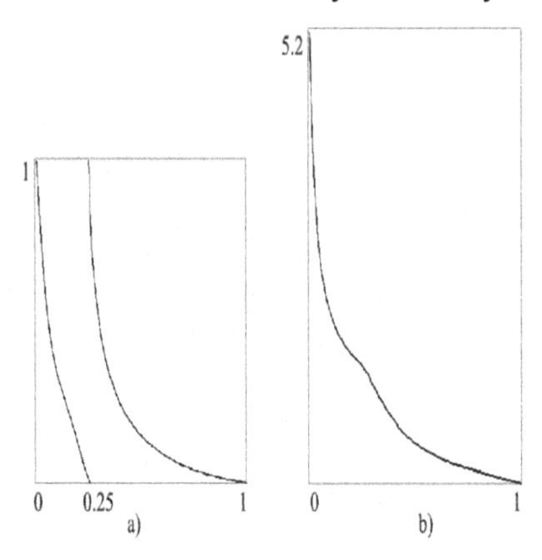

Figure 3.7 *Map τ_1 and an approximation to the invariant density of random map T.*

of asset prices on specified sets. Existence, computations and properties of invariant measures are important tools for the study of density functions for random maps (we have developed such tools in this chapter [see Sections 3.4 – 3.6 and Chapter 4]. (More general binomial models using position dependent random maps are presented in Chapter 4 and we show in that chapter that position dependent random maps are useful tools for the study of both stocks and derivative securities).

Classical binomial models are discrete time models of an asset price or a rate in which at each period there are just two possibilities: an up move and a down move. For example, if the value of one share of a stock at time $t = n$ is $S(n)$ then the value $S(n+1)$ at time $t = n+1$ may be either $S(n+1) = u \cdot S(n)$ in the up move case or $S(n+1) = d \cdot S(n)$ in the down move case where u and d are some constant nonzero factors. Binomial models are simple but very useful models that are widely used in finance. The current stock value $S(n)$ is known to all investors, but the future value $S(n+1)$ is uncertain. The difference $S(n+1) - S(n)$ as a fraction of the initial value represents the return on the stock, $K_S = \frac{S(n+1)-S(n)}{S(n)}$, which is also uncertain.

Example 3.9 *Let* $S(0) = 30$ *dollars and* $S(1) = \begin{cases} 33 & \text{with probability } \frac{3}{5} \\ 26 & \text{with probability } \frac{2}{5} \end{cases}$

The return on stock will then be $K_S = \begin{cases} \frac{33-30}{30} = 10\% & \text{with probability } \frac{3}{5} \\ \frac{26-30}{30} = -7.5\% & \text{with probability } \frac{2}{5} \end{cases}$

Now we consider a risk free asset such as an amount in a bank account which earns a fixed interest. As an alternative to keeping money in a bank account, investors may choose to invest in bonds [11]. The current value $B(n)$ of the risk free asset is known to all investors. The future value $B(n+1)$ of the risk free asset is also known.

The difference $B(n+1)-B(n)$ as a fraction of the initial value represents the return on the risk free asset, $K_B = \frac{B(n+1)-B(n)}{B(n)}$. Assume that $S(t) > 0, B(t) > 0$. Now we consider a portfolio (x,y) of x shares of the above stock and y units of risk free asset. Let $V(t)$ be the value of the portfolio (x,y) at time t; that is, $V(t) = xS(t) + yB(t)$. The difference $V(n+1)-V(n)$ as a fraction of the initial value represents the return K_V on the portfolio, $K_V = \frac{V(n+1)-V(n)}{V(n)}$, which is also uncertain.

Example 3.10 *Given the stock price in Example 3.9. Let the risk free asset price be* $B(0) = 100$ *dollars and* $B(1) = 112$ *dollars. The return on the risk free asset is* $K_B = \frac{B(1)-B(n)}{B(0)} = \frac{12}{100} = 12\%$. *The value of the portfolio at time* 0 *with* $x = 20$ *and* $y = 15$ *is* $V(0) = xS(0) + yB(0) = 20 \times 30 + 15 \times 100 = 2100$. *The time* 1 *value of this portfolio is* $V(1) = \begin{cases} 20 \times 33 + 15 \times 112 = 2,340 & \text{with probability } \frac{3}{5} \\ 20 \times 26 + 15 \times 112 = 2,200 & \text{with probability } \frac{2}{5} \end{cases}$

The return on the portfolio will be $K_V = \begin{cases} \frac{2,340-2100}{2100} = 16.19\% & \text{with probability } \frac{3}{5} \\ \frac{2,220-2100}{2100} = 9.52\% & \text{with probability } \frac{2}{5} \end{cases}$

Modern financial markets contain many other instruments besides stocks and bonds. Some of these instruments are called derivatives, because their value depends on the value of another instrument on the market. Option is the most popular derivative. A European option is a contract giving its holder the right to buy or sell the underlying security at a specified date.

3.8.1 *One period binomial model for stock option*

In this section, we present one period binomial option model. Our presentation is based on the presentation in [22]. Let $S(0)$ be the value of a risky asset (stock) at time $t = 0$. At time 1, there are just two possibilities: an up move and a down move. We call them up state and down state, and denote them by $S(1,\uparrow)$ and $S(1,\downarrow)$ respectively. Without loss of generality we assume that $S(1,\uparrow) > S(1,\downarrow)$. Now we assume that

$$S(1,\downarrow) \leq (1+r)S(0) \leq S(1,\uparrow), \tag{3.8.1}$$

where r is the interest rate, or another name is the risk free rate. The assumption is because that if nothing changes, after one period, the risky asset price is $(1+r)S(0)$ However, the price can go up or down. The inequality just shows that if the price goes up, it is higher than the original price, and if the price goes down, it is lower than the original price.

Consider a call option with the above underlying asset S with strike price K at the expiry date $T = 1$. Let $X(0)$ denote the current ($t = 0$) value of the option and $X(1)$ denote the value of the option at expiry $T = 1$. $X(1)$ is the price of the call option at time $T = 1$, or the payoff to the option contract at time $T = 1$. $X(1)$ is either zero, or the difference between the spot price and forward price (or strike price). $X(1)$ has two choices: $X(1,\uparrow)$ and $X(1,\downarrow)$, where:

$$\begin{aligned} X(1,\uparrow) &= \max[S(1,\uparrow)-K,0] \\ X(1,\downarrow) &= \max[S(1,\downarrow)-K,0] \end{aligned} \tag{3.8.2}$$

Suppose we form a portfolio containing H_1 amount of underlying risky asset (stock) S and the dollar amount H_0 in risk free asset in such a way that

$$X(1) = (1+r)H_0 + H_1 S(1). \tag{3.8.3}$$

From equation (3.8.3) we get,

$$\begin{aligned} X(1,\uparrow) &= H_0(1+r) + H_1 S(1,\uparrow) \\ X(1,\downarrow) &= H_0(1+r) + H_1 S(1,\downarrow) \end{aligned} \tag{3.8.4}$$

From equation (3.8.4) we obtain

$$X(1,\uparrow) - X(1,\downarrow) = H_1[S(1,\uparrow) - S(1,\downarrow)] \tag{3.8.5}$$

and thus,

$$H_1 = \frac{X(1,\uparrow) - X(1,\downarrow)}{S(1,\uparrow) - S(1,\downarrow)} \tag{3.8.6}$$

From equation (3.8.4) we obtain

$$X(1,\downarrow) = H_0(1+r) + \frac{X(1,\uparrow) - X(1,\downarrow)}{S(1,\uparrow) - S(1,\downarrow)} S(1,\downarrow) \tag{3.8.7}$$

Therefore,

$$\begin{aligned} -H_0(1+r) &= \frac{X(1,\uparrow) - X(1,\downarrow)}{S(1,\uparrow) - S(1,\downarrow)} S(1,\downarrow) - X(1,\downarrow) \\ &= \frac{X(1,\uparrow)S(1,\downarrow) - X(1,\downarrow)S(1,\downarrow) - X(1,\downarrow)S(1,\uparrow) + X(1,\downarrow)S(1,\downarrow)}{S(1,\uparrow) - S(1,\downarrow)} \\ &= \frac{X(1,\uparrow)S(1,\downarrow) - X(1,\downarrow)S(1,\uparrow)}{S(1,\uparrow) - S(1,\downarrow)} \end{aligned}$$

Let $R = 1 + r$. Then,

$$H_0 = \frac{X(1,\uparrow)S(1,\downarrow) - X(1,\downarrow)S(1,\uparrow)}{[S(1,\uparrow) - S(1,\downarrow)](1+r)} = \frac{X(1,\uparrow)S(1,\downarrow) - X(1,\downarrow)S(1,\uparrow)}{[S(1,\uparrow) - S(1,\downarrow)]R} \tag{3.8.8}$$

The current (at time 0) value of the portfolio containing H_1 amount of underlying risky asset S and the dollar amount H_0 in risk free asset is $H_0 + H_1 S(0)$. If there are to be no risk free arbitrage opportunities, then by law of one price, we get

$$H_0 = H_0 + H_1 S(0). \tag{3.8.9}$$

Using equation (3.8.7) and equation (3.8.8) we obtain:

$$\begin{aligned} X(0) &= H_0 + H_1 S(0) \\ &= \frac{X(1,\uparrow)S(1,\downarrow) - X(1,\downarrow)S(1,\uparrow)}{[S(1,\uparrow) - S(1,\downarrow)]R} + \frac{X(1,\uparrow) - X(1,\downarrow)}{S(1,\uparrow) - S(1,\downarrow)} S(0) \\ &= \frac{1}{R}[X(1,\uparrow)\frac{S(0)R - -S(1,\downarrow)}{S(1,\uparrow) - S(1,\downarrow)} + X(1,\downarrow)\frac{S(1,\uparrow) - S(0)R}{S(1,\uparrow) - S(1,\downarrow)}] \end{aligned} \quad (3.8.10)$$

Let

$$\pi = \frac{-S(1,\downarrow) + S(0)R}{S(1,\uparrow) - S(1,\downarrow)}. \quad (3.8.11)$$

Then

$$X(0) = \frac{1}{R}[\pi X(1,\uparrow) + (1-\pi)X(1,\downarrow)] \quad (3.8.12)$$

Equation (3.8.12) is the Cox–Ross–Rubinstein (CRR) option pricing model in one period. π in equation (3.8.11) is called the risk neutral probability, and it is always in $[0,1]$. From equation (3.8.2), $RS(0) \leq S(1,\uparrow)$. Therefore, $-S(1,\downarrow) + S(0)R \leq S(1,\uparrow) - S(1,\downarrow)$ which implies that π cannot be greater than 1. Since both the denominator and numerator in equation (3.8.11) are greater than or equal to zero, π cannot be negative.

Let the price of the risky asset S go up by a constant factor u with probability p and go down by a constant factor d, that is, $S(1,\uparrow) = uS(0), S(1,\downarrow) = dS(0)$. Then

$$\begin{aligned} \pi &= \frac{S(0)R - S(1,\downarrow)}{S(1,\uparrow) - S(1,\downarrow)} \\ &= \frac{S(0)R - S(0)d}{S(0)u - S(0)d} \\ &= \frac{R-d}{u-d} \end{aligned}$$

and

$$1 - \pi = \frac{u-d-R+d}{u-d} = \frac{u-R}{u-d}$$

From equation (3.8.12) we obtain,

$$X(0) = \frac{1}{R}[\frac{R-d}{u-d}X(1,\uparrow) + \frac{u-R}{u-d}X(1,\downarrow)] \quad (3.8.13)$$

Example 3.11 *Suppose there is a risky asset and its current value is $S(0) = 1000$. One year later there are two possible prices, either $S(1) = 1020$ or $S(1) = 1300$. Consider a call option with strike price at time 1 is $K = 1200$. Given that the spot rate is 10% during this year, then we can calculate u, d, R and π:*

$$u = \frac{S(1,\uparrow)}{S(0)} = 1300/1000 = 1.3, d = \frac{S(1,\downarrow)}{S(0)} = 1020/1000 = 1.02, R = 1 + 0.1 = 1.1$$

$$\pi = \frac{R-d}{u-d} = \frac{R-d}{u-d} = \frac{1.1-1.02}{1.3-1.02} = 1.2857$$

$$X(0) = \frac{1}{R}[\pi X(1,\uparrow) + (1-\pi)X(1,\downarrow)] = \frac{1}{1.1}(0.2857 \times 100) = 25.97$$

3.8.2 The classical binomial interest rate models and bond prices

Interest rate plays an important role in the financial markets. If the interest rate is fixed for the entire time of bonds and annuities then the valuations of bonds and annuities are simple. However, variable interest rates are a fact of modern financial markets. Valuations of annuities and coupon bearing bonds depends on these fluctuating interest rates. Valuations of bonds and annuities becomes complicated if the interest rate is variable. If at every time step the interest rate either rises by a constant factor u of the interest rate at the previous period or falls by a constant factor d, then the classical binomial model is said to describe the evolution of interest rates. We can evaluate bonds and annuities using these classical models. Note that if the functions u and d are constant then there is a possibility that eventually the interest rate become zero, or interest rate increase without bound. These are unusual cases. In reality, the probability of the interest rate going up or down is not constant and may depend on current interest rate. Moreover, it is reasonable to assume that the factors u and d are functions of interest rates, $u(x) : (0,1) \to (1,\infty)$ and $d(x) : (0,1) \to (0,1)$. That is, at time t, u and d depend on the interest rate at time $t-1$.

The price P of a coupon bearing bond at time $t = 0$ with face value F, number of coupon payment periods n, coupon rate r, the redemption value C and yield rate i is

$$P = C(\frac{1}{1+i})^n + Fr\sum_{k=1}^{n}(\frac{1}{1+i})^k.$$

For example, consider a \$1200 bond with semiannual coupon with coupon rate 8% convertible semiannually which will be redeemed at \$1500. If the yield rate is 9% convertible semiannually and the redemption date is 20 years from now, then the price of the bond at time $t = 0$ is

$$P = 1500(\frac{1}{1+.045})^{40} + 1200(.04)\sum_{k=1}^{40}(\frac{1}{1+.045})^k = 1141.17.$$

In the above example if the initial coupon rate is 8% convertible semiannually and each coupon rate is 2% greater than the preceeding coupon, then the price of the bond at time $t = 0$ is

$$P = 1500(\frac{1}{1+.045})^{40} + 1200(.04)\frac{1-(\frac{1.02}{1.045})^{40}}{.045-.02} = 1449.012302.$$

On the other hand, if the coupon rate is fixed (8% convertible semiannually) and the yield rate is changed to 9% convertible semiannually for the first 10 years and 10% convertible semiannually for the next 10 years, then the price of the bond is

$$
\begin{aligned}
P &= 1500(\frac{1}{1+.045})^{20}(\frac{1}{1+.05})^{20} \\
&+ 1200(.04)\left((\frac{1}{1+.05})^{20}\sum_{k=1}^{20}(\frac{1}{1+.05})^{k}+\sum_{k=1}^{20}(\frac{1}{1+.45})^{k}\right) \\
&= 1106.826343.
\end{aligned}
$$

Thus we see that the price at time 0 of a bond changes if the coupon rate or yield rate (or both) is not constant.

Let $i(t)$ be the yield rate during the period from time $t-1$ to t and $i(t)$ an i.i.d. random variable with mean μ and variance σ. If the coupon rate r is constant for the entire term, then the price of the bond with face value F, number of coupon payment periods n, and the redemption value C is

$$
P = C\prod_{t=1}^{n}\frac{1}{1+i(t)} + Fr\sum_{t=1}^{n}\prod_{s=1}^{t}\frac{1}{1+i(s)}.
$$

Let $r(t)$ be the coupon rate during the period from time $t-1$ to t and $r(t)$ an i.i.d. random variable with mean μ_1 and variance σ_1. If the yield rate i is constant for the entire term, then the price of the bond with face value F, number of coupon payment periods n, and the redemption value C is

$$
P = C(\frac{1}{1+i})^n + \sum_{k=1}^{n}\frac{Fr(k)}{(1+i)^k}.
$$

Let $i(0)$ be the interest rate from time $t=0$ to time 1. For $t=1,2,\ldots,N-1$, let $i(t)$ be the interest from time t to time $t+1$ and let there be two possibilities for $i(t)$: the interest rate either goes up by a factor u with probability $p_u(t)$ or the rate goes down by a factor d with probability $p_d(t) = 1-p_u(t)$. That is, $i(t) := i_u(t) = u\cdot i(t-1)$ or $i(t) := i_d(t) = d\cdot i(t-1)$. Consider a coupon bond with face value F, number of coupon payment periods n, coupon rate r, the redemption value C. Assume that the yield rate follows the binomial tree $i(t)$. Let $V(t)$ be the value of the bond (that is, the present value of the future cash flows) at the beginning of the period. We define the following values at the end of the period:

$$
\begin{aligned}
R_U(t) &= \text{cash flow in a up move} \\
R_D(t) &= \text{cash flow in a down move} \\
V_U(t) &= \text{value in a up move} \\
V_D(t) &= \text{value in a down move}
\end{aligned}
$$

Then, we can use the following recursive method to evaluate bond price:

$$
\begin{aligned}
R_U(N) &= C+Fr;\\
R_D(N) &= C+Fr;\\
V_U(N) &= 0;\\
V_D(N) &= 0;\\
V_U(t) &= \frac{p_u(t)\,(R_U(t+1)+V_U(t+1))+p_d(t)\,(R_D(t+1)+V_D(t+1))}{1+i_u(t)}, t=N-1,N-2,\ldots,1;\\
V_D(t) &= \frac{p_u(t)\,(R_U(t)+V_U(t))+(1-p_d(t))\,(R_D(t)+V_D(t))}{1+i_d(t)}, t=N-1,N-2,\ldots,1;\\
R_U(t) &= R_D(t)=Fr, t=N-1,N-2,\ldots,1;\\
V(t) &= \frac{p_u(t)\,(R_U(t+1)+V_U(t+1))+p_d(t)\,(R_D(t+1)+V_D(t+1))}{1+i(t)}, t=N-1,N-2,\ldots,1;\\
V(0) &= \frac{p_u(1)\,(R_U(1)+V_U(1))+p_d(1)\,(R_D(1)+V_D(1))}{1+i(0)};
\end{aligned}
$$

In the following example, we explain the above recursive method for stochastic interest and bond evaluation for a binomial model.

Example 3.12 *Consider the following classical binomial interest rate: the yield rate at time* $n=0$ *is* 8% *convertible semiannually,* $u=1.2, d=\frac{1}{1.2}, p_u=0.6, p_d=1-p_u=0.4$. *Then the yield rate at time* $n=1,2,3$ *follows the following binomial tree*

$$
i(0)=.04
\begin{cases}
\nearrow^{p_u=0.6,u=1.2}_{\longrightarrow}\; i(1)=0.048
\begin{cases}
\nearrow^{p_u=0.6,u=1.2}_{\longrightarrow}\; i(2)=0.0579
\begin{cases}
\nearrow^{p_u=0.6,u=1.2}_{\longrightarrow}\; i(3)=0.06912\\
\searrow^{p_d=0.4,d=1/1.2}_{\longrightarrow}\; i(3)=0.048
\end{cases}\\
\searrow^{p_d=0.4,d=1/1.2}_{\longrightarrow}\; i(2)=0.03996
\begin{cases}
\nearrow^{p_u=0.6,u=1.2}_{\longrightarrow}\; i(3)=0.048\\
\searrow^{p_d=0.4,d=1/1.2}_{\longrightarrow}\; i(3)=0.0333
\end{cases}
\end{cases}\\
\searrow^{p_d=0.4,d=1/1.2}_{\longrightarrow}\; i(1)=0.0333
\begin{cases}
\nearrow^{p_u=0.6,u=1.2}_{\longrightarrow}\; i(2)=0.03996
\begin{cases}
\nearrow^{p_u=0.6,u=1.2}_{\longrightarrow}\; i(3)=0.047952\\
\searrow^{p_d=0.4,d=1/1.2}_{\longrightarrow}\; i(3)=0.0333
\end{cases}\\
\searrow^{p_d=0.4,d=1/1.2}_{\longrightarrow}\; i(2)=0.02775
\begin{cases}
\nearrow^{p_u=0.6,u=1.2}_{\longrightarrow}\; i(3)=0.0333\\
\searrow^{p_d=0.4,d=1/1.2}_{\longrightarrow}\; i(3)=0.023125
\end{cases}
\end{cases}
\end{cases}
\qquad (3.8.14)
$$

For each starting yield rate we have all possible paths the yield rate might take. A typical path is $0.04 \to 0.04800 \to 0.03996 \to 0.048$, *that is,*

$$
\begin{aligned}
&i_u(0)=i_d(0)=0.04\\
&i_u(1)=0.04800\ (p_u=.6),\ \ i_d(1)=0.0333\ (p_d=.4)\\
&i_d(2)=0.03996\ (p_d=.4),\ \ i_u(2)=0.0576\ (p_u=.6)\\
&i_u(3)=0.04800\ (p_u=.6),\ \ i_d(3)=0.0333\ (p_d=.4)
\end{aligned}
$$

If we consider a \$1000 *par-value semiannual coupon bond with fixed coupon rate* 7.6% *and maturity date one and half years from now, then*

$$
\begin{aligned}
R_U(3) &= 1000+1000(.038)=1038\\
R_D(3) &= 1000+1000(.038)=1038\\
V_U(3) &= 0
\end{aligned}
$$

$$
\begin{aligned}
V_D(3) &= 0 \\
R_U(2) &= 38.00 \\
R_D(2) &= 38.00 \\
R_U(1) &= 38.00 \\
R_D(1) &= 38.00 \\
V_U(2) &= \frac{p_u(2+1)\,(V_U(2+1)+R_U(2+1))+p_d(2+1)\,(V_D(2+1)+R_D(2+1))}{1+i_u(2)} \\
&= \frac{0.6\,(1038+0)+0.4\,(1038+0))}{1+0.0576} = 981.47 \\
V_D(2) &= \frac{p_u(2+1)\,(V_U(2+1)+R_U(2+1))+p_d(2+1)\,(V_D(2+1)+R_D(2+1))}{1+i_d(2)} \\
&= \frac{0.6\,(1038+0)+0.4\,(1038+0))}{1+0.04} = 998.08
\end{aligned}
$$

$$
\begin{aligned}
V_U(1) &= \frac{p_u(1+1)\,(V_U(1+1)+R_U(1+1))+p_d(1+1)\,(V_D(1+1)+R_D(1+1))}{1+i_u(1)} \\
&= \frac{0.6\,(981.47+38)+0.4\,(998.08+38))}{1+0.048} = 979.1152376 \\
V_D(1) &= \frac{p_u(1+1)\,(V_U(1+1)+R_U(1+1))+p_d(2+1)\,(V_D(1+1)+R_D(1+1))}{1+i_d(1)} \\
&= \frac{0.6\,(981.47+38)+0.4\,(998.08+38))}{1+0.03333} = 993.0443908 \\
V(0) &= \frac{p_u(0+1)\,(V_U(0+1)+R_U(0+1))+p_d(0+1)\,(V_D(0+1)+R_D(0+1))}{1+i(0)} \\
&= \frac{0.6\,(979.1152376+38)+0.4\,(993.0443908+38))}{1+0.03333} = 989.7289258
\end{aligned}
$$

In the following we show that a random map model is a useful alternative for the classical interest rate binomial model in this example.

3.8.3 *Random maps with constant probabilities as useful alternative models for classical binomial models*

Random maps with constant probabilities are special cases of position dependent random maps. In Chapter 4, we show that position dependent random maps are mathematical models for generalized binomial trees in finance. Given a starting price of a risky asset, this generalized binomial tree describes all the possible paths the asset price might take and the invariant density of random maps allows us to find the probabilities of a set of asset prices. Moreover, given predetermined stationary densities of asset prices, one can construct generalized binomial trees induced by position dependent random maps and analyze these predetermined densities using the theory of random maps. In this section, we present an example of a random map with constant probabilities and study the evolution of interest rates. We show that random maps (with constant probabilities) are useful alternative models for classical binomial models. For details see Chapter 4 and [4 in Chapter 4].

Let $u, d, p_u, p_d, \tau_u, \tau_d$ are functions on $[0,1]$ defined by

$$\begin{aligned}
u(x) &= 1.2 \\
d(x) &= \frac{1}{1.2} \\
p_u(x) &= .6 \\
p_d(x) &= 1 - p_u(x) = 0.4 \\
\tau_u(x) &= x\,u(x) \\
\tau_d(x) &= x\,d(x)
\end{aligned}$$

Consider the random map $T:[0,1] \to [0,1]$ defined by

$$T = \{\tau_u(x), \tau_d(x); p_u, p_d\}.$$

The random map T satisfies the average expanding condition: $\frac{p_u}{|u'|} + \frac{p_d}{|d'|} = \frac{.6}{1.2} + \frac{.4}{\frac{1}{1.2}} = 0.98 < 1$. By Pelikan's result (see equation (3.4.8)), the random map T admits an absolutely continuous invariant measure $\mu = f^*\lambda$ where f^* is the invariant density of T. The invariant density allow us to find the probability: $\mu\{x : T(x) \in (\delta_1, \delta_2)\} = \mu(\delta_1, \delta_2)$ This random map generates binomial interest rate paths. If the starting interest rate is $x = .20$ or (equivalent to .04 or 4%) then the above random map will generate the binomial tree in the above example (Example 3.12) as follows:

$i(0) = .20$

- $\nearrow p_u{=}0.6, u{=}1.2 \longrightarrow \tau_u(0.19980) = 0.240$
  - $\nearrow p_u{=}0.6, u{=}1.2 \longrightarrow \tau_u(0.240) = 0.2895$
    - $\nearrow p_u{=}0.6, u{=}1.2 \longrightarrow \tau_u(0.2895) = 0.34560$
    - $\searrow p_d{=}0.4, d{=}1/1.2 \longrightarrow \tau_d(0.2895) = 0.240$
  - $\searrow p_d{=}0.4, d{=}1/1.2 \longrightarrow \tau_d(0.240) = 0.19980$
    - $\nearrow p_u{=}0.6, u{=}1.2 \longrightarrow \tau_d(0.19995) = 0.1665$
    - $\searrow p_d{=}0.4, d{=}1/1.2 \longrightarrow \tau_u(0.19980) = 0.239760$
- $\searrow p_d{=}0.4, d{=}1/1.2 \longrightarrow \tau_d(0.20) = 0.1665$
  - $\nearrow p_u{=}0.6, u{=}1.2 \longrightarrow \tau_u(0.1665) = 0.19980$
    - $\nearrow p_u{=}0.6, u{=}1.2 \longrightarrow \tau_u(0.19980) = 0.239760$
    - $\searrow p_d{=}0.4, d{=}1/1.2 \longrightarrow tau_d(0.19980) = 0.1665$
  - $\searrow p_d{=}0.4, d{=}1/1.2 \longrightarrow \tau_d(0.1665) = 0.13875$
    - $\nearrow p_u{=}0.6, u{=}1.2 \longrightarrow \tau_u(0.13875) = 0.1665$
    - $\searrow p_d{=}0.4, d{=}1/1.2 \longrightarrow \tau_d(0.13875) = 0.115625$

(3.8.15)

For interest rates we re-scale the unit interval $[0,1]$ in a meaningful way:

$$0 \text{ eqiv. to } 0, .20 \text{ eqiv. to } 0.04(4\%), \ldots, 1 \text{ eqiv. to } 0.20(20\%)$$

and we obtain the following binomial tree of interest rates:

$i(0) = .04$

- $\nearrow p_u{=}0.6, u{=}1.2 \longrightarrow \tau_u(.04) = 0.048$
  - $\nearrow p_u{=}0.6, u{=}1.2 \longrightarrow \tau_u(0.048) = 0.0579$
    - $\nearrow p_u{=}0.6, u{=}1.2 \longrightarrow \tau_u(0.0579) = 0.06912$
    - $\searrow p_d{=}0.4, d{=}1/1.2 \longrightarrow \tau_d(0.0579) = 0.048$
  - $\searrow p_d{=}0.4, d{=}1/1.2 \longrightarrow \tau_d(0.048) = 0.03996$
    - $\nearrow p_u{=}0.6, u{=}1.2 \longrightarrow \tau_d(0.03996) = 0.0333$
    - $\searrow p_d{=}0.4, d{=}1/1.2 \longrightarrow \tau_u(0.03996) = 0.047952$
- $\searrow p_d{=}0.4, d{=}1/1.2 \longrightarrow \tau_d(0.04) = 0.0333$
  - $\nearrow p_u{=}0.6, u{=}1.2 \longrightarrow \tau_u(0.0333) = 0.03996$
    - $\nearrow p_u{=}0.6, u{=}1.2 \longrightarrow \tau_u(0.03996) = 0.047952$
    - $\searrow p_d{=}0.4, d{=}1/1.2 \longrightarrow tau_d(0.03996) = 0.0333$
  - $\searrow p_d{=}0.4, d{=}1/1.2 \longrightarrow \tau_d(0.0333) = 0.02775$
    - $\nearrow p_u{=}0.6, u{=}1.2 \longrightarrow \tau_u(0.02775) = 0.0333$
    - $\searrow p_d{=}0.4, d{=}1/1.2 \longrightarrow \tau_d(0.02775) = 0.023125$

(3.8.16)

A typical path from this random map is

$$.20 \to 0.240 \to .20 \to 0.1665$$

which is equivalent to

$$.04 \to 0.048 \to .04 \to 0.0333$$

which is equivalent to

$$4\% \to 4.8\% \to 4\% \to 3.33\%.$$

We can use this interest rate path for the calculation of bond prices in the way we did in the above example (Example 3.12). Random map models can be applied for the evolution of stock and other assets. In Chapter 4, we will consider generalized random map model binomial models for the evaluation of asset prices where factors u and d are not necessarily constants; they are functions of current asset prices.

References

[1] Adler, R. and Flatto, L., *Geodesic flows, interval maps and symbolic dynamics*, Bull. Amer. Math. Soc. 25 (1991), 229-334.

[2] Arnold, L., *Random dynamical systems*, Springer, Berlin, 1998.

[3] Bahsoun, W., Góra, P., and Boyarsky, A., *Markov switching for position dependent random maps with application to forecasting in financial markets*, SIAM J. Appl. Dyn. Syst. 4 (2005), no. 2, 391–406.

[4] Bahsoun, W., Góra, P., and Boyarsky, A., *Stochastic perturbations for position dependent random maps*, Stoch. Dynam. 3, no. 4 (2003), 545-557.

[5] Bahsoun, W. and Góra, P., *Weakly convex and concave random maps with position dependent probabilities*, Stoch. Anal. App. 21, no. 5 (2003), 983-994.

[6] Bahsoun, W. and Góra, P., *Position dependent random maps in one and higher dimensions*, Studia Math. 166 (2005), 271–286.

[7] Barnsley, M., *Fractals everywhere*, Academic Press, London, 1998.

[8] Boyarsky, A. and Góra, P., *Laws of chaos*, Birkhaüser, 1997.

[9] Boyarsky, A. and Góra, P., *A dynamical model for interference effects and two slit experiment of quantum physics*, Phys., Lett. A 168 (1992), 103-112.

[10] Boyarsky, Abraham, Góra, Paweł, and Islam, Md. Shafiqul, *Randomly chosen chaotic maps give rise to nearly ordered behavior*, Phys. D 210 (2005), no. 3-4.

[11] Capinski, M. and Zastawniak, T., *An introduction to financial engineering*, Springer, 2003.

[12] Chatelin, Francoise, *Eigenvalues of matrices*, John Wiley and Sons, 114-117, 1993.

[13] Dunford, N. and Schwartz, J.T., *Linear operators, Part I*, Wiley Interscience (Wiley Classics Library): Chichester, 1988.

[14] Froyland, G., *Ulam's method for random interval maps*, Nonlinearity, 12 (1999) 1029-1052.

[15] Góra, Pawel, *Countably piecewise expanding transformations without absolutely continuous invariant measure*, Proceedings of Stefan Banach Center - Semester on Dynamical Systems and Ergodic Theory, Warsaw 1986 , PWN - Polish Scientific Publishers, 1989, pp.113-117.

[16] Góra, Pawel and Boyarsky, Abraham, *Absolutely continuous invariant measures for random maps with position dependent probabilities*, Math. Anal. and Appl. 278 (2003), 225-242.

[17] Góra, Pawel and Boyarsky, Abraham, *Attainable densities for random maps*, J. Math. Anal. Appl., 317 (2006), no. 1, 257–270.

[18] Góra, P., and Boyarsky, A., *Why computers like Lebesgue measure*, Comput. Math. Applic., 16, (1988), 321-329.

[19] Góra, P., Boyarsky, A., and Islam, M.S. *Invariant densities of random maps have lower bounds on their supports*, J. Appl. Math. Stoch. Anal. 2006, Art. ID 79175, 13 pp.

[20] Góra, P., Boyarsky, A., Islam, M.S., and Bahsoun,W., *Absolutely continuous invariant measures that cannot be observed experimentally*, SIAM J. Appl. Dyn. Syst. 5 (2006), no. 1, 84–90.

[21] Halfant, M., *Analytic properties of Rényi's invariant density*, Israel J. Math. 27 (1977), 1-20.

[22] Hoek, J. van der and Elliot, R. J., *Binomial models in finance*, Springer, 2006.

[23] Hunt, B.R., *Estimating invariant measures and Lyapunov exponents*, Ergod. Theor. Dynam. Syst. 16 (1996), 735-749.

[24] Islam, M.S. *Existence, approximation and properties of absolutely continuous invariant measures for random maps*, Ph.D. Thesis, Concordia University, Canada, 2004.

[25] Islam, M.S.,Góra, P., and Boyarsky, A. *A generalization of Straube's Theorem: existence of absolutely continuous invariant measures for random maps*, J. Appl. Math. Stoch. Anal. (2005), no. 2, 133–141.

[26] Islam, Shafiqul and Góra, Pawel, *Smoothness of invariant densities for random maps*, Dynamics of Continuous, Dyn. Contin. Discrete Impuls. Syst. Ser. A Math. Anal. 17 (2010), no. 2, 249–262.

[27] Jabloński, M., Góra, P., and Boyarsky, A., *A general existence theorem for absolutely continuous invariant measures on bounded and unbounded intervals*, Nonlinear World 3 (1996), 183-200.

[28] Keller, G *Piecewise monotonic transformations and exactness* Collection: Seminar on probability, Rennes 1978 (French); Exp. no. 6, 32 pp., Univ. Rennes, Rennes, 1978.

[29] Kowalski, Z. S., *Invariant measure for piecewise monotonic transformation has a lower bound on its support*, Bull. Acad. Aolon. Sci. Ser. Sci. Math. 27 (1979b), 53-57.

[30] Lasota, A. and Yorke, J.A., *On the existence of invariant measures for piece-*

wise monotonic transformations , Trans. Amer. Math. Soc 186 (1973), 481-488.

[31] Li, T.-Y., *Finite approximation for the Frobenius-Perron operator: a solution to Ulam's conjecture*, J. Approx. Theory 17 (1976), 177-186.

[32] Morita, T. *Random iteration of one-dimensional transformations*, Osaka J. Math. (1985), 22, 489-518.

[33] Motwani, R. and Raghavan, P., *Randomized algorithms*, Cambridge University Press, 1995.

[34] Pelikan, S., *Invariant densities for random maps of the interval*, Proc. Amer. Math. Soc. 281 (1984), 813-825.

[35] Rényi, A., *Representation for real numbers and their ergodic properties*, Acta Math. Acad. Sci. Hungar. 8 (1957), 477-493, 27-30.

[36] Schenk-Hoppe, K. R., *Random dynamical systems in economics*, working paper series, ISSN 1424-0459, Institute of Empirical Research in Economics, University of Zurich, Dec 2000.

[37] Schweitzer, P. J., *Perturbation theory and finite Markov chains*, J. Appl. Probab. 5, 401-404.

[38] Slomczynski, W., Kwapien, J. and Zyczkowski, K., *Entropy computing via integration over fractal measures*, Chaos 10 (2000), 180-188.

[39] Straube, E., *On the existence of Invariant, Absolutely Continuous Measures.*, Comm. Math. Phys. 81 (1981), 27-30.

[40] Ulam, S. M., *A collection of mathematical problems*, Interscience Tracts in Pure and Applied Math., 8, Interscience, New York, 1960.

[41] Ulam, S. M. and von Neumann, J., *Random ergodic theorem*, Bull. Amer. Math. Soc. 72 (1952), 46-66.

[42] Yosida, K and Hewitt, E., *Finitely additive measures*, Trans. Amer. Math. Soc. 72 (1952), 46-66.

Chapter 4

Position Dependent Random Maps

4.1 Chapter overview

In the previous chapter, we presented random maps where the probabilities of switching from one map to another in the process of iteration are constants. In this chapter, we present more general random maps where the probabilities of switching from one map to another are position dependent. Random maps with position dependent probabilities provide a useful framework for analyzing various physical, engineering, social, and economic phenomena. There are useful techniques in the theory of position dependent random maps which can be implemented in finance. For example, position dependent random maps are mathematical models for generalized binomial trees in finance. In Chapter 3, we have presented random maps with constant probabilities for the evolution of financial securities where both up move factor and down move factor are constants and they do not depend on the current values of the securities. In this chapter, we show that position dependent random map models are more general models and we can study financial securities where both up move factor and down move factor depend on the current values of the securities. The density functions of invariant measures for position dependent random maps are useful tools for the study of long term statistical behavior of some financial securities. Position dependent random maps for piecewise C^2 expanding maps were first introduced by Góra and Boyarsky in [14]. Bahsoun and Góra [2] proved the existence of invariant measures for position dependent random maps under milder conditions. Islam, Góra, and Boyarsky [17] proved the necessary and sufficient conditions for a general class of position dependent random maps. Markov switching position dependent random maps in one dimension was presented in [3, 18]. For higher dimensional Markov switching position dependent random maps the existence of invariant measures was studied by Islam in [15]. In this chapter, we present the existence of invariant measures, methods for the approximation of densities of absolutely continuous invariant measures for position dependent random maps. Some applications of position dependent random maps in finance are also presented. Our presentation is based on [2–7, 13–18].

4.2 Random maps with position dependent probabilities

Let $(X,\mathcal{B},\lambda)$ be a measure space, where λ is an underlying measure. Let $\tau_k : X \to X$, $k = 1,2,\ldots,K$, be piecewise one-to-one non-singular transformations on a common partition $\mathcal{P}$ of $X : \mathcal{P} = \{I_1, I_2, \ldots, I_q\}$ and $\tau_{k,i} = \tau_k|_{I_i}, i = 1,2,\ldots,q, k = 1,2,\ldots,K$. We define the transition function for the random map

$$T = \{\tau_1, \tau_2, \ldots, \tau_k; p_1(x), p_2(x), \ldots, p_K(x)\}$$

as follows [14]:

$$\mathbb{P}(x,A) = \sum_{k=1}^{K} p_k(x)\chi_A(\tau_k(x)), \tag{4.2.1}$$

where A is any measurable set and $\{p_k(x)\}_{k=1}^K$ is a set of position dependent probabilities, i.e., $\sum_{k=1}^K p_k(x) = 1,\ p_k(x) \geq 0$, for any $x \in X$. We define $T(x) = \tau_k(x)$ with probability $p_k(x)$ and for any non-negative integer N, $T^N(x) = \tau_{k_N} \circ \tau_{k_{N-1}} \circ \ldots \circ \tau_{k_1}(x)$ with probability

$$p_{k_N}(\tau_{k_{N-1}} \circ \ldots \circ \tau_{k_1}(x))p_{k_{N-1}}(\tau_{k_{N-2}} \circ \ldots \circ \tau_{k_1}(x)) \ldots p_{k_1}(x).$$

The transition function $\mathbb{P}$ induces an operator $\mathbb{P}_*$ on measures on $(X,\mathcal{B})$ defined by

$$\begin{aligned}
\mathbb{P}_*\mu(A) &= \int \mathbb{P}(x,A)d\mu(x) \\
&= \sum_{k=1}^{K} \int p_k(x)\chi_A(\tau_k(x))d\mu(x) \\
&= \sum_{k=1}^{K}\sum_{i=1}^{q} \int_{\tau_{k,i}^{-1}(A)} p_k(x)d\mu(x).
\end{aligned}$$

Definition 4.1 *A measure μ is T invariant if and only if $\mathbb{P}_*\mu(A) = \mu$, that is,*

$$\mu(A) = \sum_{k=1}^{K} \int_{\tau_k^{-1}(A)} p_k(x)d\mu(x).$$

4.2.1 *The Frobenius–Perron operator*

If μ has density f with respect to λ, the $P_*\mu$ also has a density which we denote by $P_T f$. By a change of variables, we obtain

$$\begin{aligned}
\int_A P_T f(x)d\lambda(x) &= \sum_{k=1}^{K}\sum_{k=1}^{q} \int_{\tau_{k,i}^{-1}(A)} p_k(x)f(x)d\lambda(x) \\
&= \sum_{k=1}^{K}\sum_{i=1}^{q} \int_{(A)} p_k(\tau_{k,i}^{-1}(x))f(\tau_{k,i}^{-1}(x))\frac{1}{J_{\tau_{k},i}(\tau_{k,i}^{-1})}d\lambda(x),
\end{aligned}$$

where $J_{\tau_k,i}$ is the Jacobian of $\tau_{k,i}$ with respect to λ. Since this holds for any measurable set A, we obtain an almost everywhere equality:

$$(P_T f)(x) = \sum_{k=1}^{K}\sum_{k=1}^{q} p_k(\tau_{k,i}^{-1}) f(\tau_{k,i}^{-1}) \frac{1}{J_{k,i}(\tau_{k,i}^{-1})} \chi_{\tau_k(I_i)}(x) = \sum_{k=1}^{K} P_{\tau_k}(p_k f)(x), \quad (4.2.2)$$

where P_{τ_k} is the Perron-Frobenius operator corresponding to the transformation τ_k. We call P_T the Frobenius–Perron operator of the random map T.

4.2.2 Properties of the Frobenius–Perron operator

The Frobenius–Perron operator P_T of a position dependent random map has properties similar to the Frobenius–Perron operator of random maps with constant probabilities:

(a) **(Linearity)** Let α, β be constants. If $f, g \in L^1([0,1])$ and $x \in [0,1]$, then

$$\begin{aligned}
(P_T(\alpha f + \beta g))(x) &= \sum_{k=1}^{K} (P_{\tau_k}(p_k(\alpha f + \beta g)))(x) \\
&= \alpha \sum_{k=1}^{K} (P_{\tau_k}(p_k f))(x) + \beta \sum_{k=1}^{K} (P_{\tau_k}(p_k g))(x) \\
&= \alpha (P_T(f))(x) + \beta (P_T(g))(x).
\end{aligned}$$

That is, $P_T : L^1 \to L^1$ is a linear operator.

(b) **(Positivity)** Let $f \in L^1([0,1])$ and $f(x) \geq 0$ for all $x \in [0,1]$. Then, for $k = 1, 2, \ldots, K$, $g(x) = p_k(x) f(x) \geq 0$ for all $x \in [0,1]$. Therefore, $(P_{\tau_k} g)(x) \geq 0$. Thus, $P_T f = \sum_{k=1}^{K} (P_{\tau_k}(p_k f))(x) = \sum_{k=1}^{K} (P_{\tau_k} g)(x) \geq 0$, that is, if $f \geq 0$, then $P_T f \geq 0$.

(c) **(Preservation of integrals)**

$$\begin{aligned}
\int_{[0,1]} P_T f(x) d\lambda(x) &= \sum_{k=1}^{K}\sum_{k=1}^{q} \int_{\tau_{k,i}^{-1}[0,1]} p_k(x) f(x) d\lambda(x) \\
&= \sum_{k=1}^{K} \int_{[0,1]} p_k(x) f(x) d\lambda(x) \\
&= \int_{[0,1]} \left((\sum_{k=1}^{K} p_k(x)) f(x) \right) d\lambda(x) \\
&= \int_{[0,1]} f(x) d\lambda(x)
\end{aligned}$$

That is, $\int_{[0,1]} P_T f d\lambda = \int_{[0,1]} f d\lambda$

(d) **(Contraction property)** $P_T : L^1([0,1]) \to L^1([0,1])$ is a contraction: Let $f \in L^1([0,1])$. Then $f = f^+ - f^-$ and $|f| = f^+ + f^-$, where $f^+ = \max(f,0)$ and $f^- = -\min(f,0)$.

$$\begin{aligned}
\| P_T f \|_1 &= \int_{[0,1]} |P_T f| = \int_{[0,1]} |P_T(f^+ - f^-)| d\lambda \\
&= \int_{[0,1]} |(P_T f^+ - P_T f^-)| d\lambda \\
&\leq \int_{[0,1]} |P_T f^+| d\lambda + \int_{[0,1]} |P_T f^-| d\lambda \\
&= \int_{[0,1]} (P_T f^+ d\lambda + \int_{[0,1]} P_T f^- d\lambda \\
&= \int_{[0,1]} P_T f d\lambda \\
&= \int_{[0,1]} f d\lambda \\
&= \| f \|_1 .
\end{aligned}$$

That is, $\| P_T f \|_1 \leq \| f \|_1$.

(e) **(Composition property)** P_T satisfies the composition property, i.e., if T and R are two random maps on $[0,1]$, then $P_{T\circ R} = P_T \circ P_R$. In particular, for any $n \geq 1, P_T^n = P_{T^n}$. Now, we present the proof of the composition property (see [2] for more details).

Proof *Let* $T = \{\tau_1, \tau_2, \ldots, \tau_K; p_1, p_2, \ldots, p_K\}$ *and* $R = \{\bar{\tau}_1, \bar{\tau}_2, \ldots, \bar{\tau}_K; \bar{p}_1, \bar{p}_2, \ldots, \bar{p}_K\}$ *be two random maps on a common partition* $\mathcal{J} = \{J_1, J_2, \cdots, J_N\}$ *of* $[0,1]$. *Then*

$$\begin{aligned}
(P_T \circ P_R) f &= P_R \left(\sum_{k=1}^{K} P_{\tau_k}(p_k f) \right) \\
&= \sum_{l=1}^{K} \sum_{k=1}^{K} P_{\bar{\tau}_l} \left(P_{\tau_k}(p_k f) \right) \\
&= \sum_{l=1}^{K} \sum_{k=1}^{K} \sum_{i=1}^{q} P_{\bar{\tau}_l} (\bar{\tau}_{l,i}^{-1} \left(P_{\tau_k}(p_k f) \right) (\bar{\tau}_{l,i}^{-1}) \frac{1}{J_{\bar{\tau}_{l,i}(\bar{\tau}_{l,i})}} \chi_{\bar{\tau}_{l,i}(Ji)} \\
&= \sum_{l=1}^{K} \sum_{k=1}^{K} \sum_{i=1}^{q} \sum_{j=1}^{q} P_{\bar{\tau}_l} (\bar{\tau}_{l,i}^{-1} (p_k(\tau_{k,j}^{-1} \circ \bar{\tau}_{l,i}^{-1}) f(\tau_{k,j}^{-1} \circ \bar{\tau}_{l,i}^{-1}) \\
&\quad \times \frac{1}{J_{\tau_{k,j}(\tau_{k,j}^{-1} \circ \bar{\tau}_{l,i}^{-1})}} \frac{1}{J_{\bar{\tau}_{l,i}(\bar{\tau}_{l,i}^{-1})}} \chi_{\tau_{k,j}(Jj)(\bar{\tau}_{l,i}^{-1})} \chi_{\bar{\tau}_{l,i}(J_i)} \\
&= \sum_{k=1}^{K} \sum_{l=1}^{K} P_{\tau_k \circ \bar{\tau}_k} (p_k(\bar{\tau}_l) \bar{p}_k f) = P_{T\circ R} f.
\end{aligned}$$

Lemma 4.2 $P_T f^* = f^*$ *if and only if* $\mu = f^* \lambda$ *is* T *invariant.*

Proof *Assume that* $\mu(A)=\sum_{k=1}^{K}\int_{\tau_k^{-1}(A)}p_k(x)d\mu(x)$, *for any* $A\in\mathcal{B}$. *Then*

$$\begin{aligned}\int_A f^*d\lambda &= \sum_{k=1}^{K}\int_{\tau_k^{-1}(A)}p_kf^*d\lambda\\ &= \sum_{k=1}^{K}\int_A P_{\tau_k}(p_kf^*)d\lambda\\ &= \int_A\sum_{k=1}^{K}P_{\tau_k}(p_kf^*)d\lambda\\ &= \int_A P_Tf^*d\lambda.\end{aligned}$$

Therefore, $P_Tf^*=f^*$.

Conversely, assume that $P_Tf^*=f^*$ *almost everywhere. Then*

$$\begin{aligned}\mu(A)=\int_A f^*d\lambda &= \int_A P_Tf^*d\lambda\\ &= \int_A\sum_{k=1}^{K}P_{\tau_k}(p_kf^*)d\lambda\\ &= \sum_{k=1}^{K}p_k\int_{\tau^{-1}(A)}p_k(x)f^*(x)d\lambda\\ &= \sum_{k=1}^{K}\int_{\tau_k^{-1}(A)}p_kd\mu\end{aligned}$$

4.2.3 *Existence of invariant measures for position dependent random maps*

In this section, we first review some results on the existence of invariant measures for position dependent random maps and then we present necessary and sufficient conditions for the existence of invariant measures for a general class of position dependent random maps.

4.2.3.1 *Existence results of Góra and Boyarsky*

In [14], Góra and Boyarsky proved the following result for the existence of invariant measures for position dependent random maps.

Theorem 4.3 *Let* $T=\{\tau_1,\tau_2,\dots,\tau_K;p_1(x),p_2(x),\dots,p_K(x)\}$ *be a position dependent random map, assume that the transformations* $\tau_k, k=1,2,\dots,K$ *are piecewise monotonic, piecewise* C^2, *and expanding on a common partition (that is, there exists a partition* $\mathcal{P}=\{P_1,I_2,\dots,I_q\}$ *such that* $\tau_{k,i}=\tau_k|_{I_i}, i=1,2,\dots,q$ *is monotonic,* C^2 *and* $|\tau_{k_i}|\geq\alpha>1$ *for some universal constant* α*), and the probabilities* $p_k(x), k=1,2,\dots K$ *are piecewise* C^1. *Then*

1. *for any* $f \in BV(I)$
$$V_I(P_T) \leq AV_I f + B \parallel f \parallel_1, \quad (4.2.3)$$
where $A = K \cdot \frac{\sum_{k=1}^{K} \beta_k}{\alpha}, B = K \cdot \frac{\sum_{k=1}^{K} \beta_k}{\alpha\delta} + \max_{1 \leq k \leq K} \sup_I |(\frac{p_k}{\tau_k'})'|, \delta = \min\{\lambda I_i : i = 1,2,\ldots,q\}, \beta_k = \sup_{x \in I} p_k(x), k = 1,2,\ldots,K.$

2. *if* $K \cdot \frac{\sum_{k=1}^{K} \beta_k}{\alpha} < 1$, *then the random map has an acim. Moreover, the Frobenius–Perron operator* P_T *is quasicompact.*

Example 4.1 *Let* $\tau : [0,1] \to [0,1]$ *be defined by*

$$\tau(x) = \begin{cases} 2x & , 0 \leq x \leq \frac{1}{2}, \\ 2 - 2x & , \frac{1}{2} < x \leq 1. \end{cases}$$

Let $\tau_1, \tau_2 : [0,1] \to [0,1]$ *be defined by* $\tau_1(x) = \tau \circ \tau(x)$ *and*

$$\tau_2(x) = \begin{cases} 3x + \frac{1}{4} & , 0 \leq x < \frac{1}{4}, \\ 3x - \frac{3}{4} & , \frac{1}{4} \leq x < \frac{9}{4}, \\ -3x + \frac{9}{4} & , \frac{1}{2} \leq x < \frac{3}{4}, \\ -3x + \frac{13}{4} & , \frac{3}{4} \leq x \leq 1 \end{cases}$$

Let $p_1, p_2 : [0,1] \to [0,1]$ *be defined by*

$$p_1(x) = \begin{cases} \frac{2}{3} & , 0 \leq x \leq \frac{1}{2}, \\ \frac{1}{3} & , \frac{1}{2} < x \leq 1. \end{cases}$$

and

$$p_2(x) = \begin{cases} \frac{1}{3} & , 0 \leq x \leq \frac{1}{2}, \\ \frac{2}{3} & , \frac{1}{2} < x \leq 1 \end{cases}$$

Consider the random map $T = \{\tau_1, \tau_2; p_1(x), p_2(x)\}$. *It can be easily shown that the random map* T *satisfies conditions of Theorem 4.3 and thus* T *has an invariant absolutely continuous measure.*

4.2.3.2 Existence results of Bahsoun and Góra

In [2], Bahsoun and Góra proved the existence of invariant measures for position dependent random maps under milder conditions.

Lemma 4.4 *[2] Let* $T = \{\tau_1, \tau_2, \ldots, \tau_K; p_1(x), p_2(x), ..., p_K(x)\}$ *be a position dependent random map, where* $\tau_k : [0,1] \to [0,1], k = 1,2,...,K$ *are piecewise one-to-one and differentiable, non-singular transformations on a common partition* $\mathcal{J} =$

$\{J_1, J_2, ..., J_q\}$ of $[0,1]$. Let $g_k(x) = \frac{p_k(x)}{|\tau_k'(x)|}, k = 1,2,...,K$. Assume that the random map T satisfies the following conditions: (i) $\sum_{k=1}^{K} g_k(x) < \alpha < 1, x \in [0,1]$; (ii) $g_k \in BV([0,1]), k = 1,2,...,K$. Then, for any $f \in BV([0,1])$, P_T satisfies the following Lasota-Yorke type inequality:

$$V_{[0,1]} P_T f \leq A V_{[0,1]} f + B \parallel f \parallel_1, \tag{4.2.4}$$

where $A = 3\alpha + \max_{1 \leq i \leq q} \sum_{k=1}^{K} V_{J_i} g_k$ and

$$B = 2\beta\alpha + \beta \max_{1 \leq i \leq q} \sum_{k=1}^{K} V_{J_i} g_k$$

with $\beta = \max_{1 \leq i \leq q} \frac{1}{\lambda(I_i)}$.

Proof *See [2].*

Let $x \in [0,1]$. Then for any $N \geq 1$ we have, $T^N(x) = \tau_{k_N} \circ \tau_{k_{N-1}} \circ \cdots \circ \tau_{k_1}(x)$ with probability

$$p_{k_N}(\tau_{k_{N-1}} \circ \cdots \circ \tau_{k_1}(x)) p_{k_{N-1}}(\tau_{k_{N-2}} \circ \cdots \circ \tau_{k_1}(x)) \ldots p_{k_1}(x).$$

For $\omega \in \{1,2,\ldots,K\}^N$, define

$$\begin{aligned} T_\omega(x) &= T^N(x), \\ p_\omega &= p_{k_N}(\tau_{k_{N-1}} \circ \cdots \circ \tau_{k_1}(x)) p_{k_{N-1}}(\tau_{k_{N-2}} \circ \cdots \circ \tau_{k_1}(x)) \ldots p_{k_1}(x), \\ g_\omega &= \frac{p_\omega}{|T_\omega'(x)|}, W_N = \max_{L \in \mathcal{J}^{(N)}} \sum_{\omega \in \{1,2,\ldots,K\}^N} V_L g_\omega \end{aligned}$$

Based on Lemma 4.4, Bahsoun and Góra proved the following Lemma for the iterates of P_T:

Lemma 4.5 *[2] Let T be a random map satisfying conditions of Lemma 11.5 and N be a positive integer such that $A_N = 3\alpha^N + W_N < 1$. Then*

$$V_{[0,1]} P_T^N f \leq A_N V_{[0,1]} f + B_N \parallel f \parallel_1, \tag{4.2.5}$$

where, $B_N = \beta_N \left(2\alpha^N + W_N\right), \beta_N = \max_{L \in \mathcal{J}^{(N)}} \frac{1}{\lambda(L)}$.

Using Lemma 4.4 and Lemma 4.5, the following theorem was proved in [2] for the existence of an absolutely continuous invariant measure of random maps:

Theorem 4.6 *[2] Let $T = \{\tau_1, \tau_2, \ldots, \tau_K; p_1(x), p_2(x), \ldots, p_K\}$ be a position dependent random map satisfying conditions of Lemma 4.4. Then, T has an invariant measure which is absolutely continuous with respect to Lebesgue measure. The operator P_T is quasicompact in $BV(I)$.*

Example 4.2 *We consider the position dependent random map*

$$T = \{\tau_1(x), \tau_2(x); p_1(x), p_2(x)\},$$

where $\tau_1, \tau_2 : [0,1] \to [0,1]$ *are defined by*

$$\tau_1(x) = \begin{cases} 3x + \frac{1}{4}, & 0 \le x < \frac{1}{4}, \\ 3x - \frac{3}{4}, & \frac{1}{4} \le x < \frac{1}{2}, \\ 4x - 2, & \frac{1}{2} \le x < \frac{3}{4}, \\ 4x - 3, & \frac{3}{4} \le x \le 1 \end{cases}$$

$$\tau_2(x) = \begin{cases} 4x, & 0 \le x < \frac{1}{4}, \\ 4x - 1, & \frac{1}{4} \le x < \frac{1}{2}, \\ 3x - \frac{3}{2}, & \frac{1}{2} \le x < \frac{3}{4}, \\ 3x - \frac{9}{4}, & \frac{3}{4} \le x \le 1, \end{cases}$$

and the position dependent probabilities $p_1, p_2 : [0,1] \to [0,1]$ *are defined by*

$$p_1(x) = \begin{cases} \frac{1}{4}, & 0 \le x < \frac{1}{4}, \\ \frac{1}{4}, & \frac{1}{4} \le x < \frac{1}{2}, \\ \frac{3}{4}, & \frac{1}{2} \le x < \frac{3}{4}, \\ \frac{3}{4}, & \frac{3}{4} \le x \le 1 \end{cases}$$

and

$$p_2(x) = \begin{cases} \frac{3}{4}, & 0 \le x < \frac{1}{4}, \\ \frac{3}{4}, & \frac{1}{4} \le x < \frac{1}{2}, \\ \frac{1}{4}, & \frac{1}{2} \le x < \frac{3}{4}, \\ \frac{1}{4}, & \frac{3}{4} \le x \le 1 \end{cases}$$

If $x \in [0, \frac{1}{4})$*, then* $\sum_{k=1}^{2} g_k(x) = \sum_{k=1}^{2} \frac{p_k(x)}{|\tau_k'(x)|} = \frac{\frac{1}{4}}{3} + \frac{\frac{3}{4}}{4} = \frac{13}{48} < 1$*. If* $x \in [\frac{1}{4}, \frac{1}{2})$*, then* $\sum_{k=1}^{2} g_k(x) = \sum_{k=1}^{2} \frac{p_k(x)}{|\tau_k'(x)|} = \frac{\frac{1}{4}}{3} + \frac{\frac{3}{4}}{4} = \frac{13}{48} < 1$*. If* $x \in [\frac{1}{2}, \frac{3}{4})$*, then* $\sum_{k=1}^{2} g_k(x) = \sum_{k=1}^{2} \frac{p_k(x)}{|\tau_k'(x)|} = \frac{\frac{3}{4}}{4} + \frac{\frac{1}{4}}{3} = \frac{13}{48} < 1$*. If* $x \in [\frac{1}{2}, \frac{3}{4})$*, then* $\sum_{k=1}^{2} g_k(x) = \sum_{k=1}^{2} \frac{p_k(x)}{|\tau_k'(x)|} = \frac{\frac{3}{4}}{4} + \frac{\frac{1}{4}}{3} =$

$\frac{13}{48} < 1$. *Moreover,* $A = 3\alpha + \max_{1\leq i\leq q} \sum_{k=1}^{K} V_{J_i} g_k = 3 \cdot \frac{13}{48} + 0 = \frac{39}{48} < 1$. *It can be easily shown that* $B = 2\beta\alpha + \beta \ \max_{1\leq i\leq q} \sum_{k=1}^{K} V_{J_i} g_k > 0$ *with* $\beta = \max_{1\leq i\leq q} \frac{1}{\lambda(J_i)}$. *The random map* $\hat{T}$ *satisfies conditions of Theorem 4.6 and thus* $\hat{T}$ *has an invariant density* f^*. *It can be easily shown by the Lasota–Yorke result [21] that both* τ_1 *and* τ_2 *have acim. Moreover,* τ_1 *and* τ_2 *are piecewise linear, expanding and Markov transformations. The matrix representation of the Frobenius–Perron operator* P_{τ_1} *is the transpose of* M_{τ_1} *where*

$$M_{\tau_1} = \begin{bmatrix} 0 & \frac{1}{3} & \frac{1}{3} & \frac{1}{3} \\ \frac{1}{3} & \frac{1}{3} & \frac{1}{3} & 0 \\ \frac{1}{4} & \frac{1}{4} & \frac{1}{4} & \frac{1}{4} \\ \frac{1}{4} & \frac{1}{4} & \frac{1}{4} & \frac{1}{4} \end{bmatrix}.$$

The matrix representation of the Frobenius–Perron operator P_{τ_2} *is the transpose of* M_{τ_2} *where*

$$M_{\tau_2} = \begin{bmatrix} \frac{1}{4} & \frac{1}{4} & \frac{1}{4} & \frac{1}{4} \\ \frac{1}{4} & \frac{1}{4} & \frac{1}{4} & \frac{1}{4} \\ \frac{1}{3} & \frac{1}{3} & \frac{1}{3} & 0 \\ \frac{1}{3} & \frac{1}{3} & \frac{1}{3} & 0 \end{bmatrix}.$$

It is easy to show that both τ_1 *and* τ_2 *have unique acim. Thus, the random map* $T = \{\tau_1(x), \tau_2(x); p_1(x), p_2(x)\}$ *also has a unique acim.*

The matrix representation of the Frobenius–Perron operator $P_T f = \sum_{k=1}^{2} P_{\tau_k}(p_k f)(x)$ *is the transpose of the matrix* M_T *where*

$$M_T = \begin{bmatrix} \frac{3}{16} & \frac{13}{16} & \frac{13}{16} & \frac{13}{16} \\ \frac{13}{16} & \frac{13}{16} & \frac{13}{16} & \frac{3}{16} \\ \frac{13}{16} & \frac{13}{16} & \frac{13}{16} & \frac{3}{16} \\ \frac{13}{16} & \frac{13}{16} & \frac{13}{16} & \frac{13}{16} \end{bmatrix}.$$

The normalized density f^* *of the unique acim of the random map* T *is the left eigenvector of the matrix* M_T *associated with the eigenvalue* 1 *(after adding the normalizing condition). In fact,* $f^* = \left[1, \frac{13}{12}, \frac{13}{12}, \frac{5}{6}\right]$ *is the unique normalized invariant density of* $\hat{T}$.

4.2.3.3 *Necessary and sufficient conditions for the existence of invariant measures for a general class of position dependent random maps*

In Chapter 3, we have presented the proof (see Theorem 3.18) of necessary and sufficient conditions for the existence of invariant measures for a general class of random maps with constant probabilities. We now state the analogous result for position dependent random maps.

Theorem 4.7 *Let (X,B,λ) be a measure space with normalized measure λ and $\tau_i : X \to X$, $i = 1,2$ be non-singular transformations. Consider the random map $T = \{\tau_1, \tau_2; p_1, p_2\}$ with position dependent probabilities p_1, p_2. Then there exists a normalized absolutely continuous (w.r.t. λ) $T-$invariant measure μ if and only if there exists $\delta > 0$ and $0 < \alpha < 1$ such that for any measurable set E and any positive integer k, $\lambda(E) < \delta$ implies*

$$\int_{\tau_1^{-1}(E)} p_1(x)d\lambda + \int_{\tau_2^{-1}(E)} p_2(x)d\lambda < \alpha;$$

$$\int_{\tau_1^{-2}(E)} p_1(x)p_1(\tau_1(x))d\lambda + \int_{\tau_2^{-1}\tau_1^{-1}(E)} p_1(x)p_2(\tau_1(x))d\lambda$$
$$+\int_{\tau_1^{-1}\tau_2^{-1}(E)} p_2(x)p_1(\tau_2(x))d\lambda + \int_{\tau_2^{-2}(E)} p_2(x)p_2(\tau_2(x))d\lambda < \alpha;$$

$\vdots$

$$\sum_{(i_1,i_2,i_3,\ldots,i_k)} \int_{\tau_{i_1}^{-1}\tau_{i_2}^{-1}\ldots\tau_{i_k}^{-1}(E)} p_{i_1}(x)p_{i_2}(\tau_{i_1}(x))\ldots p_{i_k}(\tau_{i_1}\tau_{i_2}\cdots\tau_{i_{k-1}}(x))d\lambda < \alpha.$$

Proof *The proof is analogous to the proof of Theorem 3.18 in Chapter 3.*

4.3 Markov switching position dependent random maps

The presentation of this section is based on [18]. Let $X = ([a,b],\mathcal{B},\lambda)$ be a measure space where λ is Lebesgue measure on $[a,b]$. Let $\tau_k : X \to X, k = 1,2,\ldots,K$, be piecewise one-to-one continuous non-singular transformations on a common partition $\mathcal{P}$ of $[a,b] : P = \{J_1, J_2, \ldots, J_q\}$ and $\tau_{k,i} = \tau_k|J_i$, $i = 1,2,\ldots,q$, $k = 1,2,\ldots,K$. A Markov switching position dependent random map T is a Markov process which is defined as follows: at time $n = 1$, we select a transformation τ_k randomly according to initial probabilities $p_k, k = 1,2,\ldots,K$. The probability of switching from transformation τ_k to transformation τ_l is given by $W_{k,l}$, the $(k,l)^{th}$ element of a position dependent stochastic matrix $W = W(x)$. Therefore, if we choose τ_{k_1} at time $n = 1$ when we are at position x, the Markov process at time N is given by

$$T^N(x) = \tau_{k_N} \circ \tau_{k_{N-1}} \circ \ldots \circ \tau_{k_1}(x)$$

with probability

$$W_{k_{N-1},k_N}(\tau_{k_{N-1}} \circ \ldots \circ \tau_{k_1}(x)) \cdot W_{k_{N-2},k_{N-1}}(\tau_{k_{N-2}} \circ \ldots \circ \tau_{k_1}(x)) \cdot \ldots \cdot W_{k_1,k_2}(x).$$

We assume that the probabilities $W_{k,l}(x)$ are defined on the same partition $\mathcal{P}$. Let $\Omega = \{1,2,\dots,K\}$. We define the transition function of the Markov process on $\Omega \times X$ as follows:

$$\mathbb{P}((k,x),\{l\} \times A) = W_{k,l}(x)\chi_A(\tau_k(x)),$$

where A is any measurable set and χ_A denotes the characteristic function of the set A. The random map T is the projection of the process we defined on the space X. The transition function $\mathbb{P}$ induces an operator $\mathbb{P}_*$ on measures μ on $\Omega \times X$ as follows:

$$\begin{aligned}\mathbb{P}_*\mu(\{l\} \times A) &= \int_{\Omega\times X} \mathbb{P}((k,x),\{l\} \times A)d\mu(k,x) \\ &= \int_{\Omega\times X} W_{k,l}(x)\chi_A(\tau_k(x))d\mu(k,x).\end{aligned}$$

Let ν be a measure on $\Omega \times X$ such that $\nu(\{s\} \times A) = \lambda(A)$. If μ has density f with respect to $\nu, f(s,x) = \sum_{k=1}^{K} f_k(x)\chi_{\{k\}\times X}(s,x)$, where $\sum_{k=1}^{K}\int_X f_k(x) = 1$, then $\mathbb{P}_*\mu$ also has a density which we denote by $P_T f$. By a change of variables, we obtain

$$\begin{aligned}\int_{\{l\}\times A} P_T f(s,x)d\nu(s,x) &= \sum_{k=1}^{K}\int_X W_{k,l}(x)\chi_A(\tau_k(x))f_k(x)d\lambda(x) \\ &= \sum_{k=1}^{K}\int_{\tau_k^{-1}(A)} W_{k,l}(x)f_k(x)d\lambda(x). \qquad (4.3.1)\end{aligned}$$

Using the definition of P_{τ_k}, the Frobenius–Perron operator associated with transformation τ_k (see [6] for P_{τ_k}) and (4.3.1), we obtain

$$\int_A \hat{f}_l(x)d\lambda(x) = \sum_{k=1}^{K}\int_A P_{\tau_k}(W_{k,l}f_k)(x)d\lambda(x), \qquad (4.3.2)$$

where $P_T f(s,x) = \sum_{l=1}^{K} \hat{f}_l\chi_{\{l\}\times X}(s,x)$. Since (4.3.2) is true for any $A \in \mathcal{B}$, we obtain an a.e. equality

$$\hat{f}_l(x) = \sum_{k=1}^{K} P_{\tau_k}(W_{k,l}f_k)(x). \qquad (4.3.3)$$

Thus, the density $f^*(s,x) = \sum_{l=1}^{K} f_l^*(x)\chi_{\{l\}\times X}(s,x)$ is $T-$invariant if

$$f_l^*(x) = \sum_{k=1}^{K} P_{\tau_k}(W_{k,l}f_k^*)(x). \qquad (4.3.4)$$

for $l = 1,2,\dots,K$. If we denote

$$w_l = \int_X f_l^*(x)d\lambda(x), \quad l = 1,2,\dots,K,$$

then integrating (4.3.4) with respect to λ, we obtain

$$w_l = \sum_{k=1}^{K} w_k \int_X W_{k,l}(x)\frac{f_l^*(x)}{\int_X f_k^*(x)d\lambda(x)}d\lambda(x). \qquad (4.3.5)$$

Note that, in the special case when $W_{k,l}$'s are constant, (4.3.5) reduces to $w_l = \sum_{k=1}^{K} w_k W_{k,l}$, i.e., to the case when $(w_1, w_2, \ldots, w_K)$ is a left invariant eigenvector of the matrix W.

As before, denote by $V(\cdot)$ the standard one dimensional variation of a function, and $BV([a,b])$ the space of functions of bounded variations on $[a,b]$ equipped with the norm $\| \cdot \|_{BV} = V(\cdot) + \| \cdot \|_1$, where $\| \cdot \|_1$ denotes the norm on $L^1([a,b], \mathcal{B}, \lambda)$. Let $\widehat{BV} = \prod_{k=1}^{K} BV$ denote the $K-$fold product of the space BV of functions of bounded variation and we define a norm on $\widehat{BV}$ as $\| f_1, f_2, \ldots, f_K \|_{\widehat{BV}} = \sum_{k=1}^{K} \| f_k \|_{BV}$. We also define L^1 norm on $\widehat{BV}$: $\| f_1, f_2, \ldots, f_K \|_1 = \sum_{k=1}^{K} \| f_k \|_1$. We define an operator $\widehat{P_T} : \widehat{BV} \to \widehat{BV}$ by

$$\widehat{P_T}(f_1, f_2, \ldots, f_K) = \left(\sum_{k=1}^{K} P_{\tau_k}(W_{k,1} f_k), \sum_{k=1}^{K} P_{\tau_k}(W_{k,2} f_k), \ldots, \sum_{k=1}^{K} P_{\tau_k}(W_{k,K} f_k) \right). \tag{4.3.6}$$

If $(f_1^*, f_2^*, \ldots, f_K^*)$ is fixed point of $\widehat{P_T}$, we call

$$f^* = \sum_{k=1}^{K} f_k^*$$

an invariant density of the Markov switching position dependent random map T. For more details about $\widehat{P_T}$ see [3].

Definition 4.8 [6] *We say that $\tau : [a,b] \to [a,b]$ is a Lasota–Yorke map if τ is piecewise monotone and C^2 and τ is non-singular, i.e., τ is non-singular and there exists a partition of $[a,b], a = x_0 < x_1 < \ldots < x_n = b$ such that for each $i = 0, 1, \ldots, n-1, \tau|_{(x_i, x_{i+1})}$ is monotonic and can be extended to a C^2 function on $[x_i, x_{i+1}]$.*

Lemma 4.9 *Let τ_k be a Lasota-Yorke map on $I = [0,1]$ and $W_{k,l}$ be piecewise of class C^1, for $k = 1, 2, \ldots, K$ and $l = 1, 2, \ldots, K$,. Let*

$$\alpha_l = \max_k (\sup_x \frac{2 \cdot W_{k,l}(x)}{|\tau_k'(x)|}), \; l = 1, 2, \ldots, K.$$

Then,

$$V_I(\widehat{P_T} f)_l \leq \alpha_l \sum_{k=1}^{K} V_I f_k + B_l \sum_{k=1}^{K} \|f_k\|_1, \tag{4.3.7}$$

where,

$$h_k(x) = \frac{W_{k,l}(x)}{|\tau_k'(x)|}, \delta = \min_i \lambda(J_i)$$

and

$$B_l = \frac{2}{\delta} (\max_k \sup_x h_k(x)) + (\max_k \sup_x |h_k'(x)|).$$

Proof *Since f_k is Riemann integrable, for arbitrary $\varepsilon > 0$, we can find a number θ such that for any $J_i \in \mathcal{P}$ and any partition finer than : $J_i = \cup_{p=1}^{L_i} [s_{p-1}, s_p]$ with $|s_p - s_{p-1}| < \theta$, we have*

$$\sum_{p=1}^{L_i} |f_k(s_{p-1})||s_p - s_{p-1}| \le \int_{J_i} |f_k| d\lambda + \varepsilon. \tag{4.3.8}$$

Let $0 = x_0 < x_1 < \ldots \le x_r = 1$ be such a fine partition of $I = [0,1]$.
Define $\phi_{k,i} = \tau_{k,i}^{-1}$. Let $h_k(x) = \frac{W_{k,l}(x)}{|\tau_k'(x)|}$. We have,

$$V_I(\widehat{P}_T f)_l \le \sum_{k=1}^{K} V_I P_{\tau_k}\left(W_{k,l} f_k\right). \tag{4.3.9}$$

We estimate $V_I P_{\tau_k}\left(W_{k,l} f_k\right)$:

$$\sum_{j=1}^{r} |P_{\tau_k}\left(W_{k,l} f_k\right)(x_j) - P_{\tau_k}\left(W_{k,l} f_k\right)(x_{j-1})| \tag{4.3.10}$$

$$= \sum_{j=1}^{r} |(\sum_{i=1}^{q} h_k(\phi_{k,i}(x_j)) f_k(\phi_{k,i}(x_j)) \chi_{\tau_k(J_i)}(x_j)$$
$$- \sum_{i=1}^{q} h_k(\phi_{k,i}(x_{j-1})) f_k(\phi_{k,i}(x_{j-1})) \chi_{\tau_k(J_i)}(x_{j-1}))|$$

$$\le \sum_{j=1}^{r} \sum_{i=1}^{q} |h_k(\phi_{k,i}(x_j)) f_k(\phi_{k,i}(x_j)) \chi_{\tau_k(J_i)}(x_j)$$
$$- h_k(\phi_{k,i}(x_{j-1})) f_k(\phi_{k,i}(x_{j-1})) \chi_{\tau_k(J_i)}(x_{j-1}))|. \tag{4.3.11}$$

We divide the sum on the right-hand side into three parts:
(I) the summands for which $\chi_{\tau_k(J_i)}(x_j) = \chi_{\tau_k(J_i)}(x_{j-1}) = 1$;
(II) the summands for which $\chi_{\tau_k(J_i)}(x_j) = 1$ and $\chi_{\tau_k(J_i)}(x_{j-1}) = 0$;
(III) the summands for which $\chi_{\tau_k(J_i)}(x_j) = 0$ and $\chi_{\tau_k(J_i)}(x_{j-1}) = 1$.

First, we will estimate (I).

$$\sum_{j=1}^{r} \sum_{i=1}^{q} |h_k(\phi_{k,i}(x_j)) f_k(\phi_{k,i}(x_j)) - h_k(\phi_{k,i}(x_{j-1})) f_k(\phi_{k,i}(x_{j-1}))|$$

$$\le \sum_{i=1}^{q} \sum_{j=1}^{r} |f_k(\phi_{k,i}(x_j))[h_k(\phi_{k,i}(x_j)) - h_k(\phi_{k,i}(x_{j-1}))]|$$

$$+ \sum_{i=1}^{q} \sum_{j=1}^{r} |h_k(\phi_{k,i}(x_{j-1}))[f_k(\phi_{k,i}(x_j)) - f_k(\phi_{k,i}(x_{j-1}))]|$$

$$\leq \sup_x |h_k'(x)| \sum_{i=1}^{q} \sum_{j=1}^{r} |f_k(\phi_{k,i}(x_j))[\phi_{k,i}(x_j) - \phi_{k,i}(x_{j-1})]| + (\sup_x h_k(x)) \sum_{i=1}^{q} V_{J_i} f_k$$

$$\leq \sup_x |h_k'(x)| \sum_{i=1}^{q} \left(\int_{J_i} |f_k| d\lambda(x) + \varepsilon \right) + (\sup_x h_k(x)) \sum_{i=1}^{q} V_{J_i} f_k, \text{ using } (4.3.8)$$

$$\leq \sup_x |h_k'(x)| \int_I |f_k| d\lambda(x) + (\sup_x h_k(x)) V_I f_k + q(\sup_x |h_k(x)|)\varepsilon.$$

We now consider (II) and (III) together. Notice that $\chi_{\tau_k(J_i)}(x_j) = 1$ and $\chi_{\tau_k(J_i)}(x_{j-1}) = 0$ occur only if $x_j \in \tau_l(J_i)$ and $x_{j-1} \notin \tau_l(J_i)$, i.e., if x_j and x_{j-1} are on opposite sides of an end point of $\tau_l(J_i)$, we can have at most one pair x_j, x_{j-1} like this and another pair $x_{j'} \notin \tau_l(J_i)$ and $x_{j'-1} \in \tau_l(J_i)$. Thus,

$$\sum_{i=1}^{q} (|h_k(\phi_{k,i}(x_j)) f_k(\phi_{k,i}(x_j))| + |h_k(\phi_{k,i}(x_{j'-1})) f_k(\phi_{k,i}(x_{j'-1}))|)$$

$$\leq \sup_x h_k(x) \sum_{i=1}^{q} (|f_k(\phi_{k,i}(x_j))| + |f_k(\phi_{k,i}(x_{j'-1}))|). \tag{4.3.12}$$

Since $s_i = \phi_{k,i}(x_j)$ and $r_i = \phi_{k,i}(x_{j'-1})$ are both points in J_i, we can write

$$\sum_{i=1} (|f_k(s_i)| + |f_k(r_i)|) \leq \sum_{i=1}^{q} (2|f_k(v_i)| + |f_k(v_i) - f_k(r_i)| + |f_k(v_i) - f_k(s_i)|),$$

where $v_i \in J_i$ is such that $|f_k(v_i)| \leq \frac{1}{\lambda(J_i)} \int_{I_i} |f_k| d\lambda(x)$. Thus,

$$\begin{aligned} & \sup_x h_k(x) \sum_{i=1} (|f_k(\phi_{k,i}(x_j))| + |f_k(\phi_{k,i}(x'_{j-1}))|) \\ \leq \ & \sup_x h_k(x) \sum_{i=1} \left(V_{I_i} f_k + \frac{2}{\lambda(J_i)} \int_{I_i} |f_k| d\lambda(x) \right) \\ \leq \ & \sup_x |h_k(x)| V_I f_k + \frac{2 \sup_x h_k(x)}{\delta} \int_I |f_k| d\lambda(x). \end{aligned}$$

Therefore,

$$V_I P_{\tau_k}(W_{k,l} f_k) \leq 2 \sup_x |h_k(x)| V_I f_k + (\frac{2}{\delta} (\sup_x h_k)$$
$$+ (\sup |h_k'(x)|)) \|f_k\|_1 + q(\sup_x h_k(x))\varepsilon.$$

Thus,

$$V_I(\widehat{P}_T f)_l \leq \sum_{k=1}^{K} (2 \max_k \sup_x |h_k(x)| V_I f_k + (\frac{2}{\delta} (\max_k \sup_x h_k)$$
$$+ (\max_k \sup |h_k'(x)|)) \|f_k\|_1 + q(\sup_x h_k(x))\varepsilon).$$

Since ε is arbitrarily small this proves the lemma.

Theorem 4.10 *Let τ_k be Lasota-Yorke maps and let $W_{k,l}$ be piecewise of class C^1, for $k,l = 1,2,\ldots,K$, and*

$$\alpha_l = \max_k \left(\sup_x \frac{2W_{k,l}(x)}{|\tau_k'(x)|} \right), \; l = 1,2,\ldots,K,$$

and $\sum_{l=1}^{K} \alpha_l < 1$. Then the operator $\widehat{P}_T$ is quasicompact and admits a fixed point in $\widehat{BV}$, i.e., the Markov switching random map T admits an absolutely continuous invariant measure.

Proof *The space $\widehat{BV}$ is a Banach space with norm $\|\cdot\|_{\widehat{BV}} = \sum_{k=1}^{K} \|\cdot\|_{BV}$. First, if $f = (f_1, f_2, \ldots, f_K)$ with $f_k \geq 0$, then we have*

$$\begin{aligned} \|\widehat{P}_T f\|_1 &= \sum_{l=1}^{K} \|(\widehat{P}_T f)_l\|_1 = \sum_{l=1}^{K} \int_I \sum_{k=1}^{K} P_{\tau_k}(W_{k,l} f_k) d\lambda \\ &= \sum_{k=1}^{K} \int_I \sum_{l=1}^{K} P_{\tau_k}(W_{k,l} f_k) d\lambda = \sum_{k=1}^{K} \int_I \sum_{l=1}^{K} P_{\tau_k}(f_k) d\lambda = \|f\|_1. \end{aligned}$$

For a general f it is easy to show that $\|\widehat{P}_T f\|_1 \leq \|f\|_1$. For $f \in \widehat{BV}$, by the above lemma, we obtain

$$\begin{aligned} \| \widehat{P_T f} \|_{\widehat{BV}} &= \sum_{l=1}^{K} \| (\widehat{P}_T f)_l \|_{BV} = \sum_{l=1}^{K} V_I(\widehat{P}_T f)_l + \| \widehat{P}_T f \|_1 \\ &\leq \sum_{l=1}^{K} V_I(\widehat{P}_T f)_l + \| \widehat{P}_T f \|_1 \\ &\leq \sum_{l=1}^{K} \alpha_l \left(\sum_{k=1}^{K} V_I f_k \right) + \sum_{l=1}^{K} \left(B_l \sum_{k=1}^{K} \| f_k \|_1 \right) + \| f \|_1 \\ &\leq \sum_{l=1}^{K} \alpha_l \| f \|_{\widehat{BV}} + \left(B_l + 1 - \sum_{l=1}^{K} \alpha_l \right) \cdot \| f \|_1 . \end{aligned}$$

Thus, by Ionescu-Tulcea and Marinescu Theorem [6, 14], $\widehat{P}_T$ is quasi-compact on $\widehat{BV}$ and admits a fixed point f in $\widehat{BV}$.

Example 4.3

Consider the Markov switching position dependent random map $T = \{\tau_1, \tau_2; p_1, p_2; W\}$, where τ_1, τ_2 are maps on $I = [0,1]$ defined by

$$\tau_1(x) = \begin{cases} 4x, & 0 \leq x \leq \frac{1}{4}, \\ 4x-1, & \frac{1}{4} < x \leq \frac{1}{2}, \\ 4x-2, & \frac{1}{2} < x \leq \frac{3}{4} \\ 4x-3, & \frac{3}{4} < x \leq 1 \end{cases} \qquad (4.3.13)$$

and

$$\tau_2(x) = \begin{cases} \frac{8}{3}x & , 0 \le x \le \frac{1}{3}, \\ \frac{8}{3}x - \frac{8}{9} & , \frac{1}{3} < x \le \frac{2}{3}, \\ \frac{8}{3}x - \frac{16}{9} & , \frac{2}{3} < x \le 1. \end{cases} \quad (4.3.14)$$

and W ia a stochastic switching matrix defined by

$$W = \begin{bmatrix} \frac{1}{2}x + \frac{1}{10} & \frac{9}{10} - \frac{1}{2}x \\ \frac{2}{3} & \frac{1}{3} \end{bmatrix},$$

and p_1, p_2 are initial probabilities. It is easy to show that $\alpha_1 = \frac{1}{2}$ and $\alpha_2 = \frac{9}{20}$. Hence the Markov switching random map $T = \{\tau_1, \tau_2\}$ with switching matrix W satisfies the condition of Theorem 4.10 and T has an acim.

4.4 Higher dimensional Markov switching position dependent random maps

For this section, we closely follow [15].

4.4.1 Notations and review of some lemmas

In this section we recall some lemmas from [2] and we introduce some useful notations. We include these lemmas in our situation for the convenience of the reader.

Consider the measure space $X = (X, \mathcal{B}, \lambda_n)$, where X is a bounded region in $\mathbb{R}^n$ and λ_n, is Lebesgue measure on X. Let $\tau_k : X \to X, k = 1, 2, \ldots, K$, be piecewise one-to-one continuous non-singular transformations on a common partition $\mathcal{P}$ of X : $P = \{S_1, S_2, \ldots, S_q\}$ of X and $\tau_{k,i}|S_i, i = 1, 2, \ldots, q, k = 1, 2, \ldots, K$. Suppose each S_i is a bounded closed domain having piecewise C^2 boundary of finite $(n-1)$-dimensional measure. We assume that the faces of ∂S_i meet at angles bounded uniformly away from 0. We will also assume that the entries $W_{k,l}$ are piecewise C^1 functions on the partition $\mathcal{P}$. Let $D\tau_{k,i}^{-1}(x)$ be the derivative matrix of $\tau_{k,i}^{-1}$ at x.

Let $\sup_{x\in\tau_{k,i}(S_i)} \| D\tau_{k,i}^{-1}(x) \| =: \sigma_{k,i}$ and $\sup_{x\in S_i} W_{k,l} =: \pi_{k,l,i}$. Using the smoothness of $D\tau_{k,i}^{-1}$'s and $W_{k,l}$'s we can refine the partition $\mathcal{P}$ and we obtain

$$\max_{1\le i\le q} \sum_{k=1}^{K} W_{k,l}(x) \| D\tau_{k,i}^{-1}(\tau_{k,i}(x)) \| = \sum_{k=1}^{K} \max_{1\le i\le q} \sigma_{k,i}\pi_{k,l,i}$$

The main tool of this section is the multidimensional notion of variation defined using derivatives in the distributional sense(see [13]):

$$V(f) = \int_{\mathbb{R}^n} \| Df \| = \sup\{\int_{\mathbb{R}^n} \mathrm{div}(g)d\lambda_n : g = (g_1, g_2, \ldots, g_n) \in C_0^1(\mathbb{R}^n, \mathbb{R}^n)\},$$

where $f \in L_1(\mathbb{R}^n)$ has bounded support, Df denotes the gradient of f in the distributional sense, and $C_0^1(\mathbb{R}^n, \mathbb{R}^n)\}$ is the space of continuously differentiable functions from $\mathbb{R}^n$ into $\mathbb{R}^n$ having a compact support. We will use the following properties of variation which is derived from [13]: If $f = 0$ outside a closed domain A whose boundary is Lipschitz continuous, $f|_A$ is continuous, $f|_{\text{int(A)}}$ is C^1, then

$$V(f) = \int_{\text{int(A)}} \| Df \| \, d\lambda_n + \int_{\partial A} |f| d\lambda_{n-1},$$

where λ_{n-1} is the $n-1$ dimensional measure on the boundary of A. We consider the Banach space (see [13])

$$BV(S) = \{f \in L_1(S) : V(f) < \infty\},$$

with the norm $\| f \|_{BV} = V(f) + \| f \|_1$.

Let $\widehat{BV} = \prod_{k=1}^{K} BV$ denote the $K-$fold product of the space BV and we define a norm on $\widehat{BV}$ as $\| f_1, f_2, \ldots, f_K \|_{\widehat{BV}} = \sum_{k=1}^{K} \| f_k \|_{BV}$. We also define L^1 norm on $\widehat{BV}$: $\| f_1, f_2, \ldots, f_K \|_1 = \sum_{k=1}^{K} \| f_k \|_1$. We define an operator $\widehat{P_T}$ as $\widehat{BV} \to \widehat{BV}$ by

$$\widehat{P_T}(f_1, f_2, \ldots, f_K) = \left(\sum_{k=1}^{K} P_{\tau_k}(W_{k,1} f_k), \sum_{k=1}^{K} P_{\tau_k}(W_{k,2} f_k), \ldots, \sum_{k=1}^{K} P_{\tau_k}(W_{k,K} f_k) \right). \tag{4.4.1}$$

If $(f_1^*, f_2^*, \ldots, f_K^*)$ is fixed point of $\widehat{P_T}$, we call

$$f^* = \sum_{k=1}^{K} f_k^*$$

an invariant density of the Markov switching position dependent random map T. For more details about $\widehat{P_T}$, see [3, 18].

We recall the following lemmas from [2].

Lemma 4.11 *[2] Consider $S_i \in \mathcal{P}$. Let x be a point in ∂S_i and $y = \tau_k(x)$ a point in $\partial(\tau_k(S_i))$. Let $J_{k,i}$ be the Jacobian of $\tau_k|_{S_i}$ at x and $J_{k,i}^0$ be the Jacobian of $\tau_k|_{\delta S_i}$ at x. Then $\frac{J_{k,i}^0}{J_{k,i}} \leq \sigma_{k,i}$*

Fix $1 \leq i \leq q$. Let Z denote the set of singular points of ∂S_i. Let us construct for any $x \in Z$ the largest cone with vertex at x which lies completely in S_i. Let $\theta(x)$ denote the vertex angle of this cone. Define $\beta(S_i) = \min_{x \in Z} \theta(x)$. Since the faces of δS_i meet at angles bounded away from 0, we have $\beta(S_i) > 0$. Let $\alpha(S_i) = \pi/2 + \beta(S_i)$ and $a(S_i) = |\cos(\alpha(S_i))|$.

Now we will construct C^1 field segments $L_y, y \in \partial S_i$, every L_y being a central ray of a regular cone contained in S_i, with vertex angle at y greater than or equal to $\beta(S_i)$.

We start at points $y \in Z$ where the minimal angle $\beta(S_i)$ is attained, defining L_y

to be the central rays of the largest regular cones contained in S_i. Then we extend this field of segments to the C^1 field we want, making L_y short enough to avoid overlapping. Let $\delta(y)$ be the length of $L_y, y \in \partial S_i$. By the compactness of ∂S_i we have $\delta(S_i) = \inf_{y \in \partial S_i} \delta y > 0$. Now, we shorten the L_y of our field, making them all of length $\delta(S_i)$.

Lemma 4.12 *[2] For any $S_i, i = 1,2,\ldots,q$, if F is a C^1 function on S_i, then*

$$\int_{\partial S_i} f(y) d\lambda_{n-1}(y) \leq \frac{1}{a(S_i)} \left(\frac{1}{\delta(S_i)} \int_{S_i} f d\lambda_n + V_{\text{int}(S_i)}(f) \right).$$

4.4.2 *The existence of absolutely continuous invariant measures of Markov switching position dependent random maps in $\mathbb{R}^n$*

Notations and setup of this section are similar to those in the previous section and we assume that Markov switching higher dimensional random map $T = \{\tau_1, \tau_2, \ldots, \tau_K; W\}$ satisfy conditions in the previous section.

The following lemmas are the key lemmas for proving our main result (Theorem 4.15).

Lemma 4.13 *Let $T = \{\tau_1, \tau_2, \ldots, \tau_K; W\}$ be a higher dimensional Markov switching position dependent random map and $(\widehat{P}_T f)_l$ be the l-th component of $\widehat{P}_T f$. Then,*

$$V((\widehat{P}_T f)_l) \leq \sigma_l (1 + \frac{1}{a}) \sum_{k=1}^{K} V(f_k) + \left(M_l + \frac{\sigma}{a\delta} \right) \sum_{k=1}^{K} \| f_k \|_1$$

where, $a = \min\{a(S_i) : i = 1,2,\ldots,q\} > 0, \delta = \min\{\delta(S_i) : i = 1,2,\ldots,q\} > 0$, $M_{(k,l),i} = \sup_{x \in S_i} \left(DW_{k,l}(x) - \frac{DJ_{k,i}}{J_{k,i}} W_{k,l}(x) \right), M_l = \sum_{k=1}^{K} \max_{1 \leq i \leq q} M_{(k,l),i}$, and $\sigma_l = \sum_{k=1}^{K} \max_{1 \leq i \leq q} \sigma_{k,i} \pi_{k,l,i}$.

Proof *We have $V((\widehat{P}_T f)_l) \leq \sum_{k=1}^{K} V(P_{\tau_k}(W_{k,l} f_k))$.*
To estimate $V(P_{\tau_k}(W_{k,l} f_k))$, let $F_{k,l,i} = \frac{f_k(\tau_{k,i}^{-1}) W_{k,l}(\tau_{k,i}^{-1})}{J_{k,i}(\tau_{k,i}^{-1})}, R_{k,i}(S_i), i = 1,2,\ldots,q, k = 1,2,\ldots,K$. Then

$$\int_{\mathbb{R}^n} \| DP_{\tau_k}(W_{k,l} f_k) \| d\lambda_n \leq \sum_{i=1}^{q} \int_{\mathbb{R}^n} \| D(F_{k,l,i} \chi_{R_i}) \| d\lambda_n$$

$$\leq \sum_{i=1}^{q} (\int_{\mathbb{R}^n} \| D(F_{k,l,i}) \chi_{R_i} \| d\lambda_n + \int_{\mathbb{R}^n} \| F_{k,l,i} D(\chi_{R_i}) \| d\lambda_n).$$

Now, for the first integral we have

$$\begin{aligned}
\int_{\mathbb{R}^n} \| D(F_{k,l,i})\chi_{R_i} \| d\lambda_n &= \int_{\mathbb{R}_i} \| D(F_{k,l,i}) \| d\lambda_n \\
&\leq \int_{R_i} \| D(f_k(\tau_{k,i}^{-1}))\frac{W_{k,l}(\tau_{k,i}^{-1})}{J_{k,i}(\tau_{k,i}^{-1})} \| d\lambda_n \\
&+ \int_{R_i} \| f_k(\tau_{k,i}^{-1})D\left(\frac{W_{k,l}(\tau_{k,i}^{-1})}{J_{k,i}(\tau_{k,i}^{-1})}\right) \| d\lambda_n \\
&\leq \int_{R_i} \| Df_k(\tau_{k,i}^{-1}) \| \| D\tau_{k,i}^{-1} \| \frac{W_{k,l}(\tau_{k,i}^{-1})}{J_{k,i}(\tau_{k,i}^{-1})} \| d\lambda_n \\
&+ \int_{R_i} \| f_k(\tau_{k,i}^{-1}) \| \frac{M_{k,l,i}}{J_{k,i}(\tau_{k,i}^{-1})} d\lambda_n \\
&\leq \sigma_{k,i}\pi_{k,l,i} \int_{S_i} \| Df_k \| d\lambda_n + M_{k,l,i} \int_{S_i} \| f_k \| d\lambda_n
\end{aligned}$$

For the second integral,

$$\begin{aligned}
\int_{\mathbb{R}^n} \| F_{k,l,i}D(\chi_{R_i}) \| d\lambda_n &= \int_{\partial R_i} |f_k(\tau_{k,i}^{-1})| \frac{W_{k,l}(\tau_{k,i}^{-1})}{J_{k,i}(\tau_{k,i}^{-1})} d\lambda_{n-1} \\
&= \int_{\partial S_i} |f_k| W_{k,l} \frac{J_{k,i}^0}{J_{k,i}} d\lambda_{n-1}.
\end{aligned}$$

Using Lemma 4.11 and Lemma 4.12 we get,

$$\begin{aligned}
\int_{\mathbb{R}^n} \| F_{k,l,i}D(\chi_{R_i}) \| d\lambda_n &\leq \sigma_{k,i}\pi_{k,l,i} \int_{\partial S_i} |f_k| d\lambda_{n-1} \\
&\leq \frac{\sigma_{k,i}\pi_{k,l,i}}{a} V_{S_i}(f_k) + \frac{\sigma_{k,i}\pi_{k,l,i}}{a\delta} \int_{S_i} |f_k| d\lambda_{n-1}.
\end{aligned}$$

Summing over i, we obtain

$$\begin{aligned}
V(P_{\tau_k}(W_{k,l}f_k)) &\leq (\max_{1\leq i\leq q} \sigma_{k,i}\pi_{k,l,i})(1+\frac{1}{a})V(f_k) \\
&+ \left(\max_{1\leq i\leq q} M_{k,l,i} + \frac{\max_{1\leq i\leq q}\sigma_{k,i}\pi_{k,l,i}}{a\delta}\right) \| f_k \|_1 .
\end{aligned}$$

Summing over k, we obtain

$$V((\widehat{P}_T f)_l) \leq \sigma_l(1+\frac{1}{a}) \sum_{k=1}^{K} V(f_k) + \left(M_l + \frac{\sigma_l}{a\delta}\right) \sum_{k=1}^{K} \| f_k \|_1$$

Lemma 4.14 *Let $T = \{\tau_1, \tau_2, \dots, \tau_K; W\}$ be a higher dimensional Markov switching position dependent random map. Then*

$$\begin{aligned} \| \widehat{P_T f} \|_{\widehat{BV}} &\leq \left((1+\frac{1}{a}) \sum_{l=1}^{K} \sigma_l \right) \| f \|_{\widehat{BV}} \\ &+ \left(\sum_{l=1}^{K} \left(M_l + \frac{\sigma_l}{a\delta} \right) + 1 - (1+\frac{1}{a}) \sum_{l=1}^{K} \sigma_l \right) \| f \|_1 \end{aligned}$$

Proof *The space $\widehat{BV}$ is a Banach space with norm $\| \cdot \|_{\widehat{BV}} = \sum_{k=1}^{K} \| \cdot \|_{BV}$. First, if $f = (f_1, f_2, \dots, f_K)$ with $f_k \geq 0$, then we have*

$$\begin{aligned} \|\widehat{P}_T f\|_1 &= \sum_{l=1}^{K} \|(\widehat{P}_T f)_l\|_1 = \sum_{l=1}^{K} \int_{\mathbb{R}^n} \sum_{k=1}^{K} P_{\tau_k}(W_{k,l} f_k) d\lambda_n \\ &= \sum_{k=1}^{K} \int_{\mathbb{R}^n} \sum_{l=1}^{K} P_{\tau_k}(W_{k,l} f_k) d\lambda_n = \sum_{k=1}^{K} \int_{\mathbb{R}^n} P_{\tau_k}(f_k) d\lambda_n \\ &= \sum_{k=1}^{K} \|P_{\tau_k}(f_k)\| \leq \sum_{k=1}^{K} \|f_k\| = \|f\|_1. \end{aligned}$$

For a general f it is easy to show that $\|\widehat{P}_T f\|_1 \leq \|f\|_1$. For $f \in \widehat{BV}$, by the above lemma, we obtain

$$\begin{aligned} \| \widehat{P_T f} \|_{\widehat{BV}} &= \sum_{l=1}^{K} \| (\widehat{P}_T f)_l \|_{BV} = \sum_{l=1}^{K} \left(V(\widehat{P}_T f)_l + \| (\widehat{P}_T f)_l \|_1 \right) \\ &= \sum_{l=1}^{K} V(\widehat{P}_T f)_l + \sum_{l=1}^{K} \| (\widehat{P}_T f)_l \|_1 = \sum_{l=1}^{K} V(\widehat{P}_T f)_l + \| \widehat{P}_T f \|_1 \\ &\leq \sum_{l=1}^{K} \left(\sigma_l (1+\frac{1}{a}) \sum_{k=1}^{K} V(f_k) + \left(M_l + \frac{\sigma}{a\delta} \right) \sum_{k=1}^{K} \| f_k \|_1 \right) + \| \widehat{P}_T f \|_1 \\ &\leq (1+\frac{1}{a}) \sum_{k=1}^{K} (\| f_k \|_{BV} - \| f_k \|_1) \sum_{l=1}^{K} \sigma_l + \sum_{l=1}^{K} \left(M_l + \frac{\sigma_l}{a\delta} \right) \| f \|_1 + \| f \|_1 \\ &= (1+\frac{1}{a}) (\| f \|_{\widehat{BV}} - \| f \|_1) \sum_{l=1}^{K} \sigma_l + \sum_{l=1}^{K} \left(M_l + \frac{\sigma_l}{a\delta} \right) \| f \|_1 + \| f \|_1 \\ &= \left((1+\frac{1}{a}) \sum_{l=1}^{K} \sigma_l \right) \| f \|_{\widehat{BV}} + \left(\sum_{l=1}^{K} \left(M_l + \frac{\sigma_l}{a\delta} \right) + 1 - (1+\frac{1}{a}) \sum_{l=1}^{K} \sigma_l \right) \| f \|_1 \end{aligned}$$

Theorem 4.15 *Let $T = \{\tau_1, \tau_2, \dots, \tau_K; W\}$ be a higher dimensional Markov switching position dependent random map. If*

$$\left(1 + \frac{1}{a} \right) \sum_{l=1}^{K} \sigma_l < 1,$$

then T preserves a measure which is absolutely continuous with respect to Lebesgue measure.

Proof *This follows by the above lemma and standard techniques (see [6]).*

Example 4.4 *Let $X = I^2$ and $\mathcal{P} = \{S_1, S_2, \ldots, S_{81}\}$ be a partition of X, where*

$$S_i = \{(x_1,x_2) : \frac{i-1}{9} < x_1 < \frac{i}{9}, 0 \leq x_2 < \frac{1}{9}\},$$
$$S_{i+9} = \{(x_1,x_2) : \frac{i-1}{9} < x_1 < \frac{i}{9}, \frac{1}{9} \leq x_2 < \frac{2}{9}\},$$
$$S_{i+18} = \{(x_1,x_2) : \frac{i-1}{9} < x_1 < \frac{i}{9}, \frac{2}{9} \leq x_2 < \frac{3}{9}\},$$
$$S_{i+27} = \{(x_1,x_2) : \frac{i-1}{9} < x_1 < \frac{i}{9}, \frac{3}{9} \leq x_2 < \frac{4}{9}\},$$
$$S_{i+36} = \{(x_1,x_2) : \frac{i-1}{9} < x_1 < \frac{i}{9}, \frac{4}{9} \leq x_2 < \frac{5}{9}\},$$
$$S_{i+45} = \{(x_1,x_2) : \frac{i-1}{9} < x_1 < \frac{i}{9}, \frac{5}{9} \leq x_2 < \frac{6}{9}\},$$
$$S_{i+54} = \{(x_1,x_2) : \frac{i-1}{9} < x_1 < \frac{i}{9}, \frac{6}{9} \leq x_2 < \frac{7}{9}\},$$
$$S_{i+63} = \{(x_1,x_2) : \frac{i-1}{9} < x_1 < \frac{i}{9}, \frac{7}{9} \leq x_2 < \frac{8}{9}\},$$
$$S_{i+72} = \{(x_1,x_2) : \frac{i-1}{9} < x_1 < \frac{i}{9}, \frac{8}{9} \leq x_2 \leq 1\},$$

where $i = 1, 2, \ldots, 9$. Let,

$$A = \{(x_1,x_2) : \frac{8}{9} \leq x_1 \leq 1, 0 \leq x_2 \leq 1\},$$

$$B = \{(x_1,x_2) : 0 \leq x_1 \leq \frac{1}{9}, 0 \leq x_2 \leq 1\}.$$

See Figures 4.1–Figure 4.4. Consider the Markov switching position dependent random map $T = \{\tau_1, \tau_2; W\}$ where $\tau_1, \tau_2 : I^2 \to I^2$ such that

$$\tau_1(S_i) = I^2 \setminus A, \tau_2(S_i) = I^2 \setminus B, i = 1, 2, \ldots, 81,$$

and the position dependent stochastic matrix W,

$$W = \begin{bmatrix} W_{1,1}(x) & W_{1,2}(x) \\ W_{2,1}(x) & W_{2,2}(x) \end{bmatrix},$$

where

$$W_{1,1}|_{S_j} = \begin{cases} \frac{1}{3} & \text{if j is odd} \\ \frac{2}{3} & \text{if j is even} \end{cases}, \quad W_{1,2}|_{S_j} = \begin{cases} \frac{2}{3} & \text{if j is odd} \\ \frac{1}{3} & \text{if j is even} \end{cases}$$

$$W_{2,1}|_{S_j} = \begin{cases} \frac{1}{4} & \text{if j is odd} \\ \frac{3}{4} & \text{if j is even} \end{cases}, \quad W_{2,2}|_{S_j} = \begin{cases} \frac{3}{4} & \text{if j is odd} \\ \frac{1}{4} & \text{if j is even} \end{cases}$$

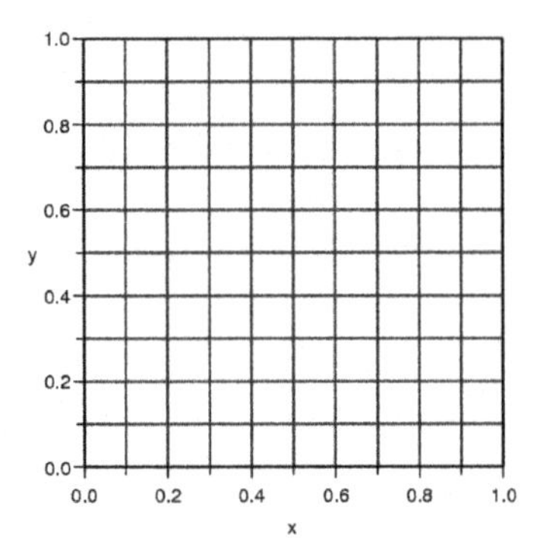

Figure 4.1 *The partitions* $\{S_i, i = 1,2,\ldots,81\}$ *of* $I^2 = I \times I$.

For the partition $\mathcal{P} = \{S_1, S_2, \ldots, S_{81}\}$ *of* I^2*, we have* $a = 1$. *The derivative matrix of* $\tau_{1,i}^{-1}$ *is* $\begin{bmatrix} 1/8 & 0 \\ 0 & 1/9 \end{bmatrix}$ *and the derivative matrix of* $\tau_{2,i}^{-1}$ *is* $\begin{bmatrix} 1/8 & 0 \\ 0 & 1/9 \end{bmatrix}$. *Therefore, the Euclidean matrix norm* $\|D\tau_{1,i}^{-1}\|$ *is* $\sqrt{145}/72$ *and the Euclidean matrix norm* $\|D\tau_{2,i}^{-1}\|$ *is* $\sqrt{145}/72$. *Thus,*

$$\sum_{l=1}^{2} \sigma_l \leq 2/3\sqrt{145}/72 + 2/3\sqrt{145}/72 + 3/4\sqrt{145}/72 + 3/4\sqrt{145}/72 \approx 0.473859045$$

and

$$\left(1 + \frac{1}{a}\right) \sum_{l=1}^{2} \sigma_l < 2(0.473859045) \approx 0.94771809 < 1.$$

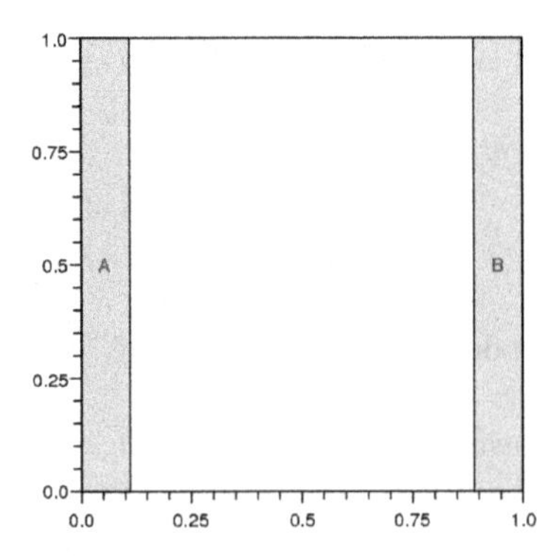

Figure 4.2 *The set* $A \subseteq I^2$ *and the set* $B \subseteq I^2$.

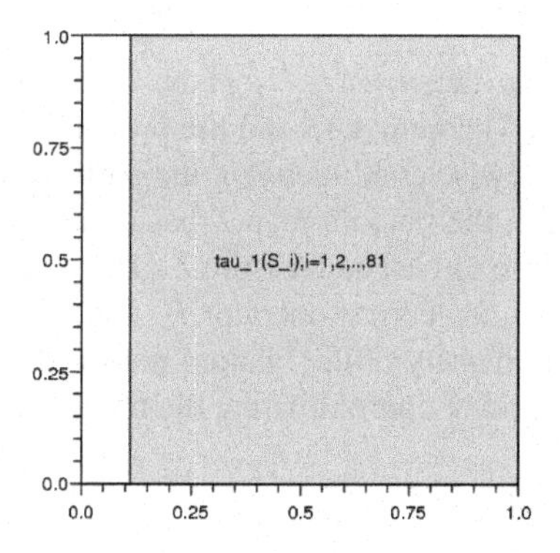

Figure 4.3 $\tau_1(S_i), i = 1,2,\ldots,81$ *of* $S_i, i = 1,2,\ldots,81$.

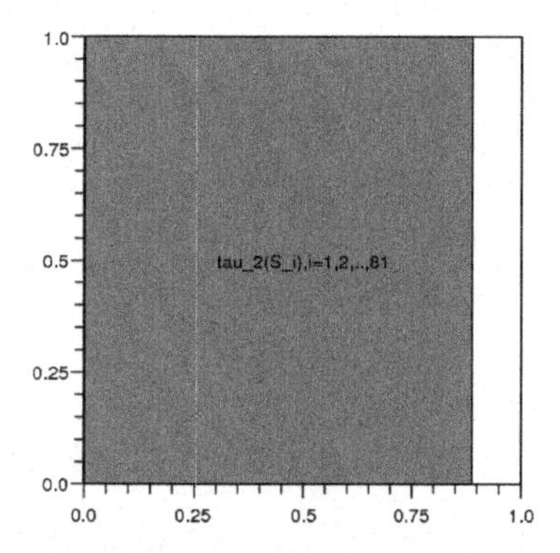

Figure 4.4 $\tau_2(S_i), i = 1,2,\ldots,81$ *of* $S_i, i = 1,2,\ldots,81$.

Therefore, by Theorem 4.15, the Markov switching position dependent random map T *admits an absolutely continuous invariant measure.*

4.5 Approximation of invariant measures for position dependent random maps

In the previous sections, we have seen that the Frobenius–Perron operator is one of the key tools for the existence of invariant measures for position dependent random maps. In this section, we describe finite dimensional numerical methods for the Frobenius–Perron operator of position dependent random maps for the approximation of invariant measures.

4.5.1 *Maximum entropy method for position dependent random maps*

Let $T = \{\tau_1, \tau_2, \ldots, \tau_K; p_1(x), p_2(x), \ldots, p_K\}$ be a position dependent random map satisfying conditions of Theorem 4.15 and the probabilities $p_1(x), p_2(x), \ldots, p_K$ are position dependent piecewise continuous on the partition $\mathcal{J} = \{J_1, J_2, \cdots, J_N\}$ of I. Hence by Theorem 4.15, the random map T has an absolutely continuous invariant measure μ with respect to Lebesgue measure λ. The invariant density f^* of μ is the fixed point of the Frobenius–Perron operator P_T defined in (4.2.2). We assume that f^* is a unique invariant density of the random map T. In this section we describe a maximum entropy method of approximating the fixed point of the operator P_T .

Let $\mathcal{D}$ be the set of all densities, that is,

$$D = \{f \in L^1(0,1) \text{ such that } \mathrm{f} \geq 0 \text{ and } \| \mathrm{f} \|_1 = \int_0^1 \mathrm{f(x)d}\lambda(\mathrm{x}) = 1\}.$$

The Boltzmann entropy [5, 19] of $f \geq 0$ is defined by

$$H(f) = -\int_I f(x) \log f(x) d\lambda(x). \tag{4.5.1}$$

For properties of H see [5, 19]. Using the Gibbs inequality

$$u - u\log u \leq v - u\log v, u, v \geq 0, \tag{4.5.2}$$

it can be shown that

$$\int_I f(x) \log f(x) d\lambda(x) \geq \int_I f(x) \log g(x) d\lambda(x) \ \forall f, g \in \mathcal{D}. \tag{4.5.3}$$

The above inequality in (4.5.3) leads to the following optimization problem [9]:

$$\max H(f) \text{ such that } \mathrm{f} \in \mathcal{D} \text{ and } \int_I \mathrm{f(x)g_n(x)d}\lambda(\mathrm{x}) = \mathrm{m_n},\ 1 \leq \mathrm{n} \leq \mathrm{N}, \tag{4.5.4}$$

where $\{g_1, g_2, ..., g_N\} \subset L^\infty(I)$ and $m_1, m_2, \cdots, m_N$ are real constants.

Proposition 4.16 [19] *Suppose that* $a_1, a_2, \cdots, a_N$ *are real numbers such that the function*

$$f(x) = \frac{e^{\sum_{n=1}^N a_n g_n(x)}}{\int_0^1 e^{\sum_{n=1}^N a_n g_n(x)} \lambda(x)} \tag{4.5.5}$$

satisfies the constraints in (4.5.4), that is,

$$\frac{\int_0^1 g_i(x) e^{\sum_{n=1}^N a_n g_n(x)} d\lambda(x)}{\int_0^1 e^{\sum_{n=1}^N a_n g_n(x)} \lambda(x)},\ i = 1, 2, \ldots, N. \tag{4.5.6}$$

Then f *solves the maximum entropy problem (4.5.4).*

Lemma 4.17 [9] *Suppose* $f \in L^1(0,1)$ *satisfies*

$$\int_I x^n f(x) d\lambda(x) = 0, \ n = 0, 1, \cdots.$$

Then $f = 0$.

Proposition 4.18 *Let* P_T *be the Frobenius–Perron operator defined in (4.2.2) of a random map* $T = \{\tau_1, \tau_2, \cdots, \tau_K; p_1(x), p_2(x), \ldots, p_K\}$ *with position dependent probabilities. Then,* $f^* \in \mathcal{D}$ *is a fixed point of* P_T *if and only if*

$$\int_I \left[x^n - \sum_{k=1}^{K} p_k(x)(\tau_k(x))^n \right] f^*(x) d\lambda(x) = 0, \ n = 1, 2, \cdots. \tag{4.5.7}$$

Proof *Suppose that* $f^* \in \mathcal{D}$ *is a fixed point of the Frobenius–Perron operator* P_T *, that is,*

$$(P_T f^*)(x) = f^*(x). \tag{4.5.8}$$

Then for $n = 1, 2, \cdots,$

$$\begin{aligned}
\int_I (f^*(x) x^n) d\lambda(x) &= \int_I (P_T f^*)(x) x^n d\lambda(x) \\
&= \int_I \sum_{k=1}^{K} (P_{\tau_k}(p_k f^*))(x) x^n d\lambda(x) \\
&= \sum_{k=1}^{K} \int_I (p_k f^*)(x) U_{\tau_k}(x^n) d\lambda(x) \\
&= \sum_{k=1}^{K} \int_I p_k(x) f^*(x) (\tau_k(x))^n d\lambda(x) \\
&= \sum_{k=1}^{K} \int_I (p_k(x))(\tau_k(x))^n f^*(x) d\lambda(x)
\end{aligned}$$

Thus,

$$\int_I \left[x^n - \sum_{k=1}^{K} p_k(x)(\tau_k(x))^n \right] f^*(x) d\lambda(x) = 0, \ n = 1, 2, \cdots.$$

Conversely, suppose f^* *satisfy (4.5.7), that is,*

$$\int_I (x^n f^*(x)) \, d\lambda(x) = \int_I f^*(x) \sum_{k=1}^{K} p_k(x)(\tau_k(x))^n d\lambda(x).$$

Now,

$$\begin{aligned}
\int_I (f^*(x)x^n)d\lambda(x) &= \int_I f^*(x)\sum_{k=1}^{K} p_k(x)(\tau_k(x))^n d\lambda(x) \\
&= \int_I f^*(x)\sum_{k=1}^{K} p_k(x)U_{\tau_k}(x^n)d\lambda(x) \\
&= \sum_{k=1}^{K}\int_I f^*(x)p_k(x)U_{\tau_k}(x^n)d\lambda(x) \\
&= \sum_{k=1}^{K}\int_I (P_{\tau_k}(p_k f^*))(x)x^n d\lambda(x) \\
&= \int_I \sum_{k=1}^{K}(P_{\tau_k}(p_k f^*))(x)x^n d\lambda(x) \\
&= \int_I (P_T f^*)(x)x^n d\lambda(x).
\end{aligned}$$

Thus,

$$\int_I (f^*(x)-(P_T f^*)(x))x^n d\lambda(x) = 0,\ n = qw1,\cdots.$$

Moreover,

$$\int_I (f^*(x)-(P_T f^*)(x))x^0 d\lambda(x) = \int_I (f^*(x)-(P_T f^*)(x))d\lambda(x) = 0.$$

Thus,

$$\int_I [f^*(x)-(P_T f^*)(x)]\,x^n d\lambda(x) = 0,\ n = 0,1,2,\cdots.$$

By Lemma 4.17, $f^*(x)-(P_T f^*)(x) = 0$. *This proves that*

$$(P_T f^*)(x) = f^*(x).$$

From the above proposition it is easy to see that the fixed point problem (4.5.8) of the Frobenius–Perron operator P_T for the position dependent random map T is equivalent to homogeneous moment problem (4.5.7). We propose the following maximum entropy method for the fixed point problem (4.5.8) of the Frobenius–Perron operator P_T for the position dependent random map T.

Algorithm: Choose N and solve the maximum entropy problem:

$$\max H(f) \text{ such that}$$

$$f \in \mathcal{D} \text{ and } \int_0^1 \left[x^n - \sum_{k=1}^{K} p_k(x)(\tau_k(x))^n\right] f(x)d\lambda(x) = 0, \tag{4.5.9}$$

where $n = 1,2,\cdots,N$ to get an approximate solution f_N of the Frobenius–Perron equation (4.5.8). In the following theorem we prove that there are constants $a_1, a_2, \cdots, a_N$ such that

$$\int_0^1 \left[x^i - \sum_{k=1}^{K} p_k(x)(\tau_k(x))^i\right] e^{\sum_{n=1}^{N} a_n[x^n - \sum_{k=1}^{K} p_k(x)(\tau_k(x))^n]} dx = 0, \; i = 1, 2, \cdots, N \tag{4.5.10}$$

which implies that, by Proposition 4.16, the corresponding density

$$f_N(x) = \frac{e^{\sum_{n=1}^{N} a_n[x^n - \sum_{k=1}^{K} p_k(x)(\tau_k(x))^n]}}{\int_0^1 e^{\sum_{n=1}^{N} a_n[x^n - \sum_{k=1}^{K} p_k(x)(\tau_k(x))^n]} dx} \tag{4.5.11}$$

is a solution to (4.5.9).

Theorem 4.19 *Suppose $f^*(x) > 0$ on I and*

$$T = \{\tau_1, \tau_2, \cdots, \tau_K; p_1(x), p_2(x), \cdots, p_K(x)\}$$

be a random map with position dependent probabilities $p_1(x), p_2(x), \cdots, p_K(x)$ such that the set

$$\{x - \sum_{k=1}^{K} p_k(x)\tau_k(x), x^2 - \sum_{k=1}^{K} p_k(x)(\tau_k(x))^2, \cdots, x^N - \sum_{k=1}^{K} p_k(x)(\tau_k(x))^N\}$$

are linearly independent. Then there exist constants $a_1, a_2, \cdots, a_N$ such that for each N the density f_N in (4.5.11) is a solution to the maximum entropy problem in (4.5.9).

Proof *For $a = (a_1, a_2, \cdots, a_N) \in \mathbb{R}^N$, define*

$$Z(a) = \int_0^1 e^{\sum_{n=1}^{N} a_n[x^n - \sum_{k=1}^{K} p_k(x)(\tau_k(x))^n]} dx \tag{4.5.12}$$

and

$$G(a) = \log Z(a) = \log\{\int_0^1 e^{\sum_{n=1}^{N} a_n[x^n - \sum_{k=1}^{K} p_k(x)(\tau_k(x))^n]} dx\}. \tag{4.5.13}$$

Then,

$$\frac{\partial G}{\partial a_i} = \frac{\int_0^1 [x^i - \sum_{k=1}^{K} p_k(x)(\tau_k(x))^i]\, e^{\sum_{n=1}^{N} a_n[x^n - \sum_{k=1}^{K} p_k(x)(\tau_k(x))^n]} dx}{\int_0^1 e^{\sum_{n=1}^{N} a_n[x^n - \sum_{k=1}^{K} p_k(x)(\tau_k(x))^n]} d\lambda(x)}$$

for each $i = 1, 2, \cdots, N$. Thus the constraints in (4.5.10) are equivalent to

$$\text{grad G} = \left(\frac{\partial \text{G}}{\partial \text{a}_1}, \frac{\partial \text{G}}{\partial \text{a}_2}, \cdots, \frac{\partial \text{G}}{\partial \text{a}_\text{N}}\right) = 0.$$

Moreover,

$$G(a) = \log \int_0^1 e^r \Pi_N(x) dx,$$

where $a = r\alpha, r = \| a \|_2$ *is the Euclidean length of a,* $\alpha = (\alpha_1, \alpha_2, \cdots, \alpha_N)$ *is the directional cosine of a and* $\Pi_N(x) = \sum_{n=1}^N \alpha_n \left[x^n - \sum_{k=1}^K p_k(x)(\tau_k(x))^n\right]$. *For a given* α*, suppose that* $\Pi_N(x) \le 0, x \in [0,1]$ *almost everywhere. Since* $\Pi_N \ne 0$ *and* $f^*(x) > 0$ *on* $[0,1]$ *by assumption, it follows that*

$$\int_0^1 \Pi_N(x) f^*(x) < 0$$

which contradicts the fact that

$$\sum_{n=1}^N \alpha_n \int_0^1 \left[x^n - \sum_{k=1}^K p_k(x)(\tau_k(x))^n\right] f^*(x) d\lambda(x) = 0$$

by Eq. (4.5.7). Thus, $\Pi_N(x) > 0$ *on some subinterval of* $[0,1]$ *from which*

$$\lim_{\|a\|_2 \to \infty} G(a) = \infty. \tag{4.5.14}$$

Therefore G is coercive. Therefore $G : \mathbb{R}^N \to \mathbb{R}$ *achieves its global minimizer at which its gradient is zero. This shows that there exist constants* $a_1, a_2, \cdots, a_N$ *such that* f_N *in (4.5.11) is a solution to the maximum entropy problem in problem (4.5.9).*

We consider a random map T with position dependent probabilities $p_1(x), p_2(x), \ldots, p_K(x)$ such that the condition of the above Theorem 4.19 is satisfied for each N. Hence the maximum entropy method is well posted. Let

$$F(a) = (F_1(a), F_2(a), \cdots, F_N(a)) = 0, \tag{4.5.15}$$

where

$$F_i(a_1, a_2, \cdots, a_N) = \int_0^1 \left[x^i - \sum_{k=1}^K p_k(x)(\tau_k(x))^i\right] e^{\sum_{n=1}^N a_n [x^n - \sum_{k=1}^K p_k(x)(\tau_k(x))^n]} dx,$$

$i = 1, 2, \cdots, N$. The main work in the maximum entropy algorithm is solving the nonlinear equations

$$F(a) = (F_1(a), F_2(a), \cdots, F_N(a)) = 0. \tag{4.5.16}$$

4.5.1.1 *Convergence of the maximum entropy method for random map*

Theorem 4.20 *Let* $T = \{\tau_1, \tau_2, \ldots, \tau_K; p_1(x), p_2(x), \ldots, p_K(x)\}$ *be a random map with position dependent probabilities. Suppose that the unique fixed point* f^* *of the Frobenius–Perron operator* P_T *of the random map T satisfies the condition that* $H(f^*) > -\infty$. *Then* $\lim_{N \to \infty} f_N = f^*$ *weakly, where* f_N *is defined in (4.5.11).*

Proof f^* *is the unique solution of Eq. (4.5.7) in* $\mathcal{D}$ *because* f^* *is the unique solution of Eq. (4.5.8) in* $\mathcal{D}$. *Thus* f^* *is the unique solutions to Eq. (4.5.8) with* $N = \infty$. *The space* $L^1(0,1)$ *is a locally convex topological vector space under the weak topology. The level sets of H are weakly compact by Lemma 2.1 (iii) in [9]. Hence the theorem follows from Lemma 4.1 in [9].*

Lemma 4.21 $\lim_{N\to\infty} H(f_N) = H(f^*)$.

Proof *The set* $\{H(f_N)\}$ *is non-increasing and lower bounded by* $H(f^*)$. *Suppose that* $H(f_N) \geq \beta$ *for all* N *with* $\beta > -\infty$. *Let*

$$E = \{f \in \mathcal{D} : H(f) \geq \beta\}.$$

By Lemma 2.1 (iii) in [9] E *is weakly compact. Moreover,* $\{f_N\} \subset E$ *and there is a subsequence* $\{f_{N_k}\}$ *of* $\{f_N\}$ *such that* f_{N_k} *converges to* $\bar{f}$ *for some* $\bar{f} \in E$. *Thus,* $H(\bar{f}) \geq \beta$. *Let*

$$F_N = D \cap \{x - \sum_{k=1}^{K} p_k(x)\tau_k(x), x^2 - \sum_{k=1}^{K} p_k(x)(\tau_k(x))^2, \ldots, x^N - \sum_{k=1}^{K} p_k(x)(\tau_k(x))^N\}^+,$$

where

$$A^+ = \{g \in L^1(0,1) : \int_0^1 f(x)g(x)d\lambda(x) = 0, \ \forall f \in A\}$$

is the annihilator of $A \subset L^1(0,1)$. *Then* F_N *is weakly closed. By similar arguments in [9], it can be shown that* $\bar{f} \in \cap_{N=1}^{\infty} F_N$. *Thus,* $H(f^*) \geq H(\bar{f}) \geq \beta$ *and thus the theorem follows.*

Theorem 4.22 *Let* $T = \{\tau_1, \tau_2, \ldots, \tau_K; p_1(x), p_2(x), \ldots, p_K(x)\}$ *be a random map with position dependent probabilities. Suppose that the unique fixed point* f^* *of the Frobenius–Perron operator* P_T *of the random map* T *satisfies the condition that* $H(f^*) > -\infty$. *Then* $\lim_{N\to\infty} \| f_N - f^* \|_1 = 0$.

Proof f_N *converges weakly to* f^* *by Theorem 4.20 and* $H(f_N) \to H(f^*) > \infty$ *by Lemma 4.21. The proof of the theorem follows from the standard consequences of an inequality in [5] (Lemma 2.2 (b)) and Lemma 2.5 in [5]).*

4.5.2 *Invariant measures of position dependent random maps via interpolation*

Let $T = \{\tau_1, \tau_2, \ldots, \tau_K; p_1(x), p_2(x), ..., p_K(x)\}$ be a position dependent random map, where $\tau_k : [0,1] \to [0,1], k = 1,2,...,K$ are piecewise one-to-one and differentiable, non-singular transformations on a common partition $\mathcal{J} = \{J_1, J_2, ..., J_q\}$ of $[0,1]$. Let $g_k(x) = \frac{p_k(x)}{|\tau_k'(x)|}, k = 1,2,...,K$. Assume that the random map T satisfies the following conditions: (i) $\sum_{k=1}^{K} g_k(x) < \alpha < 1, x \in [0,1]$; (ii) $g_k \in BV([0,1]), k = 1,2,...,K$. Moreover, we assume that the position dependent random map $T = \{\tau_1, \tau_2, \ldots, \tau_K; p_1(x), p_2(x), \ldots, p_K(x)\}$ satisfies the following assumptions:

(A) there exists $A = 3\alpha + \max_{1\leq i\leq q} \sum_{k=1}^{K} V_{J_i} g_k < 1$ and $B = 2\beta\alpha + \beta \max_{1\leq i\leq q} \sum_{k=1}^{K} V_{J_i} g_k > 0$

with $\beta = \max_{1\leq i\leq q} \frac{1}{\lambda(I_i)}$ such that $\forall f \in BV([0,1])$

$$V_{[0,1]} P_T f \leq A V_{[0,1]} f + B \| f \|_1 .$$

(B) T has a unique acim μ with density f^*.

For the existence of absolutely continuous invariant measures for T see Section 4.2.3.2. Note that the invariant density f^* of the unique acim μ is the fixed point of the Frobenius–Perron operator P_T. In the following we describe a piecewise linear approximation scheme for f^* via interpolation and we present the proof of convergence of our approximation scheme.

Let $\mathcal{P}^{(n)} = \{I_1, I_2, \dots, I_n\}$ be a partition of $I = [0,1]$ into n subintervals $I_i = [x_i, x_{i+1}], i = 0,1,2,\dots n$ of equal length $h = \frac{1}{n}$, where $x_i = ih$. Let Δ_n be the linear spline space over the interval $[0,1]$ with the above nodes and the following canonical basis:

$$\phi_i(x) = g(\frac{x - x_i}{h}), i = 0,1,2,\dots,n, \tag{4.5.17}$$

where the function $g : \mathbb{R} \to [0,1]$ is defined by

$$g(x) = \begin{cases} 1 + x, & -1 \le x \le 0, \\ 1 - x, & 0 \le x \le 1, \\ 0, & \text{otherwise.} \end{cases} \tag{4.5.18}$$

It is easy to show (see [10]) that Δ_n is a subspace of $L^1([0,1]) \cap L^\infty([0,1])$ with dimension $n+1$ and the canonical basis $\{\phi_i\}_{i=0}^n$ has the following properties:

(a) $\phi_i(x_i) = 1$ and $\phi_i(x_j) = 0$ for $j \neq i$.

(b) $\| \phi_i \| = \frac{1}{n}$ for $i = 1,2,\dots n-1$, $\| \phi_0 \| = \| \phi_n \| = \frac{1}{2n}$.

(c) $\phi_i \ge 0$ for $i = 0,1,2,\dots,n$ and $\sum_{i=0}^n \phi_i(x) = 1$.

(d) Suppose $f \in \Delta_n$. Then $f = \sum_{i=0}^n \alpha_i \phi_i(x)$ if and only if $f(x_i) = \alpha_i$ for $i = 0,1,2,\dots,n$.

For each $i = 0,1,\dots,n$, let L_i be a small interval around the node x_i. Consider the following generalized interpolation operator $Q_n : L^1([0,1]) \to \Delta_n$ defined by:

$$Q_n f(x_i) = \frac{1}{\lambda(L_i)} \int_{L_i} f(x), \ 0 \le i \le n. \tag{4.5.19}$$

The following lemma was proved in [10]:

Lemma 4.23 *1.* $\forall f \in L^1([0,1]), \lim_{n\to\infty} \| Q_n f - f \| = 0$.
2. $\forall f \in BV([0,1])$, $V_{[0,1]} Q_n f \le V_{[0,1]} f$, $\forall n$.

Now we are ready to define a sequence of finite dimensional approximations of the

Frobenius–Perron operator P_T. For each $n \geq 1$, let $P_n : L^1([0,1]) \to L^1([0,1])$ be an operator defined by

$$P_n f = Q_n \circ P_T f, \tag{4.5.20}$$

where P_T is the Frobenius–Perron operator of the position dependent random map T.

Lemma 4.24 *Let T be a random map satisfying conditions (A) – (B). Then, for any integer $n \geq 1, P_n$ has a fixed point $f_n \in \Delta_n$ and f_n is of bounded variation.*

Proof *Let $f \in BV([0,1])$. Then for each $n \geq 1$, $P_n f = Q_n \circ P_T f_n$. By Lemma 4.23,*

$$V_{[0.1]} P_n f = V_{[0,1]} Q_n \circ P_T f \leq V_{[0,1]} P_T f.$$

By Lemma 11.5,

$$V_{[0,1]} P_n f = V_{[0,1]} Q_n \circ P_T f \leq V_{[0,1]} P_T f \leq A V_{[0,1]} f + B \parallel f \parallel_1 .$$

The lemma follows by standard technique in [2] (see also [6]).

In the following we construct the matrix representation of P_n which will be useful for numerical examples in this section.

The matrix representation of P_n : For $i = 0, 1, 2 \ldots, n$,

$$\begin{aligned}
P_n \phi_i(x) &= Q_n \circ P_T \phi_i(x) \\
&= Q_n \circ \sum_{k=1}^{K} \left(P_{\tau_k}(p_k \phi_i) \right)(x) \\
&= \sum_{j=0}^{n} \left(\frac{1}{\lambda(L_j)} \int_{L_j} \sum_{k=1}^{K} \left(P_{\tau_k}(p_k \phi_i) \right)(x) d\lambda \right) \phi_j(x) \\
&= \sum_{j=0}^{n} \left(\frac{1}{\lambda(L_j)} \int_{\tau_k^{-1}(L_j)} \sum_{k=1}^{K} (p_k \phi_i)(x) d\lambda \right) \phi_j(x) \\
&= \sum_{j=0}^{n} \left(\frac{1}{\lambda(L_j)} \sum_{k=1}^{K} \int_{\tau_k^{-1}(L_j)} (p_k \phi_i)(x) d\lambda \right) \phi_j(x) \\
&= \sum_{j=0}^{n} \left(\sum_{k=1}^{K} \frac{1}{\lambda(L_j)} \int_{\tau_k^{-1}(L_j)} (p_k(x) \phi_i(x)) \, d\lambda \right) \phi_j(x)
\end{aligned}$$

For $i = 0, 1, 2, \ldots, n$, let

$$m_{ij} = \sum_{k=1}^{K} m_{\tau_k, ij}, \; j = 0, 1, 2, \ldots, n, \tag{4.5.21}$$

where

$$m_{\tau_k, ij} = \frac{1}{\lambda(L_j)} \int_{\tau_k^{-1}(L_j)} p_k(x) \phi_i(x) d\lambda, \; j = 0, 1, 2, \ldots, n \tag{4.5.22}$$

For computational simplicity, let $n = l \times N$, where l is a positive even integer, N is the number of pieces of the component maps on $[0,1]$ for the random map T and let the probabilities $p_1(x), p_2(x), \ldots, p_k(x)$ be position dependent piecewise constant on the partition $\mathcal{J} = \{J_1, J_2, \ldots\ldots, J_N\}$ of I, that is,

$$p_k = \left[p_{k,1}, p_{k,2}, \ldots, p_{k,N}\right], k = 1,2\ldots,K, p_{k,i} = p_k|_{J_i}.$$

Note that if $l(q-1)+1 \le j \le lq, q = 1,2,\ldots,N$, then we have $p_k(x) = p_{k,q}$. The matrix

$$\mathbb{M}_n = (m_{ij})_{i,j=0}^{n} \tag{4.5.23}$$

is the representation of the operator P_n and $\mathbb{M}_n$ is non-negative.

Let $c = (c_0, c_1, \ldots, c_n)$ be a normalized left eigenvector of $\mathbb{M}_n$ corresponding to the dominant eigenvalue of $\mathbb{M}_n$. Define the piecewise linear approximate density as follows:

$$f_n = \sum_{i=0}^{n} c_i \phi_i. \tag{4.5.24}$$

Lemma 4.25 *Let $T = \{\tau_1, \tau_2, \ldots, \tau_K; p_1(x), p_2(x), \ldots, p_K(x)\}$ be a position dependent random map satisfying conditions of $(A)-(B)$. Let f_n be the sequence of approximate densities defined in (4.5.24). Then the sequence $\{V_I f_n\}$ is uniformly bounded.*

Proof

$$V_I f_n = V_I P_n f_n = V_I Q_n P_T f_n \le V_I P_T f_n \le A V_I f_n + B \parallel f_n \parallel_1,$$

where $0 < A < 1$, and $B > 0$. Since $0 < A < 1$, we have $V_I f_n \le \frac{B}{1-A} \parallel f_n \parallel_1 < \infty$. Thus, the sequence $\{V_I f_n\}$ is uniformly bounded.

Theorem 4.26 *Let $T = \{\tau_1, \tau_2, \ldots, \tau_K; p_1(x), p_2(x), \ldots, p_K(x)\}$ be a random map with position dependent probabilities. Assume that the random map T satisfies conditions of $(A)-(B)$ and T has a unique acim with density f^*. Let f_n be the sequence of approximate densities of f^* from the interpolation method described above. Then* $\lim_{n\to\infty} \parallel f_n - f^* \parallel_1 = 0.$.

Proof *From Lemma 4.25, the sequence $\{f_n\}_{n\ge 1}$ is uniformly bounded in variation. By Helly's Theorem $\{f_n\}_{n\ge 1}$ is pre-compact. Let $\{f_{n_k}\}$ be a subsequence of $\{f_n\}_{n\ge 1}$ and f_{n_k} converges to $g \in L^1([0,1])$. Now,*

$$\begin{aligned} \parallel P_T g - g \parallel_1 &= \parallel g - f_{n_k} + f_{n_k} - Q_{n_k} \circ P_T f_{n_k} + Q_{n_k} \circ P_T f_{n_k} \\ &\quad - Q_{n_k} \circ P_T g + Q_{n_k} \circ P_T g - P_T g \parallel_1 \\ &\le \parallel g - f_{n_k} \parallel_1 + \parallel f_{n_k} - Q_{n_k} \circ P_T f_{n_k} \parallel_1 \\ &\quad + \parallel Q_{n_k} \circ P_T f_{n_k} - Q_{n_k} \circ P_T g \parallel_1 + \parallel Q_{n_k} \circ P_T g - P_T g \parallel_1 . \end{aligned}$$

$\parallel g - f_{n_k} \parallel_1 \to 0$ *as* $n \to \infty$ *because* $f_{n_k} \to g$. $\parallel f_{n_k} - Q_{n_k} \circ P_T f_{n_k} \parallel_1 = 0$ *because* $Q_{n_k} \circ P_T f_{n_k} = P_{n_k} f_{n_k} = f_{n_k}$. $\parallel Q_{n_k} \circ P_T f_{n_k} - Q_{n_k} \circ P_T g \parallel_1 \to 0$ *because* $Q^{n_k} \circ P_T f_{n_k} - Q_{n_k} \circ P_T g = Q_{n_k} \circ P_T (f_{n_k} - g)$ *and* $(f_{n_k} - g) \to 0$ *as* $n \to \infty$. *Finally,* $\parallel Q_{n_k} \circ P_T g -$

$P_T g \|_1 = 0$ because $Q_{n_k} h \to h$ for all $h \in L^1([0,1])$. Thus, $P_T g = g$. By the uniqueness of the fixed point of P_T we have shown that all the convergent subsequences of $\{f_n\}$ converge to the same density $g = f^*$. This shows that $\lim_{n\to\infty} f_n = f^*$.

Example 4.5 *In the following, we consider a position dependent random map T satisfying (A)–(B) with unique invariant density f^* and we apply our interpolation scheme described in the previous section. Moreover, we find the L^1 norms $\| f^* - f_n \|_1$, for some $n \geq 1$ where f_n is an approximation of f^*. We compare the interpolation scheme with Ulam's approximation method.*

We consider the position dependent random map $T = \{\tau_1(x), \tau_2(x); p_1(x), p_2(x)\}$ where $\tau_1, \tau_2 : [0,1] \to [0,1]$ are defined by

$$\tau_1(x) = \begin{cases} 3x + \frac{1}{4}, & 0 \leq x < \frac{1}{4}, \\ 3x - \frac{3}{4}, & \frac{1}{4} \leq x < \frac{1}{2}, \\ 4x - 2, & \frac{1}{2} \leq x < \frac{3}{4}, \\ 4x - 3, & \frac{3}{4} \leq x \leq 1 \end{cases}$$

$$\tau_2(x) = \begin{cases} 4x, & 0 \leq x < \frac{1}{4}, \\ 4x - 1, & \frac{1}{4} \leq x < \frac{1}{2}, \\ 3x - \frac{3}{2}, & \frac{1}{2} \leq x < \frac{3}{4}, \\ 3x - \frac{9}{4}, & \frac{3}{4} \leq x \leq 1, \end{cases}$$

and the position dependent probabilities $p_1, p_2 : [0,1] \to [0,1]$ are defined by

$$p_1(x) = \begin{cases} \frac{1}{4}, & 0 \leq x < \frac{1}{4}, \\ \frac{1}{4}, & \frac{1}{4} \leq x < \frac{1}{2}, \\ \frac{3}{4}, & \frac{1}{2} \leq x < \frac{3}{4}, \\ \frac{3}{4}, & \frac{3}{4} \leq x \leq 1 \end{cases}$$

and

$$p_2(x) = \begin{cases} \frac{3}{4}, & 0 \leq x < \frac{1}{4}, \\ \frac{3}{4}, & \frac{1}{4} \leq x < \frac{1}{2}, \\ \frac{1}{4}, & \frac{1}{2} \leq x < \frac{3}{4}, \\ \frac{1}{4}, & \frac{3}{4} \leq x \leq 1 \end{cases}$$

If $x \in [0,\frac{1}{4})$, *then* $\sum_{k=1}^{2} g_k(x) = \sum_{k=1}^{2} \frac{p_k(x)}{|\tau_k'(x)|} = \frac{\frac{1}{4}}{3} + \frac{\frac{3}{4}}{4} = \frac{13}{48} < 1$. *If* $x \in [\frac{1}{4},\frac{1}{2})$, *then* $\sum_{k=1}^{2} g_k(x) = \sum_{k=1}^{2} \frac{p_k(x)}{|\tau_k'(x)|} = \frac{\frac{1}{4}}{3} + \frac{\frac{3}{4}}{4} = \frac{13}{48} < 1$. *If* $x \in [\frac{1}{2},\frac{3}{4})$, *then* $\sum_{k=1}^{2} g_k(x) = \sum_{k=1}^{2} \frac{p_k(x)}{|\tau_k'(x)|} = \frac{\frac{3}{4}}{4} + \frac{\frac{1}{4}}{3} = \frac{13}{48} < 1$. *If* $x \in [\frac{1}{2},\frac{3}{4})$, *then* $\sum_{k=1}^{2} g_k(x) = \sum_{k=1}^{2} \frac{p_k(x)}{|\tau_k'(x)|} = \frac{\frac{3}{4}}{4} + \frac{\frac{1}{4}}{3} = \frac{13}{48} < 1$. *Moreover,* $A = 3\alpha + \max_{1\le i\le q} \sum_{k=1}^{K} V_{J_i} g_k = 3\cdot\frac{13}{48} + 0 = \frac{39}{48} < 1$. *It can be easily shown that* $B = 2\beta\alpha + \beta \max_{1\le i\le q} \sum_{k=1}^{K} V_{J_i} g_k > 0$ *with* $\beta = \max_{1\le i\le q} \frac{1}{\lambda(J_i)}$. *Thus, the random map* T *satisfies condition (A).*

It can be easily shown by the Lasota–Yorke result [21] that both τ_1 *and* τ_2 *have acim. Moreover,* τ_1 *and* τ_2 *are piecewise linear, expanding, and Markov transformations. The matrix representation of the Frobenius–Perron operator* P_{τ_1} *is the transpose of* M_{τ_1} *where*

$$M_{\tau_1} = \begin{bmatrix} 0 & \frac{1}{3} & \frac{1}{3} & \frac{1}{3} \\ \frac{1}{3} & \frac{1}{3} & \frac{1}{3} & 0 \\ \frac{1}{4} & \frac{1}{4} & \frac{1}{4} & \frac{1}{4} \\ \frac{1}{4} & \frac{1}{4} & \frac{1}{4} & \frac{1}{4} \end{bmatrix}.$$

The matrix representation of the Frobenius–Perron operator P_{τ_2} *is the transpose of* M_{τ_2} *where*

$$M_{\tau_2} = \begin{bmatrix} \frac{1}{4} & \frac{1}{4} & \frac{1}{4} & \frac{1}{4} \\ \frac{1}{4} & \frac{1}{4} & \frac{1}{4} & \frac{1}{4} \\ \frac{1}{3} & \frac{1}{3} & \frac{1}{3} & 0 \\ \frac{1}{3} & \frac{1}{3} & \frac{1}{3} & 0 \end{bmatrix}.$$

It is easy to show that both τ_1 *and* τ_2 *have unique acim. Thus, the random map* $T = \{\tau_1(x), \tau_2(x); p_1(x), p_2(x)\}$ *also has a unique acim (see Proposition 1 in [14]) and thus* T *satisfies condition (B).*

The matrix representation of the Frobenius–Perron operator $P_T f = \sum_{k=1}^{2} P_{\tau_k}(p_k f)(x)$ *is the transpose of the matrix* M_T *where*

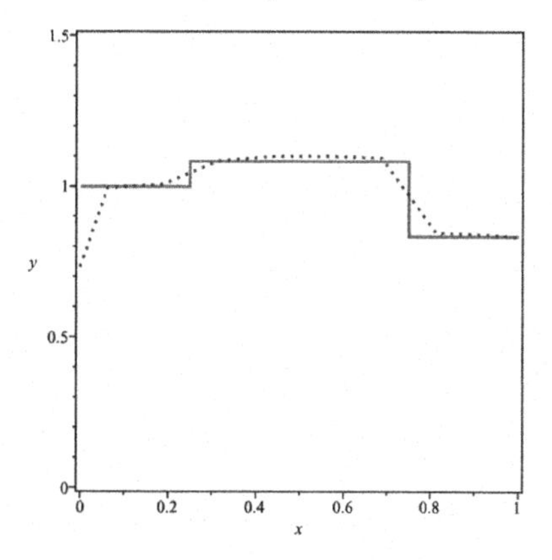

Figure 4.5 *Interpolation method for the random map T: The actual density function f^* (solid curve) and piecewise linear approximate density function f_n (dotted curve) with $n = 16$.*

$$M_T = \begin{bmatrix} \frac{3}{16} & \frac{13}{16} & \frac{13}{16} & \frac{13}{16} \\ \frac{13}{16} & \frac{13}{16} & \frac{13}{16} & \frac{3}{16} \\ \frac{13}{16} & \frac{13}{16} & \frac{13}{16} & \frac{3}{16} \\ \frac{13}{16} & \frac{13}{16} & \frac{13}{16} & \frac{13}{16} \end{bmatrix}.$$

The normalized density f^* of the unique acim of the random map T is the left eigenvector of the matrix M_T associated with the eigenvalue 1 (after adding the normalizing condition). In fact, $f^* = \left[1, \frac{13}{12}, \frac{13}{12}, \frac{5}{6}\right]$. The L^1-norm $\| f_n - f^* \|_1$ is measured (with Maple 15) to estimate the convergence of the approximate density f_n to the actual density f^* for our piecewise linear method via interpolation.

n	$\| f_n - f^* \|_1$
4	0.061819
16	0.025930
32	0.016538
64	0.009769

In the following figures we have plotted the actual density and approximate densities for interpolation method and Ulam's method.

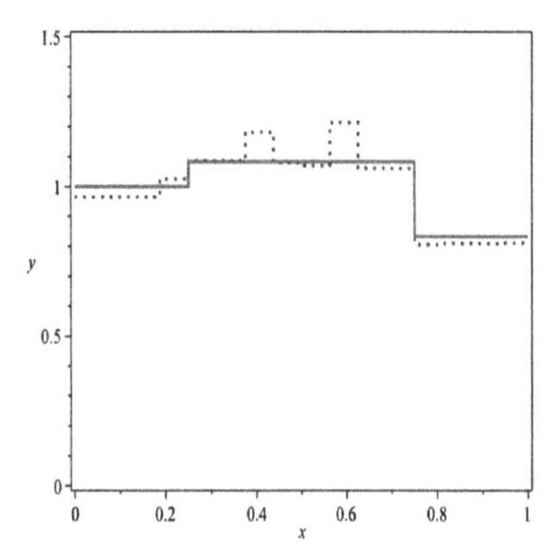

Figure 4.6 *Ulam's method for the random map T: The actual density function f^* and piecewise constant approximate density function f_n with $n = 16$.*

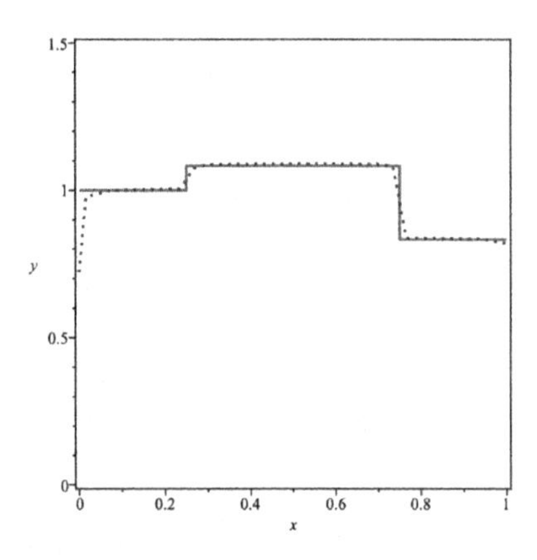

Figure 4.7 *Interpolation method for the random map T: The actual density function f^* (solid curve) and piecewise linear approximate density function f_n (dotted curve) with $n = 64$.*

4.6 Applications in finance

In Chapter 3, we considered the classical binomial models induced by random maps with constant probabilities and applied classical binomial model for binomial stock prices, options, interest rates, and other financial instruments. In a classical binomial model, there are just two possibilities for the value of a financial instrument: an up move by an up factor u or a down move by a down factor d. Note that if the factors u and d are constant then there is a possibility that eventually the value of the financial instrument becomes zero, or it increases without bound. These are unusual cases. In reality, the probability of the value of a financial instrument going up or down is not constant and may depend on current value. Moreover, it is reasonable to assume that the factors u and d are functions of current value. That is, at time t, u and d depend on the value of the financial instrument at time $t-1$.

4.6.1 *Generalized binomial model for stock prices*

In this section, we construct generalized binomial models induced by position dependent random maps and study the evolution of stock prices. Our construction of generalized binomial model is based on [4]. We consider the following assumptions:

- We consider options with one stock S. Let $S(n)$ be the price of the stock at time n.
- There is no transaction cost.
- At each period there are two possibilities: the stock price may go up by a factor u or it may go down by a factor d where the factors u and d are functions of the prices [4]: $u(x):(0,1)\to(1,\infty)$ and $d(x):(0,1)\to(0,1)$; i.e., at time n, u, and d depend on the price of stock at time $n-1$. The probabilities p_u, p_d are price dependent.

Given the factors $u(x), d(x)$ and the probabilities p_u, p_d at time $n=0$, we construct a random map.

$T=\{\tau_u(x),\tau_d(x);p_u(x),p_d(x)\}$ as follows: at time $n+1$, consider the up price to be $\tau_u(S(n))$with probability $p_u(S(n))$ and the down price to be $\tau_d(S(n))$ with probability $p_d(S(n))$. Also, $S(n+1)=u(S(n))\cdot S(n)$ or $S(n+1)=d(S(n))\cdot S(n)$. Therefore, the transformations τ_u, τ_d are given by the following formula:

$$\tau_u(x)=u(x)\cdot x, \tau_d(x)=d(x)\cdot x.$$

For each starting price $S(0)$ at time $n=0$, the random map will generate a stock price trajectory. Moreover, we can extract the up stock price $S_u(n)$, down price $S_d(n)$ for $n=1,2,\ldots N$. If the position dependent random map T satisfies conditions of Theorem 4.6, then by Theorem 4.6 T admits an absolutely continuous invariant measure $\mu=f^*\lambda$, where f^* is the invariant density of T. The invariant density allows us to find the probability: $\mu\{x:T(x)\in(\delta_1,\delta_2)\}\subset[0,1]=\mu(\delta_1,\delta_2)$. For more information see [4].

Now we are ready to use the above generalized binomial model for asset price to construct option price formula. Our construction is similar to the construction of Cox, Ross, and Rubinstein [8].

4.6.2 *Call option prices using one period generalized binomial models*

Let $C(0)$ be the current price of the call, $C_u(1), C_d(1)$ be the prices of the call at the end of the period in an up and down move, respectively; let $S_u(1), S_d(1)$ be the stock prices of share S at the end of the period in an up and down move, respectively. If the exercise price of the stock is K, then

$$\begin{aligned} C_u(1) &= \max\{S_u(1)-K,0\}, \text{ with probability } \mathrm{p_u(S(0))} \\ C_d(1) &= \max\{S_d(1)-K,0\}, \text{ with probability } \mathrm{p_d(S(0))}. \end{aligned}$$

We want to find $C(0)$. We consider the portfolio of amount A_0 of risk free bonds with interest rate r and Δ_0 number of shares of the underlying stock of the option. At the end of the period, the value of the portfolio will be

$$\begin{aligned} &\Delta_0 S_u(1)+(1+r)A_0, \text{ with probability } \mathrm{p_u(S(0))} \\ &\Delta_0 S_d(1)+(1+r)A_0, \text{ with probability } \mathrm{p_d(S(0))}. \end{aligned}$$

Replication requires that

$$\begin{aligned} C_u(1) &= \Delta_0 S_u(1)+(1+r)A_0 \\ C_d(1) &= \Delta_0 S_d(1)+(1+r)A_0 \end{aligned}$$

The above is a system of linear equations of unknown Δ_0 and A_0. Solving the above system we obtain,

$$\Delta_0 = \frac{C_u(1)-C_d(1)}{S_u(1)-S_d(1)}$$

$$A_0 = \frac{S_u(1)C_d(1)-S_d(1)C_u(1)}{(1+r)(S_u(1)-S_d(1))}$$

If there are no risk free arbitrage opportunities, then the current value $C(0)$ of the call option must be the current value of the hedging portfolio $\Delta_0 S(0)+A_0$. Therefore,

$$\begin{aligned} C(0) &= \Delta_0 S(0)+A_0 = \frac{C_u(1)-C_d(1)}{S_u(1)-S_d(1)}S(0)+\frac{S_u(1)C_d(1)-S_d(1)C_u(1)}{(1+r)(S_u(1)-S_d(1))} \\ &= \frac{C_u(1)-C_d(1)}{u(S(0))S(0)-d(S(0))S(0)}S(0)+\frac{u(S(0))S(0)C_d(1)-d(S(0))S(0)C_u(1)}{(1+r)(u(S(0))S(0)-d(S(0))S(0))} \\ &= \frac{C_u(1)-C_d(1)}{u(S(0))-d(S(0))}+\frac{u(S(0))C_d(1)-d(S(0))C_u(1)}{(1+r)(u(S(0))-d(S(0)))} \\ &= \frac{1+r-d(S(0))}{u(S(0))-d(S(0))}\frac{C_u(1)}{1+r}+\frac{u(S(0))-1-r}{u(S(0))-d(S(0))}\frac{C_d(1)}{1+r} \\ &= \frac{qC_u(1)+(1-q)C_d(1)}{1+r}, \end{aligned}$$

where

$$q=\frac{1+r-d(S(0))}{u(S(0))-d(S(0))}, \quad 1-q=\frac{u(S(0))-1-r}{u(S(0))-d(S(0))}.$$

Thus,

$$C(0)=\frac{qC_u(1)+(1-q)C_d(1)}{1+r} \text{ if } \frac{\mathrm{qC_u(1)+(1-q)C_d(1)}}{1+\mathrm{r}} > \mathrm{S(0)-K}$$

and

$$C(0)=S(0)-K \text{ if } \frac{\mathrm{qC_u(1)+(1-q)C_d(1)}}{1+\mathrm{r}} \leq \mathrm{S(0)-K}. \tag{4.6.1}$$

Example 4.6 *Let $u, d, p_u, p_d, \tau_u, \tau_d$ are functions on $[0,1]$ defined by*

$$u(x) = \begin{cases} \frac{9}{4}, & x \in [0, \frac{1}{3}) \\ \frac{5}{4}, & x \in [\frac{1}{3}, \frac{2}{3}) \\ \frac{3}{4} + \frac{1}{4x}, & x \in [\frac{2}{3}, 1] \end{cases},$$

$$d(x) = \begin{cases} \frac{7}{16}, & x \in [0, \frac{1}{3}) \\ \frac{2}{3}, & x \in [\frac{1}{3}, \frac{2}{3}) \\ \frac{3}{2} - \frac{1}{2x}, & x \in [\frac{2}{3}, 1] \end{cases},$$

$$p_u(x) = \begin{cases} 0.75, & x \in [0, \frac{1}{3}) \\ 0.8, & x \in [\frac{1}{3}, \frac{2}{3}) \\ 0.25, & x \in [\frac{2}{3}, 1] \end{cases},$$

$$p_d(x) = 1 - p_u(x).$$

Consider the random map $T : [0,1] \to [0,1]$ defined by

$$T = \{\tau_u(x), \tau_d(x); p_u(x), p_d(x)\},$$

where

$$\tau_u(x) = \begin{cases} \frac{9}{4}x, & x \in [0, \frac{1}{3}) \\ \frac{5x}{4}, & x \in [\frac{1}{3}, \frac{2}{3}) \\ \frac{3x}{4} + \frac{1}{4}, & x \in [\frac{2}{3}, 1] \end{cases},$$

$$\tau_d(x) = \begin{cases} \frac{7}{16}x, & x \in [0, \frac{1}{3}) \\ \frac{2x}{3}, & x \in [\frac{1}{3}, \frac{2}{3}) \\ \frac{3x}{2} - \frac{1}{2}, & x \in [\frac{2}{3}, 1] \end{cases},$$

If $x \in [0, \frac{1}{3})$, then

$$\sup_x \frac{p_u(x)}{|\tau_u'(x)|} + \sup_x \frac{p_d(x)}{|\tau_d'(x)|} = \frac{1}{3} + \frac{4}{7} = \frac{19}{21} < 1.$$

If $x \in [\frac{1}{3}, \frac{2}{3})$, then

$$\sup_x \frac{p_u(x)}{|\tau_u'(x)|} + \sup_x \frac{p_d(x)}{|\tau_d'(x)|} = \frac{16}{25} + \frac{6}{25} = \frac{94}{100} < 1.$$

If $x \in [\frac{2}{3}, 1]$, *then*

$$\sup_x \frac{p_u(x)}{|\tau_u'(x)|} + \sup_x \frac{p_d(x)}{|\tau_d'(x)|} = \frac{1}{3} + \frac{1}{2} = \frac{5}{6} < 1.$$

Thus, we see that the random map T satisfies the hypothesis of Theorem 4.6 and therefore, by Theorem 4.6, the random map T admits an absolutely continuous invariant measure $\mu = f^*\lambda$, where f^* is the invariant density of T. The invariant density allows us to find the probability: $\mu\{x : T(x) \in (\delta_1, \delta_2)\} = \mu(\delta_1, \delta_2)$. This random map generates binomial stock price paths. For example, if the starting stock price at $n = 0$ is $x = 0.25$, then the stock prices at time $n = 1, 2, 3$ are given by

$$s(0) = .25$$

$$\nearrow \overset{p_u=0.75, u=\frac{9}{4}}{\longrightarrow} \tau_u(.25) = 0.5625$$

$$\nearrow \overset{p_u=0.8, u=\frac{5}{4}}{\longrightarrow} \tau_u(0.5625) = 0.703125$$

$$\nearrow \overset{p_u=0.25, u=1.10555}{\longrightarrow} \tau_u(0.703125) = 0.7773437$$

$$\searrow \overset{p_d=0.75, d=0.78888}{\longrightarrow} \tau_d(0.703125) = 0.5546875$$

$$\searrow \overset{p_d=0.2, d=\frac{2}{3}}{\longrightarrow} \tau_d(0.5625) = 0.3750$$

$$\nearrow \overset{p_u=0.8, u=\frac{5}{4}}{\longrightarrow} \tau_u(0.3750) = 0.46875$$

$$\searrow \overset{p_d=0.2, d=\frac{2}{3}}{\longrightarrow} \tau_d(0.3750) = 0.250$$

$$\searrow \overset{p_d=0.25, d=\frac{7}{16}}{\longrightarrow} \tau_d(.25) = 0.109375$$

$$\nearrow \overset{p_u=0.75, u=\frac{9}{4}}{\longrightarrow} \tau_u(0.109375) = 0.24609375$$

$$\nearrow \overset{p_u=0.75, u=\frac{9}{4}}{\longrightarrow} \tau_u(0.24609375) = 0.55371093$$

$$\searrow \overset{p_d=0.25, d=0.\frac{7}{16}}{\longrightarrow} \tau_d(0.246093) = 0.1076601$$

$$\searrow \overset{p_d=0.25, d=\frac{7}{16}}{\longrightarrow} \tau_d(0.109375\ ) = 0.047851$$

$$\nearrow \overset{p_u=0.75, u=\frac{9}{4}}{\longrightarrow} \tau_u(0.047851) = 0.1076601$$

$$\searrow \overset{p_d=0.25, d=\frac{7}{16}}{\longrightarrow} \tau_d(0.047851) = 0.020935$$

(4.6.2)

For each starting stock price we have all possible paths the stock price might take. A typical path is

$$0.25 \to 0.56250 \to 0.70312 \to 0.55469 \to 0.69336 \to .54002.$$

where

$$\begin{aligned}
&s(0) = 0.25 \\
&s_u(1) = 0.5625(p_u = .75),\ \ s_d(1) = 0.109375(p_d = .25) \\
&s_u(2) = 0.70312(p_u = .8),\ \ s_d(2) = 0.3750(p_d = .2) \\
&s_d(3) = 0.55469(p_d = .75),\ \ i_u(3) = 0.7773437(p_u = .25)
\end{aligned}$$

Consider a European call option of one share of a stock with initial price $S(0) = 0.25$, strike price $K = 0.5$. Let the riskless interest rate be 5% and the expiry date of the option be one year from now. Thus

$$\begin{aligned}
C_u(1) &= \max\{s_u(1) - K\} = \max\{0.5625 - 0.5\} = 0.0625; \\
C_d(1) &= \max\{s_d(1) - K\} = \max\{0.109375 - 0.5\} = 0; \\
q &= \frac{1 + r - d(S(0))}{u(S(0)) - d(S(0))} = \frac{1 + .05 - d(.25)}{u(.25) - d(.25)} = \frac{1.05 - \frac{7}{16})}{\frac{9}{4} - \frac{7}{16}} = 0.3379310345.
\end{aligned}$$

We use the option price formula (4.6.1) and obtain

$$\begin{aligned}
C(0) &= \frac{qC_u(1) + (1 - q)C_d(1)}{1 + r} = \frac{0.3379310345 \cdot (0.0625) + (1 - 0.3379310345) \cdot 0}{1 + .05} \\
&= 0.2011494253.
\end{aligned}$$

4.6.3 The multi-period generalized binomial models and valuation of call options

Now we generalize our construction of one-period generalized binomial models to two-period generalized binomial models. Let $S(0)$ be the stock price at time 0. As we have seen in previous sections (Section 4.6.1 and Section 4.6.2), a position dependent random map T can generate stock prices at time $1,2$. Let $S_u(1), S_d(1)$ be the stock prices of share S at time 1 in an up and down move, respectively. Now we assume that at time 2 the stock price has two possibilities: the stock price may go up by a factor u or it may go down by a factor d where the factors u and d are functions of the prices [4], which is a function of prices: $u(x) : (0,1) \to (1,\infty)$ and $d(x) : (0,1) \to (0,1)$; i.e., at time 2, u and d depend on the price of underlying stock price at time 1. The probabilities p_u, p_d can be price dependent. Then at time 2, the possible stock prices are:

$$\begin{aligned}
S_{uu}(2) &= \tau_u(S_u(1)) = \tau_u(\tau_u(S(0)) \text{ with probability } \mathrm{p_u(S_u(1)) = p_u(\tau_u(S(0))} \\
S_{ud}(2) &= \tau_d(S_u(1)) = \tau_d(\tau_u(S(0)) \text{ with probability } \mathrm{p_d(S_u(1)) = p_d(\tau_u(S(0))} \\
S_{du}(2) &= \tau_u(S_d(1)) = \tau_u(\tau_d(S(0)) \text{ with probability } \mathrm{p_u(S_d(1)) = p_u(\tau_d(S(0))} \\
S_{dd}(2) &= \tau_d(S_d(1)) = \tau_d(\tau_d(S(0)) \text{ with probability } \mathrm{p_d(S_d(1)) = p_d(\tau_d(S(0))}
\end{aligned}$$

Similarly, at time 2 the possible call prices are:

$$\begin{aligned}
C_{uu}(2) &= \max\{0, S_{uu}(2) - K\} \text{ with probability } \mathrm{p_u(S_u(1)) = p_u(\tau_u(S(0))} \\
C_{ud}(2) &= \max\{0, S_{ud}(2) - K\} \text{ with probability } \mathrm{p_d(S_u(1)) = p_d(\tau_u(S(0))} \\
C_{du}(2) &= \max\{0, S_{du}(2) - K\} \text{ with probability } \mathrm{p_u(S_d(1)) = p_u(\tau_d(S(0))} \\
C_{dd}(2) &= \max\{0, S_{dd}(2) - K\} \text{ with probability } \mathrm{p_d(S_d(1)) = p_d(\tau_d(S(0))}
\end{aligned}$$

Applying the similar idea in the previous section we obtain

$$\begin{aligned}
C_u(1) &= \frac{qC_{uu}(1) + (1-q)C_{ud}(1)}{1+r} \\
C_d(1) &= \frac{qC_{du}(1) + (1-q)C_{dd}(1)}{1+r}
\end{aligned}$$

and

$$\begin{aligned}
C(0) &= \frac{qC_u(1) + (1-q)C_d(1)}{1+r} \\
&= \frac{q\left(\frac{qC_{uu}(2)+(1-q)C_{ud}(2)}{1+r}\right) + (1-q)\left(\frac{qC_{du}(2)+(1-q)C_{dd}(2)}{1+r}\right)}{1+r} \\
&= \frac{q^2C_{uu}(2) + q(1-q)C_{ud}(2) + q(1-q)C_{du}(2) + q^2C_{dd}(2)}{(1+r)^2} \\
&= \frac{q^2\max\{0, S_{uu}(2)-K\} + q(1-q)\left(\max\{0, S_{ud}(2)-K\} + \max\{0, S_{du}(2)-K\}\right)}{(1+r)^2} \\
&\quad + \frac{(1-q)^2\max\{0, S_{dd}(2)-K\}}{(1+r)^2}
\end{aligned}$$

The above construction of two-period generalized binomial models can be easily extended to multi-period generalized binomial models for option prices.

4.6.4 *The generalized binomial interest rate models using position dependent random maps and valuation of bond prices*

In a classical binomial model for interest rates, if the up factor u and the down factor d are constant then there is a possibility that the interest rates eventually become zero, or they increase without bound. In reality, the probability of the interest rate going up or down is not constant and may depend on the current interest rate. Moreover, it is reasonable to assume that the factors u and d are functions of interest rates, $u(x) : (0,1) \to (1,\infty)$ and $d(x) : (0,1) \to (0,1)$. That is, at time $t+1$, u and d depend on the interest rate at time t. In the following we construct a position dependent random map $T = \{\tau_u(x), \tau_d(x); p_u(x), p_d(x)\}$, where the transformation τ_u is the law which moves the interest rate up with a probability $p_u(x)$ and the transformation τ_d is the law which moves the interest rate down with a probability $p_d(x) = 1 - p_u(x)$. At time $t+1$, consider the upward interest rate to be $\tau_u(i(t))$ with probability $p_u(i(t))$ and the downward interest rate to be $\tau_d(i(t))$ with probability $p_d(i(t))$. Moreover, $i(t+1) = u(i(t)) \cdot i(t)$ or $i(t+1) = d(i(t)) \cdot i(t)$. Therefore, the transformation τ_u and τ_d are given by

$$\tau_u(x) = u(x) \cdot x \text{ and } \tau_d(x) = d(x) \cdot x.$$

This position dependent random map generates all possible paths the interest rate can take and we can use these paths to calculate bond prices. Now we present an example.

Example 4.7 *Let $u, d, p_u, p_d, \tau_u, \tau_d$ are functions on $[0,1]$ defined by*

$$u(x) = \begin{cases} 1 + \frac{1}{e^x}, & x \in [0, \frac{1}{3}) \\ \frac{5}{4}, & x \in [\frac{1}{3}, \frac{2}{3}) \\ \frac{3}{4} + \frac{1}{4x}, & x \in [\frac{2}{3}, 1] \end{cases},$$

$$d(x) = \begin{cases} e^x - \frac{1}{2}, & x \in [0, \frac{1}{3}) \\ \frac{2}{3}, & x \in [\frac{1}{3}, \frac{2}{3}) \\ \frac{3}{2} - \frac{1}{2x}, & x \in [\frac{2}{3}, 1] \end{cases},$$

$$p_u(x) = \begin{cases} 0.8, & x \in [0, \frac{1}{3}) \\ 0.75, & x \in [\frac{1}{3}, \frac{2}{3}) \\ 0.25, & x \in [\frac{2}{3}, 1] \end{cases},$$

$$p_d(x) = 1 - p_u(x).$$

Consider the random map $T : [0,1] \to [0,1]$ *defined by*

$$T = \{\tau_u(x), \tau_d(x); p_u(x), p_d(x)\},$$

where

$$\tau_u(x) = \begin{cases} x + \frac{x}{e^x}, & x \in [0, \frac{1}{3}) \\ \frac{5x}{4}, & x \in [\frac{1}{3}, \frac{2}{3}) \\ \frac{3x}{4} + \frac{1}{4}, & x \in [\frac{2}{3}, 1] \end{cases},$$

$$\tau_d(x) = \begin{cases} xe^x - \frac{x}{2}, & x \in [0, \frac{1}{3}) \\ \frac{2x}{3}, & x \in [\frac{1}{3}, \frac{2}{3}) \\ \frac{3x}{2} - \frac{1}{2}, & x \in [\frac{2}{3}, 1] \end{cases},$$

If $x \in [0, \frac{1}{3})$, *then*

$$\sup_x \frac{p_u(x)}{|\tau_u'(x)|} + \sup_x \frac{p_d(x)}{|\tau_d'(x)|} = 0.5413864422 + 0.4000000000 = 0.9413864422 < 1.$$

If $x \in [\frac{1}{3}, \frac{2}{3})$, *then*

$$\sup_x \frac{p_u(x)}{|\tau_u'(x)|} + \sup_x \frac{p_d(x)}{|\tau_d'(x)|} = 0.6000000000 + 0.3750000000 = 0.9750000000 < 1.$$

If $x \in [\frac{2}{3}, 1]$, *then*

$$\sup_x \frac{p_u(x)}{|\tau_u'(x)|} + \sup_x \frac{p_d(x)}{|\tau_d'(x)|} = 0.3333333333 + 0.5000000000 = 0.8333333333 < 1.$$

Thus, we see that the random map T *satisfies the hypothesis of Theorem 4.6 and therefore, by Theorem 4.6, the random map* T *admits an absolutely continuous invariant measure* $\mu = f^*\lambda$, *where* f^* *is the invariant density of* T. *The invariant density allows us to find the probability:* $\mu\{x : T(x) \in (\delta_1, \delta_2)\} = \mu(\delta_1, \delta_2)$. *This random map generates binomial interest rate paths. For example, if the starting interest rate is* $x = 0.15$ *(equivalent to* $0.03 = 3\%$*) then a typical path from this random map is*

$$0.15 \to 0.27910 \to .49024 \to 0.32683 \to 0.56254 \to 0.70318$$

which is equivalent to

$$0.03 \to 0.05582 \to 0.098084 \to 0.065366 \to 0.112508 \to 0.140636$$

which is equivalent to

$$3\% \to 5.582\% \to 9.8084\% \to 6.5366\% \to 11.2508\% \to 14.0636\%.$$

In the following we explain the above generalized binomial model induced by the position dependent random maps T and we use the generalized binomial interest rate path to bond evaluation: If the yield rate at time $n = 0$ is 3% (equivalently 0.15) convertible semiannually, then the yield rate at time $n = 1,2,3$ is given by

$$
\begin{array}{llll}
 & & & \nearrow^{p_u=0.75,u=\frac{5}{4}}\!\!\longrightarrow \tau_u(0.49022)=0.612775 \\
 & & \nearrow^{p_u=0.8,u=2.321939531}\!\!\longrightarrow \tau_u(0.27910)=0.49022 & \\
 & & & \searrow^{p_d=0.25,d=\frac{2}{3}}\!\!\longrightarrow \tau_d(0.49022)=0.3268133 \\
 & \nearrow^{p_u=0.8,u=2.161834243}\!\!\longrightarrow \tau_u(.15)=0.27910 & & \\
 & & & \nearrow^{p_u=0.8,u=2.257857655}\!\!\longrightarrow \tau_u(0.229410)=0.41179266 \\
 & & \searrow^{p_d=0.2,d=0.821939531}\!\!\longrightarrow \tau_d(0.27910)=0.229410 & \\
 & & & \searrow^{p_d=0.2,d=0.757857655}\!\!\longrightarrow \tau_d(0.229410)=0.1738608 \\
i(0)=.15 & & & \\
 & & & \nearrow^{p_u=0.8,u=2.208243920}\!\!\longrightarrow \tau_u(0.189168)=0.3427253 \\
 & & \nearrow^{p_u=0.8,u=2.104369960}\!\!\longrightarrow \tau_u(0.099275)=0.189168 & \\
 & & & \searrow^{p_d=0.2,d=0.708243920}\!\!\longrightarrow \tau_d(0.189168)=0.1339770 \\
 & \searrow^{p_d=0.2,d=0.6618}\!\!\longrightarrow \tau_d(.15)=0.099275 & & \\
 & & & \nearrow^{p_u=0.8,u=2.061834423}\!\!\longrightarrow \tau_u(0.059998)=0.1165021015 \\
 & & \searrow^{p_d=0.2,d=0.604369960}\!\!\longrightarrow \tau_d(0.099275)=0.059998 & \\
 & & & \searrow^{p_d=0.2,d=0.561834423}\!\!\longrightarrow \tau_d(0.059998)=0.03370894171
\end{array}
\tag{4.6.3}
$$

After re-scaling we obtain

$$
\begin{array}{llll}
 & & & \nearrow i(3)=0.122555 \\
 & & \nearrow i(2)=0.098044 & \\
 & & & \searrow i(3)=0.06536266 \\
 & \nearrow i(1)=0.05582 & & \\
 & & & \nearrow i(3)=0.082358532 \\
 & & \searrow i(2)=0.045882 & \\
 & & & \searrow i(3)=0.03477216 \\
i(0)=.03 & & & \\
 & & & \nearrow i(3)=0.06854506 \\
 & & \nearrow i(2)=0.0378336 & \\
 & & & \searrow i(3)=0.0267954 \\
 & \searrow i(1)=0.019855 & & \\
 & & & \nearrow i(3)=0.00233004 \\
 & & \searrow i(2)=0.0119996 & \\
 & & & \searrow i(3)=0.0067416
\end{array}
\tag{4.6.4}
$$

For each starting yield rate we have all possible paths the yield rate might take. A typical path is as follows:

$$
\begin{aligned}
& i_u(0) = i_d(0) = 0.03 \\
& i_u(1) = 0.05582\ (p_u = .8),\ \ i_d(1) = 0.019855\ (p_d = .2) \\
& i_u(2) = .09804\ (p_u = .8),\ \ i_d(2) = 0.045882\ (p_d = .2) \\
& i_d(3) = 0.0653626\ (p_u = .75),\ \ i_d(3) = 0.102256\ (\text{prob} = .25)
\end{aligned}
$$

If we consider a \$1000 par-value semiannual coupon bond with fixed coupon rate 7.6% and maturity date one and half years from now, then

$$
\begin{aligned}
R_U(3) &= 1000 + 1000(.038) = 1038 \\
R_D(3) &= 1000 + 1000(.038) = 1038 \\
V_U(3) &= 0 \\
V_D(3) &= 0 \\
V_U(2) &= \frac{p_u(2)\left(R_U(2+1) + V_U(2+1)\right) + p_d(2)\left(R_D(2+1) + V_D(2+1)\right)}{1 + i_u(2)} \\
&= \frac{0.75\,(1038 + 0) + 0.25\,(1038 + 0))}{1 + 0.098048} = 945.3138661 \\
V_D(2) &= \frac{p_u(2)\left(R_U(2+1) + V_U(2+1)\right) + p_d(2)\left(R_D(2+1) + V_D(2+1)\right)}{1 + i_d(2)} \\
&= \frac{0.75\,(1038 + 0) + 0.25\,(1038 + 0))}{1 + 0.045878} = 992.4675727
\end{aligned}
$$

$$
\begin{aligned}
R_U(2) &= 38 \\
R_D(2) &= 38 \\
V_U(1) &= \frac{p_u(1)\,(R_U(1+1)+V_U(1+1))+p_d(1)\,(R_D(1+1)+V_D(1+1))}{1+i_u(1)} \\
&= \frac{0.8\,(38+945.3138661)+0.2\,(38+992.4675727))}{1+0.05582} = 940.2593316 \\
V_D(1) &= \frac{p_u(1)\,(R_U(1+1)+V_U(1+1))+p_d(1)\,(R_D(1+1)+V_D(1+1))}{1+i_d(1)} \\
&= \frac{0.8\,(38+945.3138661)+0.2\,(38+992.4675727))}{1+0.019854} = 973.4183594 \\
V(0) &= \frac{p_u(1)\,(R_U(1)+V_U(1))+p_d(1)\,(R_D(1)+V_D(1))}{1+i(0)} \\
&= \frac{0.8\,(38+940.2593316)+0.2\,(38+973.4183594))}{1+0.03} = 956.2049875
\end{aligned}
$$

References

[1] Bahsoun, W. and Góra, P., *Position dependent random maps in one and higher dimensions*, Studia Math., 166, 271-286, 2005.

[2] Bahsoun, W., Góra, P., and Boyarsky, A., *Markov switching for position dependent random maps with application to forecasting in financial markets*, SIAM J. Appl. Dyn. Syst. 4, no. 2, 391–406, 2005.

[3] Bahsoun, W., Góra, P., Mayoral, S., and Morales, M., *Random dynamics and finance: constructing implied binomial trees from predetermined stationary density*, Appl. Stochastic Models Bus. Ind., 23, 181-212, 2007.

[4] Barnsley, M., *Fractals everywhere*, Academic Press, London, 1998.

[5] Borwein, J. M. and Lewis, A.S., *Convergence of the best entropy estimates*, SIAM J. Optim. 1(2), 191-205, 1991.

[6] Boyarsky, A. and Gora, P., *Laws of chaos: invariant measures and dynamical systems in one dimension*, Birkhauser, 1997.

[7] Boyarsky, A. and Góra, P., *A dynamical model for interference effects and two slit experiment of quantum physics*, Phys., Lett. A 168, 103-112, 1992.

[8] Cox, J. C., Ross, S.A., and Rubinstein, M., *Option pricing: a simplified approach*, Journal of Financial Economics, 7, 229-264, 1979.

[9] Ding, J., *A maximum entropy method for solving Frobenius–Perron equations*, Appl. Math. Comp., 93, 155-168, 1998.

[10] Ding, J. and Rhee, N. H., *Approximations of Frobenius–Perron operators via interpolations*, Nonlinear Analysis, 57, 831–842, 2004.

[11] Dunford, N. and Schawartz, J.T., *Linear operators, Part I*, Wiley Interscience (Wiley Classics Library): Chichester, 1988.

[12] Froyland, G., *Ulam's method for random interval maps*, Nonlinearity, 12, 1029-1052, 1999.

[13] Giusti, E., *Minimal surfaces and functions of bounded variation*, Monographs in Mathematics, 80. Birkhuser Verlag, Basel, 1984.

[14] Góra, Pawel and Boyarsky, Abraham, *Absolutely continuous invariant measures for random maps with position dependent probabilities*, Math. Anal. and Appl. 278, 225-242, 2003.

[15] Islam, M. Shafiqul, *Invariant measures for higher dimensional Markov switching position dependent random maps*, International Journal of Bifurcation and Chaos, Volume 19, Issue 1, 2009.

[16] Islam, M. Shafiqul, *Maximum entropy method for position dependent random maps,* International Journal of Bifurcation and Chaos,Volume 21, Issue 6 (June 2011) DOI No: 10.1142/S0218127411029458.

[17] Islam, M. Shafiqul, Gora, P., and Boyarsky, A., *A generalization of Straube's Theorem: existence of absolutely continuous invariant measures for random maps*, J. Appl. Math. Stoch. Anal., no. 2, 133-141, 2005.

[18] Islam, M. Shafiqul, Góra, P., and Boyarsky, A., *Approximation of absolutely continuous invariant measures for Markov switching position dependent random maps*, Int. J. Pure Appl. Math. 25, no. 1, 51-78, 2005.

[19] Lasota, A. and Mackey, M. C., *Chaos, fractals, and noise. Stochastic aspects of dynamics*, Applied Mathematical Sciences 97, Springer-Verlag, New York, 1994.

[20] Lasota, A. and Yorke, J.A., *On the existence of invariant measures for piecewise monotonic transformations*, Trans. Amer. Math. Soc. 186, 481-488, 1973.

[21] Li, T.-Y., *Finite approximation for the Frobenius–Perron operator: a solution to Ulam's conjecture*, J. Approx. Theory. 17, 177-186, 1976.

[22] Morita, T., *Random iteration of one-dimensional transformations*, Osaka J. Math., 22, 489-518, 1985.

[23] Pelikan, S., *Invariant densities for random maps of the interval*, Proc. Amer. Math. Soc. 281, 813-825, 1984.

[24] Schenk-Hoppe, K. R., *Random dynamical systems in economics*, working paper series, ISSN 1424-0459, Institute of Empirical Research in Economics, University of Zurich, Dec. 2000.

[25] Slomczynski, W., Kwapien, J., and Zyczkowski, K., *Entropy computing via integration over fractal measures*, Chaos 10, 180-188, 2000.

[26] Straube, E., *On the existence of invariant, absolutely continuous measures*, Comm. Math. Phys. 81, 27-30, 1981.

[27] Ulam, S. M., *A collection of mathematical problems*, Interscience Tracts in Pure and Applied Math. 8, Interscience, New York, 1960.

[28] Yosida, K. and Hewitt, E., *Finitely additive measures*, Trans. Amer. Math. Soc. 72, 46-66, 1952.

Chapter 5

Random Evolutions as Random Dynamical Systems

5.1 Chapter overview

In mathematical language, an RE is a solution of a stochastic operator integral equation in a Banach space. The operator coefficients of such equations depend on random parameters. The random evolution (RE), in physical language, is a model for a dynamical system whose state of evolution is subject to random variations. Such systems arise in many branches of science, e.g., random Hamiltonian and Shrödinger's equations with random potential in quantum mechanics, Maxwell's equation with a random reflective index in electrodynamics, transport equation, storage equation, etc. There is a lot of applications of REs in financial and insurance mathematics [11]. One of the recent applications of RE is associated with geometric Markov renewal processes which are regime-switching models for a stock price in financial mathematics, which will be studied intensively in the next chapters. Another recent application of RE is a semi-Markov risk process in insurance mathematics [11]. The REs are also examples of more general mathematical objects such as multiplicative operator functionals (MOFs) [7, 10], which are random dynamical systems in Banach space. The REs can be described by two objects: 1) operator dynamical system $V(t)$ and 2) random process x_t. Depending on structure of $V(t)$ and properties of the stochastic process x_t we have different kinds of REs: continuous, discrete, Markov, semi-Markov, etc. In this chapter we deal with various problems for REs, including martingale property, asymptotical behavior of REs, such as averaging, merging, diffusion approximation, normal deviations, averaging and diffusion approximation in redusible phase space for x_t, rate of convergence for limit theorems for REs.

5.2 Multiplicative operator functionals (MOF)

Multiplicative operator functionals (MOF) are a generalization of real multiplicative functionals from the probabilistic point of view and a stochastic analogue of semigroup of operators from the functional analysis point of view.

Let $(B,\mathcal{B},\|\cdot\|)$ be a separable Banach space and $L(B)$ be a space of linear bounded operators on B. Let x_t be a Markov process in a measurable space $(X,\mathcal{X})$ and let $\mathcal{F}_t := \sigma\{x_s; 0 \leq s \leq t\}$ be a sigma-algebra generated by the process x_t and let

$(\Omega, \mathcal{F}, P)$ be a probability space.

The **continuous time multiplicative operator functional (MOF)** [7,10] of $(x_t, L(B))$ is called a map $(t,\omega) \to V(t,\omega)$ from $R_+ \times \Omega$ to $L(B)$ such that:

1) $\omega \to V(t,\omega)f$ is an $\mathcal{F}_t$-measurable, $\forall t \in R_+, \forall f \in B$;

2) $t \to V(t,\omega)$ is right-continuous a.s., $\forall f \in B$;

3) $V(t+s,\omega)f = V(t,\omega)\theta_t V(s,\omega)f$ a.s., $\forall s,t \in R_+, \forall f \in B$, where θ_t is a shift operator ;

4) $V(0,\omega)f = f$ a.s., $\forall f \in B$.

Example 1. The *additive functional of a Markov process* x_t is a function α_t such that:

1) α_t is an $\mathcal{F}_t$-measurable;

2)

$$\alpha_s + \theta_s \alpha_t = \alpha_{t+s},$$

$0 \le s \le t$,

3) $\alpha_0 = 0$ a.s., where θ_s is a shift operator. One of the many examples of multiplicative functionals is the following one:

$$\alpha_t := \int_0^t a(x_u)du,$$

where $a(x)$ is a measurable function on X. Let $\Gamma(t)$ be a right-continuous semigroup of operators on B. Then

$$V(t,\omega)f := \Gamma(\alpha_t)f$$

is a right-continuous MOF of (x_t, B).

Example 2. The *multiplicative functional of a Markov process* x_t is a function ϕ_t such that:

1) ϕ_t is an $\mathcal{F}_t$-measurable;

2)

$$\phi_s \times \theta_s \phi_t = \phi_{t+s},$$

$0 \le s \le t$,

3) $\phi_0 = 1$ a.s.,

where θ_s is a shift operator.

Let α_t be an additive functional as in Example 1. Then

$$\phi_t := e^{\int_0^t \alpha(x_s)ds}$$

is a multiplicative functional. If we define the following operator

$$V(t,\omega)f := e^{\int_0^t \alpha(x_s)ds} f,$$

then $V(t,\omega)$ a right-continuous MOF of (x_t, B).

Example 3. As we will see in the next section, the random evolutions are also MOF of a Markov or semi-Markov process.

The definition of MOF can also be written for discrete time case. Let $x_k, k =$

$0,1,2,\ldots$ be a Markov chain on a measurable space $(X,\mathcal{X})$ and let $\mathcal{F}_n := \sigma\{x_k; k = 0,1,2,\ldots\}$ be a sigma-algebra generated by the chain x_n.

The **discrete time multiplicative operator functional (MOF)** of $(x_n, L(B))$ is called a map $(n,\omega) \to V(n,\omega)$ from $N \times \Omega$ to $L(B)$ such that:

1) $\omega \to V(n,\omega)f$ is an $\mathcal{F}_n$-measurable, $\forall n \in N, \forall f \in B$;

2) $V(n+m,\omega)f = V(m,\omega)\theta_m V(n,\omega)f$ a.s., $\forall n, mt \in N, \forall f \in B$, where θ_m is a shift operator;

4) $V(0,\omega)f = f$ a.s., $\forall f \in B$.

Example 1. The *discrete time additive functional of a Markov chain x_n* is a function α_n such that:

1) α_n is an $\mathcal{F}_n$-measurable;

2)

$$\alpha_m + \theta_m \alpha_n = \alpha_{n+m},$$

$0 \le m \le n$,

3) $\alpha_0 = 0$ a.s., where θ_m is a shift operator.

Let $\Gamma(t)$ be a right-continuous semigroup of operators on B. Then

$$V(n,\omega)f := \Gamma(\alpha_n)f$$

is a MOF of (x_n, B).

Example 2. The *multiplicative functional of a Markov chain x_n* is a function ϕ_n such that:

1) ϕ_n is an $\mathcal{F}_n$-measurable;

2)

$$\phi_m \times \theta_m \phi_n = \phi_{n+m},$$

$0 \le m \le n$,

3) $\phi_0 = 1$ a.s.,

where θ_s is a shift operator.

Let α_n be an additive functional as in Example 1. Then

$$\phi_n := e^{\sum_0^n \alpha(x_n)}$$

is a multiplicative functional. If we define the following operator

$$V(n,\omega)f := e^{\sum_0^n \alpha(x_n)} f,$$

then $V(n,\omega)$ a MOF of (x_t, B).

Example 3. As we will see in the next section, the discrete time (or discrete) random evolutions are also discrete time MOF of a Markov chain.

5.3 Random evolutions

5.3.1 Definition and classification of random evolutions

Let $(\Omega, \mathcal{F}, \mathcal{F}_t, \mathcal{P})$ be a probability space, $t \in R_+ := [0, +\infty]$, let (X, Ξ) be a measurable phase space, and let $(B, \mathcal{B}, \|\cdot\|)$ be a separable Banach space.

Let us consider a Markov renewal process $(x_n, \theta_n; n \geq 0)$, $x_n \in X$, $\theta_n \in R_+$, $n \geq 0$, with stochastic kernel

$$Q(x,A,t) := P(x,A)G_x(t), P(x,A) := \mathcal{P}(x_{n+1} \in A/x_n = x), G_x(t) := \mathcal{P}(\theta_{n+1}/x_n = x), \tag{5.3.1}$$

$x \in X, a \in \Xi, t \in R_+$. Process $x_t := x_{\nu(t)}$ is called a semi-Markov process, where $\nu(t) := \max\{n : \tau_n \leq t\}$, $\tau_n := \sum_{k=0}^{n} \theta_k$, $x_n = x_{\tau_n}$, $\mathcal{P}\{\nu(t) < +\infty, \forall t \in R_+\} = 1$. We note, that if $G_x(t) = 1 - e^{-\lambda(x)t}$, where $\lambda(x)$ is a measurable and bounded function on X, then x_t is called a jump Markov process.

Let $\{\Gamma(x); x \in X\}$ be a family of operators on the dense subspace $B_0 \in B$, which is common domain for $\Gamma(x)$, independent of x, noncommuting and unbounded in general, such that map $\Gamma(x)f : X \to B$ is strongly $\Xi/\mathcal{B}$-measurable for all $f \in B$, $\forall t \in R_+$; also, let $\{\mathcal{D}(x,y); x,y \in X\}$ be a family of bounded linear operators on B, such that map $\mathcal{D}(x,y)f : X \times X \to B$ is $\Xi \times \Xi/\mathcal{B}$-measurable, $\forall f \in B$.

Random Evolution (RE) is defined by the solution of stochastic operator integral equation in separable Banach space B:

$$V(t)f = f + \int_0^t V(s)\Gamma(x_s)f ds + \sum_{k=1}^{\nu(t)} V(\tau_k-)[\mathcal{D}(x_{k-1},x_k) - I]f, \tag{5.3.2}$$

where I is an identity operator on B, $\tau_k- := \tau_k - 0, f \in B$.

If x_t in (5.3.1) is a Markov or semi-Markov process, then RE in (5.3.2) is called a **Markov or semi-Markov RE**, respectively.

If $\mathcal{D}(x,y) \equiv I$, $\forall x,y \in X$, then $V(t)$ in (5.3.2) is called **a continuous RE**.

If $\Gamma(x) \equiv 0$, $\forall x \in X$, is a zero operator on B, then $V(t)$ in (5.3.2) is called **a jump RE**.

RE $V_n := V(\tau_n)$ is called **a discrete RE** .

Operators $\Gamma(x)$, $x \in X$, describe a continuous component $V^c(t)$ of RE $V(t)$ in (5.3.2), and operators $\mathcal{D}(\S,\dagger)$ describe a jump component $V^d(t)$ of RE $V^d(t)$ in (5.3.2).

In such a way, RE is described by two objects: 1) operator dynamical system $V(t)$; 2) random process x_t.

We note, that it turned out to be [5, 6]

$$V(t) = \Gamma_{x_t}(t - \tau_{\nu(t)}) \prod_{k=1}^{\nu(t)} \mathcal{D}(x_{k-1},x_k)\Gamma_{x_{k-1}}(\theta_k), \tag{5.3.3}$$

where $\Gamma_x(t)$ are the semigroups of operators of t generated by the operators $\Gamma(x)$, $\forall x \in X$. We also note, that RE in (5.3.2) is usually called **a discontinuous RE.** Under above introduced conditions the solution $V(t)$ of the equation (5.3.2) is unique and can be represented by product (5.3.3), that can be proved by constructive method [5].

Remark. From the definition of random evolutions it follows that they are other examples of MOFs, as they satisfy all the conditions for MOFs.

5.3.2 Some examples of RE

Connection of RE with applied problems is explained by the generality of definition (5.3.2) of RE. It includes any homogeneous linear evolutionary system. If, for example,

$$\Gamma(x) := v(x)\frac{d}{dz}, \mathcal{D}(x,y) \equiv I, B = C^1(R),$$

then the equation (5.3.2) is a transport equation which describes the motion of a particle with random velosity $v(x_t)$. In such a way, various interpretations of operators $\Gamma(x)$ and $\mathcal{D}(x,y)$ give us many realizations of RE.

Example 1. Impulse traffic process. Let $B = C(R)$ and operators $\Gamma(x)$ and $\mathcal{D}(x,y)$ are defined by the following way:

$$\Gamma(x)f(z) := v(z,x)\frac{d}{dz}f(z), \mathcal{D}(x,y)f(z) := f(z+a(x,y)), \tag{5.3.4}$$

where functions $v(z,x)$ and $a(x,y)$ are continuous and bounded on $R \times X$ and $X \times X$, respectively, $\forall z \in R$, $\forall x,y \in X$, $f(z) \in C^1(R) := B_0$. Then the equation (5.3.2) takes the form:

$$f(z_t) = f(z) + \int_0^t v(z_s,x_s)\frac{d}{dz}f(z_s)ds + \sum_{k=1}^{v(t)}[f(z_{\tau_k}- + a(x_{k-1},x_k)) - f(z_{\tau_k}-)], \tag{5.3.5}$$

and RE $V(t)$ is defined by the relation:

$$V(t)f(z) = f(z_t)$$

$z_0 = z$. Equation (5.3.5) is a functional one for **impulse traffic process** z_t, which satisfies the equation:

$$z_t = z + \int_0^t v(z_s,x_s)ds + \sum_{k=1}^{v(t)} a(x_{k-1},x_k). \tag{5.3.6}$$

We note that impulse traffic process z_t in (5.3.6) is a realization of discontinuous RE.

Example 2. Summation on a Markov chain. Let us put $v(z,x) \equiv 0$, $\forall z \in R$, $\forall x \in X$, in (5.3.6). Then the process

$$z_t = z + \sum_{k=1}^{v(t)} a(x_{k-1},x_k) \tag{5.3.7}$$

is a summation on a Markov chain $(x_n; n \geq 0)$ and it is a realization of a jump RE. Let $z_n := z_{\tau_n}$ in (5.3.7). Then discrete process

$$z_n = z + \sum_{k=1}^{n} a(x_{k-1},x_k)$$

is a realization of a discrete RE.

Example 3. Diffusion process in random media. Let $B = C(R)$, $B_0 = C^2(R)$, $P_x(t,z,A)$ be a Markov continuous distribution function, with respect to the diffusion process $\xi(t)$, that is the solution of the stochastic differential equation in R with semi-Markov switchings:

$$d\xi(t) = \mu(\xi(t),x_t)dt + \sigma(\xi(t),x_t)dw_t, \xi(0) = z, \tag{5.3.8}$$

where x_t is a semi-Markov process independent on a standard Wiener process w_t, coefficients $\mu(z,x)$ and $\sigma(z,x)$ are bounded and continuous functions on $R \times X$. Let us define the following contraction semigroups of operators on B:

$$\Gamma_x(t)f(z) := \int_R P_x(t,z,dy)f(y), f(y) \in B, x \in X. \tag{5.3.9}$$

Their infinitesimal operators $\Gamma(x)$ have the following kind:

$$\Gamma(x)f(z) = \mu(z,x)\frac{d}{dz}f(z) + 2^{-1}\sigma^2(z,x)\frac{d^2}{dz^2}f(z), f(z) \in B_0.$$

The process $\xi(t)$ is a continuous one; that is why the operators $\mathcal{D}(\S,\dagger) \equiv \mathcal{I}, \forall x,y \in X$, are identity operators. Then the equation (5.3.2) takes the form:

$$f(\xi(t)) = f(z) + \int_0^t [\mu(\xi(s),x_s)\frac{d}{dz} + 2^{-1}\sigma^2(\xi(s),x_s)\frac{d^2}{dz^2}]f(\xi(s))ds, \tag{5.3.10}$$

and RE $V(t)$ is defined by the relation

$$V(t)f(z) = E[f(\xi(t))/x_s; 0 \le s \le t; \xi(0) = z].$$

Equation (5.3.10) is a functional one for diffusion process $\xi(t)$ in (5.3.8) in semi-Markov random media x_t. We note that diffusion process $\xi(t)$ in (5.3.8) is a realization of continuous RE.

Example 4. The Geometric Markov Renewal Process (GMRP) [12]. Let $(x_n, \theta_n)_{n\in\mathbf{Z}^+}$ be a Markov renewal process on the phase space $X \times \mathbf{R}^+$ with the semi-Markov kernel $Q(x,A,t)$ and $x(t) := x_{\nu(t)}$ be a semi-Markov process. Let $\rho(x)$ be a bounded continuous function on X such that $\rho(x) > -1$. We define a stochastic functional S_t with Markov renewal process $(x_n;\theta_n)_{n\in Z_+}$ as follows:

$$S_t := S_0 \prod_{k=0}^{\nu(t)} (1 + \rho(x_k)), \tag{5.3.11}$$

where $S_0 > 0$ is the initial value of S_t. We call the process $(S_t)_{t\in\mathbf{R}_+}$ in (5.3.11) a geometric Markov renewal process (GMRP). This process $(S_t)_{t\in\mathbf{R}_+}$ we call such by analogy with the geometric compound Poisson process

$$S_t = S_0 \prod_{k=1}^{N(t)} (1 + Y_k), \tag{5.3.12}$$

where $S_0 > 0, N(t)$ is a standard Poisson process, $(Y_k)_{k \in Z_+}$ are i. i. d. random variable, which is a trading model in many financial applications as a pure jump model (see [12] and Chapter 6).

Let $B : C_0(\mathbf{R}_+)$ be a space of continuous functions on $\mathbf{R}_+$, vanishing at infinity, and let us define a family of bounded contracting operators $D(x)$ on $C_0(\mathbf{R}_+)$:

$$D(x)f(s) := f(s(1+\rho(x)), \; x \in X, s \in \mathbf{R}_+. \tag{5.3.13}$$

With these contraction operators $D(x)$ we define the following jump semi-Markov random evolution (JSMRE) $V(t)$ of geometric Markov renewal process in (5.3.11)

$$V(t) = \prod_{k=0}^{\nu(t)} D(x_k) := D(x_{\nu(t)}) \circ D(x_{\nu(t)-1}) \circ \ldots \circ D(x_1) \circ D(x_0). \tag{5.3.14}$$

Using (5.3.13) we obtain from (5.3.14)

$$V(t)f(s) = \prod_{k=0}^{\nu(t)} D(x_k)f(s) = f(s \prod_{k=0}^{\nu(t)} (1+\rho(x_k)) = f(S_t), \tag{5.3.15}$$

where S_t is defined in (5.3.11) and $S_0 = s$.

5.3.3 *Martingale characterization of random evolutions*

The main approaches to the study of RE are martingale methods. The main idea is that process

$$M_n := V_n - I - \sum_{k=0}^{n-1} E[V_{k+1} - V_k / \mathcal{F}_k], V_0 = I, \tag{5.3.16}$$

is an $\mathcal{F}_\backslash$-martingale in B, where

$$\mathcal{F}_n := \sigma x_k, \tau_k; 0 \le k \le n, V_n := V(\tau_n),$$

E is an expectation by probability $\mathcal{P}$. Representation of the martingale M_n (see (5.3.4)) in the form of martingale-difference

$$M_n = \sum_{k=0}^{n-1} [V_{k+1} - E(V_{k+1} / \mathcal{F}_k)] \tag{5.3.17}$$

gives us the possibility to calculate the weak quadratic variation:

$$< l(M_n f) > := \sum_{k=0}^{n-1} E[l^2((V_{k+1} - V_k)f) / \mathcal{F}_k], \tag{5.3.18}$$

where $l \in B^*$, and B^* is a dual space to B, dividing points of B. The martingale method of obtaining the limit theorems for the sequence of RE is founded on the solution of the following problems: 1) weak compactness of the family of measures generated by the sequences of RE; 2) any limiting point of this family of measures is the solution of the martingale problem; 3) the solution of the martingale problem is unique. The conditions 1)-2) guarantee the existence of weakly converging subsequence, and condition 3) gives the uniqueness of the weak limit. It follows from 1-3) that consequence of RE converges weakly to the unique solution of the martingale problem. The weak convergence of RE in series scheme we obtain from the criterion of weak compactness of the processes with values in separable Banach space [5]. The limit RE we obtain from the solution of some martingale problem in the form of some integral operator equations in Banach space B. We also use the representation

$$V_{k+1} - V_k = [\Gamma_{x_k}(\theta_{k+1})\mathcal{D}(x_k, x_{k+1}) - I]V_k, V_k := V(\tau_k), \tag{5.3.19}$$

and the following expression for semigroups of operators $\Gamma_x(t)$[5]:

$$\Gamma_x(t)f = f + \sum_{k=1}^{n-1} \frac{t^k}{k!}\Gamma^k(x)f + (n-1)^{-1}\int_0^t (t-s)^n \Gamma_x(s)\Gamma^n(x) f ds, \forall x \in X, \tag{5.3.20}$$

$\forall f \in \cap_{x \in X} Dom(\Gamma^n(x))$. Taking into account (5.3.4)-(5.3.8) we obtain the limit theorems for RE. In the previous section we considered the evolution equation associated with random evolutions by using the jump structure of the semi-Markov process or jump Markov process. In order to deal with more general driving processes and to consider other applications, it is useful to re-formulate the treatment of random evolution in terms of a **martingale problem.** It has been shown by Stroock and Varadhan [9] that the entire theory of multidimensional diffusion processes (and many other continuous-parameter Markov processes) can be so formulated. Suppose that we have an evolution equation of the form:

$$\frac{df}{dt} = Gf. \tag{5.3.21}$$

The **martingale problem** is to find a Markov process $x(t), t \geq 0$, and RE $V(t)$ so that for all smooth functions

$$V(t)f(x(t)) - \int_0^t V(s)Gf(x(s))ds \quad is \quad a \quad martingale. \tag{5.3.22}$$

It is immediate that this gives the required solution. Indeed, the operator

$$f \to T(t)f := E_x[V(t)f(x(t))]$$

defines a semigroup of operators on the Banach space B, whose infinitesimal generator can be computed by taking the expectation:

$$E_x[V(t)f(x(t))] - f(x) = E_x[\int_0^t V(s)Gf(x(s))ds],$$

and

$$\lim_{t\to 0} t^{-1}[E_x[V(t)f(x(t))] - f(x)] = \lim_{t\to 0} t^{-1} E_x[\int_0^t V(s)Gf(x(s))ds] = Gf(x).$$

Remark. In case $V(t) \equiv I$-identity operator, the above reduces to the usual martingale problem for Markov process [3].

Remark. In case $B = R$ the problem reduces to the determination of a real-valued multiplicative functional, which is related to a Feynman-Kac type formula. In the case of the one-dimensional Wiener process a wide class of multiplicative functionals is provided by

$$V(t) = \exp \int_0^t a(x(s))ds + \int_0^t b(x(s))dw(s),$$

where $w(t)$ is a standard Wiener process.

Let us illustrate the martingale problem for discontinuous RE over a jump Markov process, diffusion process, etc.

Martingale problem for discontinuous RE over a jump Markov process. Let $x(t), t \geq 0$, be a conservative regular jump Markov process on a measurable state space (X, Ξ) with rate function $\lambda(x) > 0$ and a family of probability measures $P(x, dy)$. Let also $V(t)$ be a discontinuous RE in (5.3.2). For any Borel function f we have the sum:

$$f(x(t)) = f(x(0)) + \sum_{0 \leq s \leq t} [f(x(s+0)) - f(x(s-0))]. \tag{5.3.23}$$

From this we see that the product $V(t)f(x(t))$ satisfies the differential equation:

$$\frac{dV(t)f(x(t))}{dt} = V(t)\Gamma(x(t))f(x(t)), if \tau_k < t < \tau_{k+1},$$

and the jump across $t = \tau_k$ is evaluated as

$$V(t)f(x(t))|_{\tau_k-}^{\tau_k+} = V(\tau_k-)\mathcal{D}(x(\tau_k-), x(\tau_k+))f(x(\tau_k+0)) - f(x(\tau_k-0))$$

leading to the equation:

$$\begin{aligned} V(t)f(x(t)) &= f(x) + \textstyle\int_0^t V(s)\Gamma(x(s))f(x(s))ds \\ &+ \textstyle\sum_{0 \leq \tau_k \leq t} V(\tau_k-)[\mathcal{D}(x(\tau_k-), x(\tau_k+))f(x(\tau_k+)) \\ &- f(x(\tau_k-))], x(0) = x, \tau_k\pm := \tau_k \pm 0. \end{aligned} \tag{5.3.24}$$

To put this in the appropriate form of the martingale problem, we use the following identity from the theory of Markov processes: for any positive Borel-measurable function $\phi(.,.)$:

$$E_x[\sum_{0 \leq \tau_k \leq t} \phi(x(\tau_k-), x(\tau_k+))] = E_x[\int_0^t \lambda(x(s)) \int_X \phi(x(s), y)P(x(s), dy)ds]. \tag{5.3.25}$$

We note, that the difference

$$\sum_{0 \leq \tau_k \leq t} \phi(x(\tau_k-), x(\tau_k+)) - \int_0^t \lambda(x(s))(P\phi)(x(s))ds$$

is a martingale, where P is an operator generated by $P(x,A)$, $x \in X$, $A \in \Xi$. Applying this to the above computations we see that

$$V(t)f(x(t)) = f(x) + \int_0^t V(s)Gf(x(s))ds + Z(t), \tag{5.3.26}$$

where $Z(t), t \geq 0$, is a martingale and

$$Gf(x) = \Gamma(x)f + \lambda(x)\int_X [\mathcal{D}(x,y)f(y) - f(x)]P(x,dy).$$

Martingale problem for discontinuous RE over semi-Markov process. It is known, that process $(x(t),\gamma(t))$ (with $\gamma(t) := t - \tau_{\nu(t)}$ and $x(t)$ as semi-Markov process) is a Markov process in $X \times R_+$ with infinitesimal operator

$$\hat{Q} := \frac{d}{dt} + \frac{g_x(t)}{\bar{G}_x(t)}[P - I],$$

where $g_x(t) := dG_x(t)/dt$, $\bar{G}_x(t) := 1 - G_x(t)$, P is an operator generated by $P(x,A)$, $x \in X$, $A \in \Xi$, $P(x,A)$ and $G_x(t)$ are defined in (5.3.1). We note, that in the Markov case, $G_x(t) = 1 - \exp{-\lambda(x)t}$, $g_x(t) = \lambda(x)\exp{-\lambda(x)t}$, $\hat{G}_x(t) = \exp{-\lambda(x)t}$, and $g_x(t)/\hat{G}_x(t) = \lambda(x)$, $\forall x \in X$. Hence, $\hat{Q} = \lambda(x)[P - I]$ is an infinitesimal operator of a jump Markov process $x(t)$ in X. Using the reasonings in (5.3.23)-(5.3.26) of the previous example for Markov process $y(t) := (x(t),\gamma(t))$ in $X \times R_+$ we obtain that the solution of martingale problem are operator

$$Gf(x,t) = \frac{d}{dt}f(x,t) + \Gamma(x)f(x,t) + \frac{g_x(t)}{\hat{G}_x(t)}\int_X [\mathcal{D}(\S,\dagger)\{(\dagger,\sqcup) - \{(\S,\imath)]\mathcal{P}(\S,\lceil\dagger),$$

and the process $y(t)$.

Martingale problem for RE over Wiener process. Let $w(t), t \geq 0$, be the Wiener process in R^d and consider the linear stochastic equation:

$$V(t) = I + \int_0^t V(s)\Gamma_0(w(s))ds + \sum_{j=1}^d \int_0^t V(s)\Gamma_j(w(s))dw_j(s),$$

where the final term is a stochastic integral of the Ito variety and $\Gamma_0, \ldots, \Gamma_d$ are bounded operators on a Banach space B. If f is any C^2 function, Ito's formula gives

$$f(w(t)) = f(w(0)) + 2^{-1}\int_0^t \Delta f(w(s))ds + \sum_{j=1}^d \int_0^t \frac{\partial f}{\partial w_j}(w(s))dw_j(s).$$

Using the stochastic product rule

$$d(Mf) = Mdf + (dM)f + (dM)df \tag{5.3.27}$$

and rearranging terms, we have:

$$V(t)f(w(t)) = f(w(0)) + \int_0^t V(s)(2^{-1}\Delta f + \sum_{j=1}^{d} \Gamma_j \frac{\partial f}{\partial w_j} + \Gamma_0 f)(w(s))ds + Z(t),$$

where $Z(t) := \sum_{j=1}^{d} \int_0^t V(s)(\frac{\partial f}{\partial w_j}(w(s)) + \Gamma_j(w(s))f(w(s)))dw_j(s)$, which is a martingale. Therefore we have obtained the solution of the martingale problem, with the infinitesimal generator

$$Gf = 2^{-1}\Delta f(w) + \sum_{j=1}^{d} \Gamma_j(w)\frac{\partial f}{\partial w_j}(w) + \Gamma_0(w)f(w).$$

This corresponds to the stochastic solution of the parabolic system

$$\frac{\partial u}{\partial t} = Gu.$$

Martingale problem for RE over diffusion process. Let $\xi(t), t \geq 0$, be the diffusion process in R:

$$d\xi(t) = a(\xi(t))dt + \sigma(\xi(t))dw(t)$$

and consider the linear stochastic equation:

$$V(t) = I + \int_0^t V(s)\Gamma_0(\xi(s))ds + \int_0^t V(s)\Gamma_1(\xi(s))d\xi(s),$$

with the bounded operators Γ_0 and Γ_1 on B. If f is any C^2 function Ito's formula gives:

$$\begin{aligned} f(\xi(t)) &= f(\xi(0)) + \int_0^t [a(\xi(s))\frac{df(\xi(s))}{d\xi} + 2^{-1}\sigma^2(\xi(s))\frac{d^2 f(\xi(s))}{d\xi^2}]ds \\ &+ \int_0^t \frac{\partial f(\xi(s))}{\partial \xi}\sigma(\xi(s))dw(s). \end{aligned}$$

Using the stochastic product rule (5.3.27) we have:

$$V(t)f(\xi(t)) = f(\xi(0)) + \int_0^t V(s)(a\frac{df}{d\xi} + 2^{-1}\sigma^2\frac{d^2 f}{d\xi^2} + \Gamma_1\frac{df}{d\xi} + \Gamma_0 f)(\xi(s))ds + Z(t),$$

where

$$Z(t) := \int_0^t V(s)(\sigma\frac{df}{d\xi} + \Gamma_1 f)(\xi(s))dw(s),$$

which is a martingale. Therefore, we have obtained the solution of the martingale problem with the operator

$$Gf = a\frac{df}{d\xi} + 2^{-1}\sigma^2\frac{d^2 f}{d\xi^2} + \Gamma_1\frac{df}{d\xi} + \Gamma_0 f.$$

We will obtain other solutions of martingale problems for RE in the limit theorems for RE.

5.3.4 Analogue of Dynkin's formula for RE

Let $x(t), t \geq 0$, be a strongly measurable strong Markov process, let $V(t)$ be a multiplicative operator functional (MOF) of $x(t)$ [7, 10], let A be the infinitesimal operator of semigroup

$$(T(t)f)(x) := E_x[V(t)f(x(t))], \tag{5.3.28}$$

and let τ be a stopping time for $x(t)$. It is known [10], that if $Ah = g$ and $E_x\tau < +\infty$, then

$$E_x[V(\tau)h(x(\tau)) - h(x) = E_x\int_0^\tau V(t)Ah(x(t))dt. \tag{5.3.29}$$

Formula (5.3.28) is an analogue of Dynkin's formula for MOF [10]. In fact, if we set $V(t) \equiv I$-identity operator, then from (5.3.29) we obtain:

$$E_x[h(x(\tau))] - h(x) = E_x\int_0^\tau Qh(x(t))dt, \tag{5.3.30}$$

where Q is an infinitesimal operator of $x(t)$ (see (5.3.28)). Formula (5.3.30) is the well-known Dynkin's formula. Let $x(t)$, $t \geq 0$, be a continuous Markov process on (X, Ξ) and $V(t)$ be a continuous RE:

$$dV(t)/dt = V(t)\Gamma(x(t)), V(0) = I. \tag{5.3.31}$$

We note, that the function $u(t,x) := E_x[V(t)f(x(t))]$ satisfies the following equation [10]:

$$du(t,x)/dt = Qu(t,x) + \Gamma(x)u(t,x), u(0,x) = f(x), \tag{5.3.32}$$

where Q is an infinitesimal operator of $x(t)$. From (5.3.29) and (5.3.32) we obtain the **analogue of Dynkin's formula for continuous Markov RE** $V(t)$ in (5.3.31):

$$E_x[V(\tau)h(x(\tau))] - h(x) = E_x\int_0^\tau V(t)[Q + \Gamma(x(t))]h(x(t))dt. \tag{5.3.33}$$

Let $x(t), t \geq 0$, be a jump Markov process with infinitesimal operator Q and $V(t)$ be a discontinuous Markov RE in (5.3.2). In this case the function $u(t,x) := E_x[V(t)f(x(t))]$ satisfies the equation [10]:

$$du(t,x)/dt = Qu(t,x) + \Gamma(x)u(t,x) + \lambda(x)\int_X P(x,dy)[\mathcal{D}(x,y) - I]u(t,y), u(0,x) = f(x). \tag{5.3.34}$$

From (5.3.29) and (5.3.34) we obtain the **analogue of Dynkin's formula for discontinuous Markov RE** in (5.3.2):

$$\begin{aligned} E_x[V(\tau)f(x(\tau))] - f(x) &= E_x\int_0^\tau V(s)[Q + \Gamma(x(t)) \\ &\quad + \lambda(x)\int_X P(x(t),dy)(\mathcal{D}(x(t),y) - I)]f(x(t))dt. \end{aligned} \tag{5.3.35}$$

Let finally $x(t), t \geq 0$, be a semi-Markov process, and $V(t)$ be a semi-Markov random evolution in (5.3.2). Let us define the process

$$\gamma(t) := t - \tau_{\nu(t)}). \tag{5.3.36}$$

Then the process

$$y(t) := (x(t), \gamma(t)) \tag{5.3.37}$$

is a Markov process in $X \times R_+$ with infinitesimal operator [6]

$$\hat{Q} := \frac{d}{dt} + \frac{g_x(t)}{\bar{G}_x(t)}[P - I], \tag{5.3.38}$$

where $g_x(t) := dG_x(t)/dt$, $\bar{G}_x(t) := 1 - G_x(t)$, P is an operator generated by the kernel $P(x,A)$. Hence, the process $(V(t)f; x(t); \gamma(t); t \geq 0) \equiv (V(t)f; y(t); t \geq 0)$ in $B \times X \times R_+$ is a Markov process with infinitesimal operator

$$L(x) := \hat{Q} + \Gamma(x) + \frac{g_x(t)}{\bar{G}_x(t)} \int_X P(x,dy)[\mathcal{D}(x,y) - I], \tag{5.3.39}$$

where $\hat{Q}$ is defined in (5.3.38).

Let $f(x,t)$ be a function on $X \times R_+$ bounded by x and differentiable by t, and let τ be a stopping time for $y(t) = (x(t), \gamma(t))$. Then for semi-Markov RE $V(t)$ in (5.3.2) we have from (5.3.29), (5.3.36)- (5.3.39) the following **analogue of Dynkin's formula:**

$$\begin{aligned} E_y[V(\tau)f(y(\tau))] - f(y) &= E_y \int_0^\tau V(s)[\hat{Q} + \Gamma(x(t)) \\ &\quad + \frac{g_x(t)}{\bar{G}_x(t)} \int_X P(x(t),dy)[\mathcal{D}(x(t),y) - I]f(y(t))dt, \end{aligned} \tag{5.3.40}$$

where $y := y(0) = (x,0), f(y) = f(x,0)$.

5.3.5 *Boundary value problems for RE*

Let $x(t), t \geq 0$, be a continuous Markov process in semicompact state space (X, Ξ). Let $V(t)$ be a continuous Markov RE in (5.3.31), and let G be an open set satisfying the following conditions:

$$\forall x \in G, \exists U : E_x \tau_U < +\infty, U \in \Xi, \tau_U := \inf_t t : x(t) \notin U, P_x \tau_G = +\infty = 0, \forall x \in X. \tag{5.3.41}$$

If $f(x)$ is a bounded measurable function on ∂G (boundary of G) and function

$$b(x) := E_x[V(\tau_G)f(x(\tau_G))] \tag{5.3.42}$$

is continuous on X, then function $b(x)$ is the solution of the equation [10]:

$$Qb(x)+\Gamma(x)b(x)=0, \forall x\in G, \tag{5.3.43}$$

where Q is an infinitesimal operator of $x(t)$. If function

$$H(x) := E_x[\int_0^{\tau_G} V(t)g(x(t))dt] \tag{5.3.44}$$

is continuous and bounded, then this function satisfies the following equation [10]:

$$QH(x)+\Gamma(x)H(x)=-g(x), \forall x\in X. \tag{5.3.45}$$

It follows from (5.3.41)-(5.3.44) that the boundary value problem

$$QH(x)+\Gamma(x)H(x)=-g(x), H(x)|_{\partial G}=f(x) \tag{5.3.46}$$

has the following solution:

$$H(x)=E_x\int_0^{\tau_G}[V(s)g(x(s))ds]+E_x[V(\tau_G f(x(\tau_G))]. \tag{5.3.47}$$

Let $x(t), t\geq 0$, be a jump Markov process in (X,Ξ), let $V(t)$ be a discontinuous Markov RE in (5.3.2), and let conditions (5.3.41) be satisfied. It follows from (5.3.44)-(5.3.47) and from (5.3.44), that the boundary value problem

$$QH(x)+\Gamma(x)H(x)+\int_X P(x,dy)[\mathcal{D}(x,y)-I]H(y)=-g(x), H(x)|_{\partial G}=f(x)$$

has the following solution:

$$H(x)=E_x\int_0^{\tau_G} V(s)g(x(s))ds+E_x[V(\tau_G)f(x(\tau_G))].$$

5.4 Limit theorems for random evolutions

The main approach to the investigation of SMRE in the limit theorems is a martingale method.

The martingale method of obtaining the limit theorems (averaging and diffusion approximation) for the sequence of SMRE is bounded on the solution of the following problems:

1) weak compactness of the family of measures generated by the sequence of SMRE;

2) any limiting point of this family of measures is the solution of the martingale problem;

3) the solution of the martingale problem is unique.

The conditions 1–2) guarantee the existence of weakly converging subsequence, and condition 3) gives the uniqueness of a weak limit.

From 1–3) it follows that consequence of SMRE converges weakly to the unique solution of the martingale problem.

5.4.1 Weak convergence of random evolutions

A **weak convergence of SMRE** in series scheme we obtain from the criterion of weak compactness of the process with values in separable Banach space [5]. The **limit SMRE** we obtain from the solution of some martingale problem in a kind of some integral operator equations in Banach space B.

The main idea is that process

$$M_n := V_n - I - \sum_{k=0}^{n-1} E\left[V_{k+1} - V_k / \mathcal{F}_k\right], \quad V_0 = I, \tag{5.4.1}$$

is an $\mathcal{F}_n$**–martingale** in B, where

$$\mathcal{F}_n := \sigma\{x_k, \tau_k; 0 \le k \le n\}, \quad V_n := V(\tau_n),$$

E is an expectation by probability $\mathcal{P}$ on a probability space $(\Omega, \mathcal{F}, \mathcal{P})$.

Representation of the martingale M_n in the form of **martingale - differences**

$$M_n = \sum_{k=0}^{n-1} \left[V_{k+1} - E(V_{k+1} / \mathcal{F}_k)\right] \tag{5.4.2}$$

gives us the possibility to calculate the **weak quadratic variation**:

$$< l(M_n f) > := \sum_{k=0}^{n-1} E\left[l^2((V_{k+1} - V_k)f) / \mathcal{F}_k\right], \tag{5.4.3}$$

where $l \in B^*$, and B^* is a dual space to B, dividing points of B.

From (5.3.19) it follows that

$$V_{k+1} - V_k = \left[\Gamma_{x_k}(\theta_{k+1}) D(x_k, x_{k+1}) - I\right] \cdot V_k. \tag{5.4.4}$$

We note that the following expression for semigroup of operators $\Gamma_x(t)$ is fulfilled:

$$\begin{aligned} \Gamma_x(t) f &= I + \sum_{k=1}^{n-1} \frac{t^k}{k!} \Gamma^k_{(x)} f + \frac{1}{(n-1)!} \int_0^t (t-s)^n \Gamma_x(s) \Gamma^n_{(x)} f ds, \\ \forall x \in X, \quad \forall f &\in \bigcap_x Dom(\Gamma^n(x)). \end{aligned} \tag{5.4.5}$$

Taking into account (5.4.1)–(5.4.5) we obtain the above-mentioned results.

Everywhere we suppose that the following conditions be satisfied:

A) there exist Hilbert spaces H and H^* such that they are compactly embedded in Banach spaces B and B^*, respectively, $H \subset B, \quad H^* \subset B^*$, where B^* is a dual space to B, that divides points of B;

B) operators $\Gamma(x)$ and$(\Gamma(x))^*$ are dissipative on any Hilbert space H and H^*, respectively;

C) operators $D(x,y)$ and $D^*(x,y)$ are contractive on any Hilbert space H and H^*, respectively;

D) $(x_n; \quad n \geq 0)$ is a uniformly ergodic Markov chain with stationary distribution $\rho(A), \quad A \in \mathcal{X}$;

E) $m_i(x) := \int_0^\infty t^i G_x(dt)$ are uniformly integrable, $\forall i = 1,2,3$, where

$$G_x(t) := \mathcal{P}\{\omega : \theta_{n+1} \leq t/x_n = x\}; \tag{5.4.6}$$

F)

$$\int_X \rho(dx)\|\Gamma(x)f\|^k < +\infty; \int_X \rho(dx)\|PD_j(x,\cdot)f\|^k < +\infty;$$

$$\int_X \rho(dx)\|\Gamma(x)f\|^{k-1} \cdot \|PD_j(x,\cdot)f\|^{k-1} < +\infty; \quad \forall k = 1,2,3,4, f \in B, \tag{5.4.7}$$

where P is an operator generated by the transition probabilities $P(x,A)$ of Markov chain $(x_n; \quad n \geq 0)$:

$$P(x,A) := \mathcal{P}\{\omega : x_{n+1} \in A/x_n = x\}, \tag{5.4.8}$$

and $\{D_j(x,y); \quad x,y \in X, \quad j = 1,2\}$ is a family of some closed operators.

If $B := C_0(R)$, then $H := W^{l,2}(R)$ is a Sobolev space [8], and $W^{l,2}(R) \subset C_0(R)$ and this embedding is compact. For the spaces $B := L_2(R)$ and $H := W^{l,2}(R)$ it is the same.

It follows from the conditions A - B) that operators $\Gamma(x)$ and $(\Gamma(x))^*$ generate a strongly continuous contractive semigroup of operators $\Gamma_x(t)$ and $\Gamma_x^*(t), \quad \forall x \in X$, in H and H^*, respectively. From the conditions A–C it follows that SMRE $V(t)$ in (1) is a contractive operator in $H, \quad \forall t \in R_+$, and $\|V(t)f\|_H$ is a semimartingale $\forall f \in H$. In such a way, the conditions A - C) supply the following result:

SMRE $V(t)f$ is a tight process in B, namely, $\forall \Delta > 0$ there exists a compact set K_Δ:

$$\mathcal{P}\{V(t)f \in K_\Delta; 0 \leq t \leq T\} \geq 1 - \Delta. \tag{5.4.9}$$

This result follows from the Kolmogorov-Doob inequality [4] for semimartingale $\|V(t)f\|_H$ [5].

Condition (5.4.9) is the main step in proving limit theorems and rates of convergence for the sequence of SMRE in series schemes.

5.4.2 Averaging of random evolutions

Let's consider a SMRE in series scheme:

$$V_\varepsilon(t) = f + \int_0^t \Gamma(x(s/\varepsilon))V_\varepsilon(s)f\,ds + \sum_{k=1}^{\nu(t/\varepsilon)} [D^\varepsilon(x_{k-1},x_k) - I]\, V_\varepsilon(\varepsilon\tau_k-)f, \quad (5.4.10)$$

where

$$D^\varepsilon(x,y) = I + \varepsilon D_1(x,y) + 0(\varepsilon), \quad (5.4.11)$$

$\{D_1(x,y); x,y \in X\}$ is a family of closed linear operators, $\|0(\varepsilon)f\|/\varepsilon \to 0 \quad \varepsilon \to 0$, ε is a small parameter,

$$f \in B_0 := \bigcap_{x,y\in X} Dom(\Gamma^2(x)) \cap Dom(D_1^2(x,y)). \quad (5.4.12)$$

Another form for $V_\varepsilon(t)$ in (5.4.10) is:

$$V_\varepsilon(t) = \Gamma_{x(t/\varepsilon)}(t - \varepsilon\tau_{\nu(t/\varepsilon)}) \prod_{k=1}^{\nu(t/\varepsilon)} D^\varepsilon(x_{k-1},x_k)\Gamma_{k-1}(\varepsilon\theta_k). \quad (5.4.13)$$

Under conditions A - C) the sequence of SMRE $V_\varepsilon(t)f$ is tight (see (5.4.9)). $\rho - a.s.$

Under conditions $D), E), i = 2, F), k = 2, j = 1$, the sequence of SMRE $V_\varepsilon(t)f$ is weakly compact $\rho - a.s.$ in $D_B[0,+\infty)$ with limit points in $C_B[0,+\infty)$, $f \in B_0$.

Let's consider the following process in $D_B[0,+\infty)$:

$$M^\varepsilon_{\nu(t/\varepsilon)} f^\varepsilon := V^\varepsilon_{\nu(t/\varepsilon)} f^\varepsilon - f^\varepsilon - \sum_{k=0}^{\nu(t/\varepsilon)-1} E_\rho[V^\varepsilon_{k+1} f^\varepsilon_{k+1} - V^\varepsilon_k f^\varepsilon_k / \mathcal{F}_k], \quad (5.4.14)$$

where $V^\varepsilon_n := V_\varepsilon(\varepsilon\tau_n)$ (see (5.3.19)),

$$f^\varepsilon := f + \varepsilon f_1(x(t/\varepsilon)),$$

$$f^\varepsilon_k := f^\varepsilon(x_k),$$

function $f_1(x)$ is defined from the equation

$$(P-I)f_1(x) = \left[(\hat{\Gamma}+\hat{D}) - (m(x)\Gamma(x)+PD_1(x,\cdot))\right] f,$$
$$\hat{\Gamma} := \int_X \rho(dx)m(x)\Gamma(x), \quad \hat{D} := \int_X \rho(dx)PD_1(x,\cdot),$$
$$m(x) := m_1(x) \tag{5.4.15}$$

(see E), $f \in B_0$.

The process $M^{\varepsilon}_{\nu(t/\varepsilon)} f^{\varepsilon}$ is an $\mathcal{F}^{\varepsilon}_t$–martingale with respect to the σ–algebra $\mathcal{F}^{\varepsilon}_t := \sigma\{x(s/\varepsilon); 0 \le s \le t\}$.

The martingale $M^{\varepsilon}_{\nu(t/\varepsilon)} f^{\varepsilon}$ in (5.4.14) has the asymptotic representation:

$$M^{\varepsilon}_{\nu(t/\varepsilon)} f^{\varepsilon} = V^{\varepsilon}_{\nu(t/\varepsilon)} f - f - \varepsilon \sum_{k=0}^{\nu(t/\varepsilon)} (\hat{\Gamma}+\hat{D}) V^{\varepsilon}_k f + 0_f(\varepsilon), \tag{5.4.16}$$

where $\hat{\Gamma}, \hat{D}, f, f^{\varepsilon}$ are defined in (5.4.14)–(5.4.15) and

$$\|0_f(\varepsilon)\|/\varepsilon \to const \quad as \varepsilon \to 0, \quad \forall f \in B_0.$$

We've used (5.3.19), (5.3.20) *as* $n = 2$, and representation (5.4.4) and (5.4.14) in (5.4.16).

The families $l(M^{\varepsilon}_{\nu(t/\varepsilon)} f^{\varepsilon})$ and

$$l\left(\sum_{k=0}^{\nu(t/\varepsilon)} E_{\rho}[(V^{\varepsilon}_{k+1} f^{\varepsilon}_{k+1} - V^{\varepsilon}_k f^{\varepsilon}_k)/\mathcal{F}_k]\right)$$

are weakly compact for all $l \in B^*_0$ in a some dense subset from B^*. Let $V_0(t)$ be a limit process for $V_{\varepsilon}(t) as \quad \varepsilon \to 0$.

Since (see (5.4.13))

$$[V_{\varepsilon}(t) - V^{\varepsilon}_{\nu(t/\varepsilon)}] = [\Gamma_{x(t/\varepsilon)}(t - \varepsilon \tau_{\nu(t/\varepsilon)}) - I] \cdot V^{\varepsilon}_{\nu(t/\varepsilon)} \tag{5.4.17}$$

and the right-hand side in (5.4.17) tends to zero *as* $\varepsilon \to 0$, then it's clear that the limits for $V_{\varepsilon}(t)$ and $V^{\varepsilon}_{\nu(t/\varepsilon)}$ are the same , namely, $V_0(t) \quad \rho - a.s.$

The sum $\varepsilon \cdot \sum_{k=0}^{\nu(t/\varepsilon)} (\hat{\Gamma}+\hat{D}) V^{\varepsilon}_k f$ converges strongly *as* $\varepsilon \to 0$ to the integral

$$m^{-1} \cdot \int_0^t (\hat{\Gamma}+\hat{D}) V_0(s) f ds.$$

The quadratic variation of the martingale $l(M^{\varepsilon}_{\nu(t/\varepsilon)} f^{\varepsilon})$ tends to zero, and, hence,

$$M^{\varepsilon}_{\nu(t/\varepsilon)} f^{\varepsilon} \to 0 \quad as \quad \varepsilon \to 0, \quad \forall f \in B_0, \quad \forall e \in B_0^*.$$

Passing to the limit in (5.4.16) $as\varepsilon \to 0$ and taking into account all the previous reasonings we obtain that the limit process $V_0(t)$ satisfies the equation:

$$0 = V_0(t)f - f - m^{-1} \int_0^t (\hat{\Gamma} + \hat{D}) V_0(s) f ds, \qquad (5.4.18)$$

where

$$m := \int_X \rho(dx) m(x), \quad f \in B_0, \quad t \in [0, T].$$

5.4.3 *Diffusion approximation of random evolutions*

Let us consider SMRE $V_\varepsilon(t/\varepsilon)$, where $V_\varepsilon(t)$ is defined in (5.4.10) or (5.4.13), with the operators

$$D^{\varepsilon}(x,y) := I + \varepsilon D_1(x,y) + \varepsilon^2 D_2(x,y) + 0(\varepsilon^2), \qquad (5.4.19)$$

$\{Di(x,y); \quad x,y \in X, \quad i = 1,2\}$ are closed linear operators and $\|0(\varepsilon^2) f\| / \varepsilon^2 \to 0, \varepsilon \to 0$

$$\forall f \in B_0 := \bigcap_{x,y \in X} Dom(\Gamma^4(x)) \bigcap Dom(D_2(x,y)),$$

$$Dom(D_2(x,y)) \subseteq Dom(D_1(x,y)); \quad D_1(x,y) \subseteq Dom(D_1(x,y)),$$
$$\forall x,y \in X, \quad \Gamma^i(x) \subset Dom(D_2(x,y)), \quad i = \overline{1,3}. \qquad (5.4.20)$$

In such a way

$$V_\varepsilon(t/\varepsilon) = \Gamma_{x(t/\varepsilon^2)}(t/\varepsilon - \varepsilon \tau_{\nu(t/\varepsilon^2)}) \prod_{k=1}^{\nu(t/\varepsilon^2)} D^{\varepsilon}(x_{k-1}, x_k) \Gamma_{x_{k-1}}(\varepsilon, \theta_k), \qquad (5.4.21)$$

where $D^{\varepsilon}(x,y)$ are defined in (5.4.19).

Under conditions A) - C) the sequence of SMRE $V_\varepsilon(t/\varepsilon) f$ is tight (see (5.4.9)) $\rho - a.s.$

Under conditions $D), E), i = 3, F), k = 4$, the sequence of SMRE $V_\varepsilon(t/\varepsilon) f$ is weakly compact $\rho - a.s.$ in $D_B[0, +\infty)$ with limit points in $C_B[o, +\infty)$, $f \in B_0$.

Let the **balance condition** be satisfied:

$$\int_X \rho(dx)[m(x)\Gamma(x) + PD_1(x,\cdot)]f = 0, \quad \forall f \in B_0 \tag{5.4.22}$$

Let us consider the following process in $D_B[0,+\infty)$:

$$M^{\varepsilon}_{\nu(t/\varepsilon^2)} f^{\varepsilon} := V^{\varepsilon}_{\nu(t/\varepsilon^2)} f^{\varepsilon} - f^{\varepsilon} - \sum_{k=0}^{\nu(t/\varepsilon^2)-1} E_{\rho}[V^{\varepsilon}_{k+1} f^{\varepsilon}_{k+1} - V^{\varepsilon}_k f^{\varepsilon}_k / \mathcal{F}_k], \tag{5.4.23}$$

where $f^{\varepsilon} := f + \varepsilon f_1(x(t/\varepsilon^2)) + \varepsilon^2 f_2(x(t/\varepsilon^2))$, and functions f_1 and f_2 are defined from the following equations:

$$\begin{aligned}
(P-I)f_1(x) &= -[m(x)\Gamma(x) + PD_1(x,\cdot)]f, \\
(P-I)f_2(x) &= [\hat{L} - L(x)]f, \\
\hat{L} : &= \int_X \rho(dx) L(x),
\end{aligned} \tag{5.4.24}$$

$$\begin{aligned}
L(x) \quad := \quad & (m(x)\Gamma(x) + PD_1(x,\cdot))(R_0 - I)(m(x)\Gamma(x) + PD_1(x,\cdot)) + \\
& + m_2(x)\Gamma^2(x)/2 + m(x)PD_1(x,\cdot)\Gamma(x) + PD_2(x,\cdot),
\end{aligned}$$

R_0 is a potential operator of $(x_n; \quad n \geq 0)$.

The balance condition (5.4.22) and condition $\prod(\hat{L} - L(x)) = 0$ give the solvability of the equations in (5.4.24).

The process $M^{\varepsilon}_{\nu(t/\varepsilon^2)} f^{\varepsilon}$ is an $\mathcal{F}^{\varepsilon}_t$–martingale with respect to the σ–algebra $\mathcal{F}^{\varepsilon}_t := \sigma\{x(s/\varepsilon^2); 0 \leq s \leq t\}$.

This martingale has the asymptotic representation:

$$M^{\varepsilon}_{\nu(t/\varepsilon^2)} f^{\varepsilon} = V^{\varepsilon}_{\nu(t/\varepsilon^2)} f - f - \varepsilon^2 \sum_{k=0}^{\nu(t/\varepsilon^2)-1} \hat{L} V^{\varepsilon}_k f - 0_f(\varepsilon t), \tag{5.4.25}$$

where $\hat{L}$ is defined in (5.4.25) and

$$\|0_f(\varepsilon)\|/\varepsilon \to const \quad \varepsilon \to 0, \quad \forall f \in B_0.$$

We have used (5.3.19), (5.3.20) *as* $n = 3$, and representation (5.4.23) and (5.4.24) in (5.4.25).

The families $l(M^{\varepsilon}_{\nu(t/\varepsilon^2)} f^{\varepsilon})$ and $l(\sum_{k=0}^{\nu(t/\varepsilon^2)} E_{\rho}[(V^{\varepsilon}_{k+1} f^{\varepsilon}_{k+1} - V^{\varepsilon}_k f^{\varepsilon}_k)/\mathcal{F}_k])$ are weakly compact for all $l \in B_0^*$, $\quad f \in B_0$.

Set $V^0(t)$ for the limit process for $V_\varepsilon(t/\varepsilon) as \quad \varepsilon \to 0$.

From (5.4.13) we obtain that the limits for $V_\varepsilon(t/\varepsilon)$ and $V^\varepsilon_{\nu(t/\varepsilon^2)}$ are the some, namely, $V^0(t)$.

The sum $\varepsilon^2 \sum_{k=0}^{\nu(t/\varepsilon^2)} \hat{L} V_k^\varepsilon f$ converges strongly *as* $\varepsilon \to 0$ to the integral $m^{-1} \int_0^t \hat{L} V^0(s) f ds$.

Set $M^0(t)f$ be a limit martingale for $M^\varepsilon_{\nu(t/\varepsilon^2)} f^\varepsilon \quad as \quad \varepsilon \to 0$.

Then, from (5.4.24)–(5.4.25) and previous reasonings we have *as* $\varepsilon \to 0$:

$$M^0(t)f = V^0(t)f - f - m^{-1} \cdot \int_0^t \hat{L} V^0(s) f ds. \tag{5.4.26}$$

The quadratic variation of the martingale $M^0(t)f$ has the form:

$$< l(M^0(t)f) > = \int_0^t \int_X l^2(\sigma(x)\Gamma(x)V^0(s)f)\sigma(dx)ds, \tag{5.4.27}$$

where

$$\sigma^2(x) := [m_2(x) - m^2(x)]/m.$$

The solution of martingale problem for $M^0(t)$ (namely, to find the representation of $M^0(t)$ with quadratic variation (5.4.26)) is expressed by the integral over Wiener orthogonal martingale measure $W(dx, ds)$ with quadratic variation $\rho(dx) \cdot ds$:

$$M^0(t)f = \int_0^t \int_x \sigma(x)\Gamma(x)V^0(s) f W(dx, ds). \tag{5.4.28}$$

In such a way, the limit process $V^0(t)$ satisfies the following equation (see (5.4.26) and (5.4.27)):

$$V^0(t)f = f + m^{-1} \cdot \int_0^t \hat{L} \cdot V^0(s) f ds + \int_0^t \int_X \sigma(x)\Gamma(x)V^0(s) f W(dx, ds). \tag{5.4.29}$$

If the operator $\hat{L}$ generates the semigroup $U(t)$ then the process $V^0(t)f$ in (5.4.29) satisfied the equation:

$$V^0(t)f = U(t)f + \int_0^t \int_x \sigma(x)U(t-s)\Gamma(x)V^0(s) f W(dx, ds). \tag{5.4.30}$$

The **uniqueness** of the limit evolution $V_0(t)f$ in the **averaging** scheme follows from the equation (5.4.30) and the fact that if the operator $\hat{\Gamma}+\hat{D}$ (see (5.4.15)) generates a semigroup, then $V_0(t)f = \exp\{(\hat{\Gamma}+\hat{D})\cdot t\}f$ and this representation is unique.

The **uniqueness** of the limit evolution $V^0(t)f$ in a **diffusion approximation** scheme follows from the uniqueness of the solution of martingale problems for $V^0(t)f$ (see (5.4.26)–(5.4.27)) [9]. The latter is proved by **dual SMRE** in series scheme by constructing the limit equation in diffusion approximation and by using a dual identity [5].

5.4.4 Averaging of random evolutions in reducible phase space, merged random evolutions

Suppose that the following conditions hold true:

a) **decomposition** of phase space X (**reducible** phase space):

$$X = \bigcup_{u \in U} X_u, \quad X_u \bigcap X_{u'} = \emptyset, \quad u \neq u' : \tag{5.4.31}$$

where $(U, \mathcal{U})$ is a some measurable phase space (**merged** phase space);

b) Markov renewal process $(x_n^\varepsilon, \theta_n; n \geq 0)$ on $(X, \mathcal{X})$ has the **semi-Markov kernel**:

$$Q_\varepsilon(x, A, t) := P_\varepsilon(x, A) G_x(t), \tag{5.4.32}$$

where $P_\varepsilon(x,A) = P(x,A) - \varepsilon^l P_1(x,A), \quad x \in X, \quad A \in \mathcal{X}, \quad = \infty, \in;$ $P(x,A)$ are the transition probabilities of the **supporting nonperturbed** Markov chain $(x_n; n \geq 0)$;

c) the stochastic kernel $P(x,A)$ is adapted to the decomposition (5.4.31) in the following form:

$$P(x, X_u) = \begin{cases} 1, x \in X_u \\ 0, x \overline{\in} X_u, \quad u \in U; \end{cases}$$

d) the Markov chain $(x_n; n \geq 0)$ is uniformly ergodic with stationary distributions $\rho_u(B)$:

$$\rho_u(B) = \int_{X_u} P(x, B) \rho_u(dx), \quad \forall u \in U, \quad \forall B \in \mathcal{X}. \tag{5.4.33}$$

e) there is a family $\{\rho_u^\varepsilon(A); u \in U, A \in \chi, \varepsilon > 0\}$ of stationary distributions of perturbed Markov chain $(x_n^\varepsilon; n \geq 0)$;

f)

$$b(u) := \int_{X_u} \rho_u(dx) P_1(x, X_u) > 0, \quad \forall u \in U,$$

$$b(u, \Delta) := -\int_{X_u} \rho_u(dx) P_1(x, X_\Delta) > 0, \quad \forall u \overline{\in} \Delta, \quad \Delta \in U; \tag{5.4.34}$$

g) the operators $\Gamma(u) := \int_{X_u} \rho_u(dx)m(x)\Gamma(x)$ and

$$\hat{D}(u) := \int_{X_u} \rho_u(dx) \int_{X_u} P(x,dy)D_1(x,y) \tag{5.4.35}$$

are closed $\forall u \in U$ with common domain B_0, and operators $\hat{\Gamma}(u) + \hat{D}(u)$ generate the semigroup of operators $\forall u \in U$.

Decomposition (5.4.31) in a) defines the **merging** function

$$u(x) = u \quad \forall x \in Xu, \quad u \in U. \tag{5.4.36}$$

We note that σ–algebras $\mathcal{X}$ and $\mathcal{U}$ are coordinated such that

$$X_\Delta = \bigcup_{u\in\Delta} Xu, \quad \forall u \in U, \quad \Delta \in \mathcal{U}. \tag{5.4.37}$$

We set $\prod_u f(u) := \int_{X_u} \rho_u(dx)f(x)$ and $x^\varepsilon(t) := x^\varepsilon_{\nu(t/\varepsilon^2)}$.

SMRE in reducible phase space X is defined by the solution of the equation:

$$\begin{aligned} V_\varepsilon(t) &= I + \int_0^t \Gamma(x^\varepsilon(s/\varepsilon))V_\varepsilon(s)ds \\ &\quad + \sum_{k=0}^{\nu(t/\varepsilon)} [D^\varepsilon(x^\varepsilon_{k-1}, x^\varepsilon_k) - I]V_\varepsilon(\varepsilon\tau_k^-), \end{aligned} \tag{5.4.38}$$

where $D^\varepsilon(x,y)$ are defined in (5.4.11).

Let's consider the martingale

$$\begin{aligned} M^\varepsilon_{\nu(t/\varepsilon)} f^\varepsilon(x^\varepsilon(t/\varepsilon)) &:= V^\varepsilon_{\nu(t/\varepsilon)} f^\varepsilon(x^\varepsilon(t/\varepsilon)) - f^\varepsilon(x) \\ &\quad - \sum_{k=0}^{\nu(t/\varepsilon)-1} E_{\rho^\varepsilon_u}[V^\varepsilon_{k+1} f^\varepsilon_{k+1} - V^\varepsilon_k f^\varepsilon_k / \mathcal{F}^\varepsilon_k], \end{aligned} \tag{5.4.39}$$

where

$$\begin{aligned} \mathcal{F}^\varepsilon_n &:= \sigma\{x^\varepsilon_k, \theta_k; 0 \le k \le n\}, \\ f^\varepsilon(x) &:= \hat{f}(u(x)) + \varepsilon f^1(x), \quad \hat{f}(u) := \int_{X_u} \rho_u(dx) f(x), \end{aligned} \tag{5.4.40}$$

$$\begin{aligned} (P-I)f_1(x) &= [-(m(x)\Gamma(x) + PD_1(x,\cdot)) + \hat{\Gamma}(u) \\ &\quad + \hat{D}(u) + (\Pi_u - I)P_1]\hat{f}(u), \end{aligned} \tag{5.4.41}$$

$$f_k^\varepsilon := f^\varepsilon(x_k^\varepsilon), \quad V_n^\varepsilon := V_\varepsilon(\varepsilon\tau_n),$$

and $V_\varepsilon(t)$ is defined in (5.4.38), P_1 is an operator generated by $P_1(x,A)$ (see (5.4.32)).

The following representation is true [5]:

$$\Pi_u^\varepsilon = \Pi_u - \varepsilon^r \Pi_u P_1 R_0 + \varepsilon^{2r} \Pi_u^\varepsilon (P_1 R_0)^2, \quad r = 1,2, \tag{5.4.42}$$

where $\Pi_u^\varepsilon, \Pi_u, P_1$ are the operators generated by $\rho_u^\varepsilon, \quad \rho_u$ and $P_1(x,A)$ respectively, $x \in X, \quad A \in \mathcal{X}, \quad u \in U$.

It follows from (5.4.42) that for any continuous and bounded function $f(x)$

$$E_{\rho_u^\varepsilon} f(x) \to \varepsilon \to 0 E_{\rho_u} f(x), \quad \forall u \in U,$$

and all the calculations with $E_{\rho_u^\varepsilon}$ in this section reduce to the calculations with E_{ρ_u} in a similar way as in Section 5.4.3.

Under conditions $A - C)$ the sequence of SMRE $V_\varepsilon(t)f$ in (5.4.38), $f \in B_0$, is tight $\rho_u - a.s., \quad \forall u \in U$.

Under conditions $D), E), i = 2, F), k = 2, j = 1$, the sequence of SMRE $V_\varepsilon(t)f$ is weakly compact $\rho_u - a.s., \quad \forall u \in U$, in $D_B[0,+\infty)$ with limit points in $C_B[0,+\infty)$.

We note that $u(x^\varepsilon(t/\varepsilon)) \to \hat{x}(t)$ *as* $\varepsilon \to 0$, where $\hat{x}(t)$ is a **merged** jump Markov process in $(U,\mathcal{U})$ with **infinitesimal operator** $\Lambda(\hat{P} - I)$,

$$\begin{aligned} \Lambda \hat{f}(u) &:= [b(u)/m(u)]\hat{f}(u), \\ \hat{P}\hat{f}(u) &:= \int_U [b(u,du')/b(u)]\hat{f}(u), \\ m(u) &:= \int_{X_u} \rho_u(dx) m(x), \end{aligned} \tag{5.4.43}$$

$b(u)$ and $b(u,\Delta)$ are defined in (5.4.34). We also note that

$$\Pi_u P_1 = \Lambda(\hat{P} - I), \tag{5.4.44}$$

where Π_u is defined in (5.4.42), P_1–in (5.4.42), Λ and $\hat{P}$–in (5.4.43).

Using (5.3.19), (5.3.20) *as* $n = 2$, and (5.4.40)–(5.4.41), (5.4.42) *as* $r = 1$, (5.4.44), we obtain the following representation:

$$M_{\nu(t/\varepsilon)}^\varepsilon f^\varepsilon(x^\varepsilon(t/\varepsilon)) = V_{\nu(t/\varepsilon)}^\varepsilon \hat{f}(u(x^\varepsilon(t/\varepsilon))) - \hat{f}(u(x)) -$$

$$\varepsilon \sum_{k=0}^{\nu(t/\varepsilon)} [m(u)\hat{\Gamma}(u) + m(u)\hat{D}(u) + m(u)\Lambda(\hat{P} - I)] V_k^\varepsilon \hat{f}(u(x_k^\varepsilon)) + 0_f(\varepsilon), \tag{5.4.45}$$

where $\|0_f(\varepsilon)\|/\varepsilon \to const \quad \varepsilon \to 0, \quad \forall f \in B_0$. Since the third term in (5.4.45) tends to the integral

$$\int_0^t [\Lambda(\hat{P}-I)+\hat{\Gamma}(\hat{x}(s))+\hat{D}(\hat{x}(s))] \times \hat{V}_0(s)\hat{f}(\hat{x}(s))ds$$

and the quadratic variation of the martingale $l(M^{\varepsilon}_{\nu(t/\varepsilon)} f^{\varepsilon}(x^{\varepsilon}(t/\varepsilon)))$ tends to zero *as* $\varepsilon \to 0($ and, hence, $M^{\varepsilon}_{\nu(t/\varepsilon)} f^{\varepsilon}(x^{\varepsilon}(t/\varepsilon)) \to 0, \varepsilon \to 0), \quad \forall l \in B_0^*$, then we obtain from (5.4.45) that the limit evolution $\hat{V}_0(t)$ satisfies equation:

$$\hat{V}_0(t)\hat{f}(\hat{x}(t)) = \hat{f}(u) + \int_0^t [\Lambda(\hat{P}-I)+\hat{\Gamma}(\hat{x}(s))+\hat{D}(\hat{x}(s))]\hat{V}_0(s)\hat{f}(\hat{x}(s))ds. \quad (5.4.46)$$

RE $\hat{V}_0(t)$ is called a **merged** *RE* in an averaging scheme.

5.4.5 Diffusion approximation of random evolutions in reducible phase space

Let us consider SMRE $V_{\varepsilon}(t/\varepsilon)$ with expansion (5.4.19), where $V_{\varepsilon}(t)$ is defined in (5.4.38), and conditions $A) - F)(as \quad i=3, k=4, j=1,2)$ and conditions $a) - f)(e=2)$ are satisfied.

Let the balance condition

$$\int_{X_u} \rho_u(dx)[m(x)\Gamma(x)+PD_1(x,\cdot)]f = 0, \quad \forall u \in U, \quad (5.4.47)$$

be also satisfied and operator

$$L(u) := \int_{X_u} \rho_u(dx)L(x)/m(u), \quad (5.4.48)$$

generates the semigroup of operators, where $L(x)$ is defined in (5.4.24) and $m(u)$ in (5.4.43).

Let us also consider the martingale

$$\begin{aligned} M^{\varepsilon}_{\nu(t/\varepsilon^2)} f^{\varepsilon}(x^{\varepsilon}(t/\varepsilon^2)) &= V^{\varepsilon}_{\nu(t/\varepsilon^2)} f^{\varepsilon}(x^{\varepsilon}(t/\varepsilon^2)) - f^{\varepsilon}(x) \\ &- \sum_{k=0}^{\nu(t/\varepsilon^2)} E\rho^{\varepsilon}_u[V^{\varepsilon}_{k+1}f^{\varepsilon}_{k+1} - V^{\varepsilon}_k f^{\varepsilon}_k/\mathcal{F}^{\varepsilon}_k], \quad (5.4.49) \end{aligned}$$

where

$$\begin{aligned} f^{\varepsilon}(x) &:= \hat{f}(u(x)) + \varepsilon f^1(x) + \varepsilon^2 f^2(x), \\ (P-I)f^1(x) &= [m(x)\Gamma(x)+PD_1(x,\cdot)]\hat{f}(u), \\ (P-I)f^2(x) &= [m(u)L(u)-L(x)+(\Pi_u - I)P_1]\hat{f}(u), \quad (5.4.50) \end{aligned}$$

where $L(u)$ is defined in (5.4.48).

From the balance condition (5.4.47) and from the condition

$$\Pi_u[L(u)-L(x)+(\Pi_u-I)P_1]=0$$

it follows that functions $f^i(x), i=1,2$, are defined unique.

Set $\hat{V}^0(t)$ for the limit of $V_\varepsilon(t/\varepsilon)$ *as* $\varepsilon\to 0$. From (5.4.17) we obtain that the limits for $V_\varepsilon(t/\varepsilon)$ and $V^\varepsilon_{\nu(t/\varepsilon^2)}$ are the same, namely, $\hat{V}^0(t)$.

Weak compactness of $V_\varepsilon(t/\varepsilon)$ is analogical to the one in Section 2.3 with the use of (5.4.32) *as* $l=2$ and (5.4.41) *as* $r=2$. That is why all the calculations in Section 5.4.3 we use in this section replacing E_{ρ_u} by $E_{\rho_u^\varepsilon}$ that reduce to the rates by E_{ρ_u} *as* $\varepsilon\to 0$.

Using (5.3.19), (5.3.20) *as* $n=3$, and representations (5.4.19) and (5.4.49)–(5.4.50) we have the following representation for $M^\varepsilon f^\varepsilon$:

$$\begin{aligned} M^\varepsilon_{\nu(t/\varepsilon^2)}f^\varepsilon &= V^\varepsilon_{\nu(t/\varepsilon^2)}\hat{f}(u(x^\varepsilon(t/\varepsilon^2)))-\hat{f}(u)(x)-\varepsilon^2 \\ &\sum_{k=0}^{\nu(t/\varepsilon^2)}[m(u)L(u(x_k^\varepsilon)+\Pi_uP_1]V_k^\varepsilon\hat{f}(u(x_k^\varepsilon))+0_f(\varepsilon), \end{aligned} \tag{5.4.51}$$

where $L(u)$ is defined in (5.4.48), $\|0_f(\varepsilon)\|/\varepsilon\to const$ $\varepsilon\to 0$. The sum in (5.4.51) converges strongly *as* $\varepsilon\to 0$ to the integral

$$\int_0^t[\Lambda(\hat{P}-I)+L(\hat{x}(s))]\hat{V}^0(s)\hat{f}(\hat{x}(s))ds, \tag{5.4.52}$$

because of the relation (5.4.43), where $\hat{x}(t)$ is a jump Markov process in (U,U) with infinitesimal operator $\Lambda(\hat{P}-I)$, $\hat{x}(0)=u\in U$.

Let $\hat{M}^0(t)f$ be a limit martingale for

$$M^\varepsilon_{\nu(t/\varepsilon^2)}f^\varepsilon(x^\varepsilon(t/\varepsilon^2)) \quad as \quad \varepsilon\to 0.$$

In such a way from (5.4.46)–(5.4.51) we have the equation *as* $\varepsilon\to 0$:

$$\begin{aligned} \hat{M}^0(t)\hat{f}(\hat{x}(t)) &= \hat{V}^0(t)\hat{f}(\hat{x}(t))-\hat{f}(u) \\ &-\int_0^t[\Lambda(\hat{P}-I)+L(\hat{x}(s))]\hat{V}^0(s)\hat{f}(\hat{x}(s))ds. \end{aligned} \tag{5.4.53}$$

The quadratic variation of the martingale $\hat{M}^0(t)$ has the form:

$$< l(\hat{M}^0(t)\hat{f}(u)) >= \int_0^t \int_{X_u} l^2(\sigma(x,u)\Gamma(x)\hat{V}^0(s)\hat{f}(u))\rho_u(dx)ds, \quad (5.4.54)$$

where

$$\sigma^2(x,u) := [m_2(x) - m^2(x)]/m(u).$$

The solution of the martingale problem for $\hat{M}^0(t)$ is expressed by integral:

$$\hat{M}^0(t)\hat{f}(\hat{x}(t)) = \int_0^t \hat{W}(ds,\hat{x}(s))\hat{V}^0(s)\hat{f}(\hat{x}(s)), \quad (5.4.55)$$

where

$$\hat{W}(t,u)f := \int_{X_u} W_{\rho_u}(t,dx)\sigma(x,u)\Gamma(x)f.$$

Finally, from (5.4.52)–(5.4.54) it follows that the limit process $\hat{V}^0(t)$ satisfies the following equation:

$$\begin{aligned} \hat{V}^0(t)\hat{f}(\hat{x}(t)) &= \hat{f}(u) + \int_0^t [\Lambda(\hat{P}-I) + L(\hat{x}(s))]\hat{V}^0(s)\hat{f}(\hat{x}(s))ds \\ &+ \int_0^t \hat{W}(ds,\hat{x}(s))\hat{V}^0(s)\hat{f}(\hat{x}(s)). \end{aligned} \quad (5.4.56)$$

RE $\hat{V}^0(t)$ in (5.4.56) is called a **merged** *RE* in a diffusion approximation scheme. If the operator $\hat{U}^0(t)$ be a solution of Cauchy problem:

$$\begin{cases} d\hat{U}^0(t)dt = \hat{U}^0(t)L(\hat{x}(t)) \\ \hat{U}^0(0) = I, \end{cases}$$

then the operator process $\hat{V}^0\hat{f}(\hat{x}(t))$ satisfies equation:

$$\begin{aligned} \hat{V}^0(t)\hat{f}(\hat{x}(t)) &= \hat{U}^0(t)\hat{f}(u) + \int_0^t \hat{U}^0(t-s)\Lambda(\hat{P}-I)\hat{V}^0(s)\hat{f}(\hat{x}(s))ds \\ &+ \int_0^t \hat{U}^0(t-s)\hat{W}(ds,\hat{x}(s))\hat{V}^0(s)\hat{f}(\hat{x}(s)). \end{aligned} \quad (5.4.57)$$

The uniqueness of the limit *RE* $\hat{V}^0(t)$ is established by dual SMRE.

5.4.6 Normal deviations of random evolutions

The averaged evolution obtained in averaging and merging schemes can be considered as the first approximation to the initial evolution. The diffusion approximation of the SMRE determines the second approximation to the initial evolution, since the first approximation under balance condition — the averaged evolution — appears to be trivial.

Here we consider the **double approximation** to the SMRE — the averaged and the diffusion approximation — provided that the balance condition fails. We introduce the **deviation process** as the normalized difference between the initial and averaged evolutions. In the limit we obtain the **normal deviations** of the initial SMRE from the averaged one.

Let us consider the SMRE $V_\varepsilon(t)$ in (5.4.10) and the averaged evolution $V_0(t)$ in (5.4.18). Let's also consider the deviation of the initial evolution $V_\varepsilon(t)f$ from the averaged one $V_0(t)f$:

$$W_\varepsilon(t)f := \varepsilon^{-1/2} \cdot [V_\varepsilon(t) - V_0(t)]f, \quad \forall f \in B_0. \tag{5.4.58}$$

Taking into account the equations (5.4.10) and (5.4.58) we obtain the relation for $W_\varepsilon(t)$:

$$\begin{aligned} W_\varepsilon(t)f &= \varepsilon^{-1/2} \int_0^t (\Gamma(x(s/\varepsilon)) - \hat{\Gamma})V_\varepsilon(s)fds \\ &+ \int_0^t \hat{\Gamma} W_\varepsilon(s)fds + \\ &\varepsilon^{-1/2}[V_\varepsilon^d(t) - \int_0^t \hat{D} \cdot V_0(s)ds]f, \quad \forall f \in B_0, \end{aligned} \tag{5.4.59}$$

where

$$V_\varepsilon^d(t)f := \sum_{k=1}^{\nu(t/\varepsilon)} [D^\varepsilon(x_{k-1}, x_k) - I]V_\varepsilon(\varepsilon\tau_k^-)f,$$

and $\hat{\Gamma}, \hat{D}$ are defined in (5.4.15).

If the process $W_\varepsilon(t)f$ has the weak limit $W_0(t)f \quad as \quad \varepsilon \to 0$ then we obtain:

$$\int_0^t \hat{\Gamma} W_\varepsilon(s)fds \to \int_0^t \hat{\Gamma} W_0(s)fds, \varepsilon \to 0. \tag{5.4.60}$$

Since the operator $\Gamma(x) - \hat{\Gamma}$ satisfies to the balance condition

$$(\Pi(\Gamma(x) - \hat{\Gamma})f = 0),$$

then the diffusion approximation of the first term in the right-hand side of (5.4.59) gives:

$$\varepsilon^{-1/2}\int_0^t e((\Gamma(x(s/\varepsilon))-\hat{\Gamma})f)ds \to l(\sigma_1 f)w(t), \varepsilon\to 0 \tag{5.4.61}$$

where

$$\begin{aligned} l^2(\sigma_1 f) &= \int_X \rho(dx)[m(x)l((\Gamma(x)-\hat{\Gamma})f)(R_0-I)m(x)l((\Gamma(x)-\hat{\Gamma})f) \\ &\quad +2^{-1}\cdot m_2(x)l^2((\Gamma(x)-\hat{\Gamma})f)]/m, \end{aligned}$$

$\forall l \in B_0$, $w(t)$ is a standard Wiener process.

Since $\prod(PD_1(x,\cdot)-\hat{D})f=0$, then the diffusion approximation of the third term in the right-hand side of (5.4.59) gives the following limit:

$$\varepsilon^{-1/2}\cdot l(V_\varepsilon^d(t)f-\int_0^t \hat{D}V_0(s)fds)\to l(\sigma_2 f)\cdot w(t), \varepsilon\to 0, \tag{5.4.62}$$

where

$$l^2(\sigma_2 f) := \int_X \rho(dx)l((PD_1(x,\cdot)-\hat{D})f)(R_0-I)\cdot l((PD_1(x,\cdot)-\hat{D})f).$$

The passage to the limit in the representation (5.4.59) as $\varepsilon\to 0$ by encountering (5.4.60)–(5.4.62) arrives at the equation for $W_0(t)f$:

$$W_0(t)f=\int_0^t \hat{\Gamma}W_0(s)fds+\sigma f w(t), \tag{5.4.63}$$

where the variance operator σ is determined from the relation:

$$l^2(\sigma f) := l^2(\sigma_1 f)+l^2(\sigma_2 f), \quad \forall l\in B_0, \quad \forall l\in B_0^*, \tag{5.4.64}$$

where operators σ_1 and σ_2 are defined in (5.4.61) and (5.4.62), respectively.

Double approximation of the SMRE has the form:

$$V_\varepsilon(t)f\approx V_0(t)f+\sqrt{\varepsilon}W_0(t)f$$

for small ε, which perfectly fits the standard form of the *CLT* with non zero limiting mean value.

5.4.7 Rates of convergence in the limit theorems for RE

The rates of convergence in the averaging and diffusion approximation scheme for the sequence of SMRE are considered in this section.

Averaging Scheme. The problem is to estimate the value

$$\|E_\rho[V_\varepsilon(t)f^\varepsilon(x(t/\varepsilon)) - V_0(t)f]\|, \quad \forall f \in B_0, \tag{5.4.65}$$

where $V_0(t), V_\varepsilon(t), f^\varepsilon, f$ and B_0 are defined in (5.4.18), (5.4.10), (5.4.14), (5.4.12), respectively.

We use the following representation

$$\begin{aligned} \|E_\rho[V_\varepsilon(t)f^\varepsilon(x(t/\varepsilon)) &- V_0(t)f]\| \leq \|E_\rho[V_\varepsilon(t)f - V_\varepsilon(\tau_{\nu(t/\varepsilon)})f]\| + \\ +\|E_\rho[V_\varepsilon(\tau_{\nu(t/\varepsilon)})f &- V_0(t)f]\| + \varepsilon\|E_\rho V_\varepsilon(t)f_1(x(t/\varepsilon))\| \end{aligned} \tag{5.4.66}$$

that follows from (5.4.60) and (5.4.18), (5.4.14), (5.4.12).

For the first term in the right-hand side of (5.4.66) we obtain (see (5.4.17) and (5.4.5) *as* $n = 2$):

$$\|E_\rho[V_\varepsilon(t)f - V_\varepsilon(\tau_{\nu(t/\varepsilon)}))f]\| \leq \varepsilon \cdot C_1(T,f), \quad \forall t \in [0,T], \tag{5.4.67}$$

where

$$\begin{aligned} C_1(T,f) &:= \int_X \rho(dx)[C_0(T,x,f) + C_0^2(T,x,f)], \\ C_0(T,x,f) &:= T \cdot m_2(x)\|\Gamma(x)f\|/2m, \quad \forall f \in B_0. \end{aligned}$$

For the second term in the right-hand side of (5.4.66) we have from (5.4.16) and (5.4.65) (since $E_\rho M^\varepsilon_{\nu(t/\varepsilon)}f^\varepsilon(x(t/\varepsilon)) = 0$):

$$\begin{aligned} \|E_\rho[V_\varepsilon(\tau_{\nu(t/\varepsilon)})f &- V_0(t)f]\| \leq \varepsilon\|E_\rho[V^\varepsilon_{\nu(t/\varepsilon)} - I]f_1(x(t/\varepsilon))\| \\ &+ \varepsilon\|E_\rho[\sum_{k=0}^{\nu(t/\varepsilon)-1}(\hat{\Gamma}+\hat{D})V^\varepsilon_k f - \varepsilon^{-1}m^{-1}\int_0^1(\hat{\Gamma}+\hat{D})V_0(s)fds]\| \\ &+ \varepsilon \cdot C_2(T,f), \end{aligned} \tag{5.4.68}$$

where constant $C_2(T,f)$ is expressed by the algebraic sum of $\int_X m_i(x)\|\Gamma^i(x)f\|\rho$ (dx) and

$$\int_X m_i(x)\|PD_1(x,\cdot)\cdot\Gamma^i(x)f\|\rho(dx), \quad i = 1,2, \quad f \in B_0,$$

and $\|R_0\|$, R_0 is a potential of Markov chain $(x_n; n \geq 0)$.

For the third term in the right-hand side of (5.4.66) we obtain:

$$E_\rho \|f_1(x)\| \leq 2C_3(f), \tag{5.4.69}$$

where

$$C_3(f) := \|R_0\| \cdot \int_X p(dx)[m(x)\|\Gamma(x)f\| + \|PD_1(x,\cdot)f\|].$$

Finally, from (5.4.66)–(5.4.69) we obtain the estimate of the value in (5.4.65), namely, **rate of convergence in averaging scheme for SMRE**:

$$\|E_\rho[V_\varepsilon(t)f^\varepsilon(x(t/\varepsilon)) - V_0(t)f]\| \leq \varepsilon \cdot C(T,f), \tag{5.4.70}$$

where constant $C(T,f)$ is expressed by $C_i(T,f)$, $i = \overline{1,3}$.

Diffusion Approximation. The problem is to estimate the value:

$$\|E_\rho[V_\varepsilon(t/\varepsilon)f^\varepsilon(x(t/\varepsilon^2)) - V^0(t)f]\|, \quad \forall f \in B_0, \tag{5.4.71}$$

where $V_\varepsilon(t/\varepsilon)$, f^ε, $V^0(t)$, f, B_0 are defined in (5.4.21), (5.4.23), (5.4.29), (5.4.20), respectively.

Here, we use the following representation:

$$\begin{aligned}
&\|E_\rho[V_\varepsilon(t/\varepsilon)f^\varepsilon(x(t/\varepsilon^2)) - V^0(t)f]\| \leq \|E_\rho[V_\varepsilon(t/\varepsilon)f - V_\varepsilon(\tau_{\nu(t/\varepsilon^2)})f]\| \\
&+\|E_\rho[V_\varepsilon(\tau_{\nu(t/\varepsilon^2)})f - V^0(t)f]\| + \varepsilon\|E_\rho[V_\varepsilon(t/\varepsilon)f_1(x(t/\varepsilon^2))]\| \\
&+\varepsilon^2\|E_\rho[V_\varepsilon(t/\varepsilon)f_2(x(t/\varepsilon^2))]\|,
\end{aligned} \tag{5.4.72}$$

that follows from (5.4.70) and (5.4.23), (5.4.17), respectively.

First of all we have for the fourth term in the right-hand side of (5.4.71):

$$\varepsilon^2\|E_\rho[V_\varepsilon(t/\varepsilon)f_2(x(t/\varepsilon^2))]\| \leq \varepsilon^2 \cdot 2\|R_0\| \cdot \int_X \rho(dx)\|L(x)f\| := \varepsilon^2 d_1(f), \tag{5.4.73}$$

where $L(x)$ is defined in (5.4.24).

For the third term in the right-hand side of (5.4.71) we obtain:

$$\varepsilon\|E_\rho[V_\varepsilon(t/\varepsilon)f_1(x(t/\varepsilon^2))]\| \leq \varepsilon \cdot d_2(f), \tag{5.4.74}$$

where

$$d_2(f) := 2\|R_0\| \cdot \int_X \rho(dx)[m(x)\|\Gamma(x)f\| + \|PD_1(x,\cdot)f\|], \quad f \in B_0.$$

For the first term in the right-hand side of (5.4.71) we have from (5.4.70):

$$\|E_\rho[V_\varepsilon(t/\varepsilon)f - V_\varepsilon(\tau_{\nu(t/\varepsilon^2)})f]\| \leq \varepsilon \cdot C_1(T,f), \tag{5.4.75}$$

where $C_1(T,f)$ is defined in (5.4.70).

For the second term in the right-hand side of (5.4.76) we use the asymptotic representation (5.4.25) for the martingale $M^\varepsilon_{\nu(t/\varepsilon^2)} f^\varepsilon$ and the conditions

$$E_\rho M^\varepsilon f^\varepsilon = 0, \quad E_\rho M^0(t) f = 0, \quad \forall f \in B_0 \tag{5.4.76}$$

$$\begin{aligned} \|E_\rho[V_\varepsilon(\tau_{\nu(t/\varepsilon^2)})f - V^0(t)f]\| &\leq \varepsilon\|E_\rho[V^\varepsilon(\tau_{\nu(t/\varepsilon^2)})f_1 - f_1(x)]\| + \\ &+ \varepsilon^2\|E_\rho[V_\varepsilon(\tau_{\nu(t/\varepsilon^2)})f_2 - f_2(x)]\| + \varepsilon^2\|E_\rho[\sum_{k=0}^{\nu(t/\varepsilon^2)-1} \hat{L}V_k^\varepsilon f \\ &- \varepsilon^{-2} m^{-1}\int_0^t \hat{L}V^0(s) f ds]\| + \varepsilon \cdot d_3(f), \end{aligned} \tag{5.4.77}$$

where constant $d_3(f)$ is expressed by the algebraic sum of

$$\int_X m_i(x)\|\Gamma^j(x)PD_e(x,\cdot)f\|\rho(dx), \quad i = \overline{1,3}, \quad j = \overline{0,3}, \quad e = \overline{1,2}.$$

We note that

$$\|E_\rho[\sum_{k=0}^{\nu(t/\varepsilon^2)-1} \hat{L}V_k^\varepsilon f - \varepsilon^{-2}m^{-1}\int_0^t \hat{L}V^0(s) f ds]\| \leq d_4(T,f). \tag{5.4.78}$$

Finally, from (5.4.72)–(5.4.78) we obtain the estimate of the value in (5.4.71), namely, **rate of convergence in diffusion approximation scheme for SMRE**:

$$\|E_\rho[V_\varepsilon(t/\varepsilon)f^\varepsilon(x(t/\varepsilon^2)) - V^0(t)f]\| \leq \varepsilon \cdot d(T,f), \tag{5.4.79}$$

where constant $d(T,f)$ is expressed by $d_i, i = \overline{1,4}$, and $C_1(T,f), \quad f \in B_0$.

References

[1] Arnold, L. *Random dynamical systems.* Springer-Verlag, 1998.

[2] Bhattacharya, R. and Majumdar, M. *Random dynamical systems: theory and applications.* Cambridge University Press, 2007.

[3] Dynkin, E. B. *Markov processes.* Springer-Verlag, 1991.

[4] Jacod, J. and Shiryaev, A. N. *Limit theorems for stochastic processes.* Springer-Verlag, 2010.

[5] Korolyuk, V. S. and Swishchuk A. V. *Evolution of systems in random media.* Chapman & Hall CRC, 1995.

[6] Korolyuk, V. S. and Swishchuk A. V. *Semi-Markov random evolutions.* Kluwer AP, 1995.

[7] Pinsky, M. *Lectures on random evolutions.* World Scientific Publishers, 1991.

[8] Sobolev, S. L. *Some applications of functional analysis in mathematical physics.* American Mathematical Society, 1991.

[9] Strook, D. and Varadhan, S.R.S. *Multidimensional diffusion processes.* Springer-Verlag, 1979.

[10] Swishchuk, A. V. *Random evolutions and their applications.* Kluwer AP, 1997.

[11] Swishchuk, A. V. *Random evolutions and their applications. New Trends.* Kluwer AP, 2000.

[12] Swishchuk, A. and Islam, S. *The geometric Markov renewal processes with applications to finance.* Stochastic Analysis and Applications. v. 29, n. 4, 684-705, 2010.

Chapter 6

Averaging of the Geometric Markov Renewal Processes (GMRP)

6.1 Chapter overview

We introduce the geometric Markov renewal processes as a model for a security market and study these processes in a series scheme. We consider its approximations in the form of averaged, merged, and double averaged geometric Markov renewal processes. Weak convergence analysis and rates of convergence of ergodic geometric Markov renewal processes are presented. Martingale properties, infinitesimal operators of geometric Markov renewal processes are presented and a Markov renewal equation for expectation is derived. As an application, we consider the case of two ergodic classes. Moreover, we consider a generalized binomial model for a security market induced by a position dependent random map as a special case of a geometric Markov renewal process.

6.2 Introduction

In various practical situations, the evolution of systems is influenced by an external random medium. For a fixed state of the medium, the evolution of the system is completely determined by the internal evolutionary laws, while the quantitative characteristics of the system change in accordance with the changes of the medium. At the same time, the probability laws governing the behavior of the external medium are independent of the evolution of the system. Therefore, the mathematical model of the evolution of a system placed in a random medium consists of two processes, namely, a switching process which describes the changes of the medium, and a switched process which describes the evolution of the system. A large literature exists regarding switched-switching processes and their applications [8, 9, 10, 11, 14]. A mathematical model of the evolution of a financial security market is an example of a switched process in a random medium. We are interested in introducing switched-switching processes which are generalizations of processes considered in the literature, e.g., the geometric compound Poisson processes [1] and binomial models [4]. Let $N(t)$ be a standard Poisson process, $(Y_k)_{k\in Z_+}$ be i. i. d. random variables which are independent of $N(t)$ and $S_0^* > 0$. The geometric compound Poisson processes

$$S_t^* = S_0^* \prod_{k=1}^{N(t)} (1+Y_k), \; t > 0 \tag{6.2.1}$$

is a trading model in many financial applications with pure jumps [1, 4, 11]. On the other hand, a classical binomial model [4] is a representation of a random map [8] with constant probabilities (see [3]). In this chapter, motivated by the geometric compound Poisson processes (6.2.1), we introduce the geometric Markov renewal processes (GMRP) (6.4.1) (see Section 6.4) for a security market where Markov renewal processes and semi-Markov processes are treated as switching processes. The geometric Markov renewal processes (6.4.1) will be our main trading model in further analysis. We apply Markov renewal theories and phase merging algorithms to the study of the geometric Markov renewal processes. We consider a generalized binomial model induced by position dependent random maps and show that a generalized binomial model is a special case of a geometric Markov renewal process.

In Section 6.3, we present the notation and summarize the results we shall need in the sequel. The geometric Markov renewal processes are introduced in Section 6.4; here we present the jump semi-Markov random evolution, a Markov renewal equation for expectation, infinitesimal operators, and martingale properties of the geometric Markov renewal processes. The ergodic averaged geometric Markov renewal processes and weak convergence are presented in Section 6.5. We present rates of convergence of ergodic geometric Markov renewal processes in Section 6.6. We present merged geometric Markov renewal processes in Section 6.7. In Section 6.8 we consider a generalized binomial model induced by a random map with position dependent probabilities as a special case of the geometric Markov renewal process. In Section 6.9, we present a number of applications of GMRP.

6.3 Markov renewal processes and semi-Markov processes

Let $(\Omega, \mathcal{B}, \mathcal{F}_t, \mathbf{P})$ be a standard probability space with complete filtration $\mathcal{F}_t$ and let $(x_k)_{k\in\mathbf{Z}_+}$ be a Markov chain in the phase space $(X, \mathcal{X})$ with transition probability $P(x,A)$, where $x \in X, A \in \mathcal{X}$. Let $(\theta_k)_{k\in\mathbf{Z}_+}$ is a renewal process which is a sequence of independent and identically distributed (i. i. d.) random variables with a common distribution function $F(x) := \mathbf{P}\{w : \theta_k(w) \leq x\}$. The renewal process $(\theta_k)_{k\in\mathbb{Z}_+}$ counts events and the random variables θ_k can be interpreted as lifetimes (operating periods, holding times, renewal periods) of a certain system in a random environment. From the renewal process $(\theta_k)_{k\in\mathbf{Z}_+}$ we can construct another renewal process $(\tau_k)_{k\in\mathbf{Z}_+}$ defined by

$$\tau_k := \sum_{n=0}^{k} \theta_n. \tag{6.3.1}$$

The random variables τ_k are called renewal times (or jump times). The process

$$\nu(t) := \sup\{k : \tau_k \leq t\} \tag{6.3.2}$$

is called the counting process.

Definition[11, 14] A homogeneous two-dimensional Markov chain $(x_n, \theta_n)_{n\in \mathbf{Z}_+}$ on the phase space $X \times \mathbf{R}_+$ is called a Markov renewal process (MRP) if its transition probabilities are given by the semi-Markov kernel

$$Q(x,A,t) = \mathbf{P}\{x_{n+1} \in A, \theta_{n+1} \le t | x_n = x\}, \ \forall x \in X, A \in \mathcal{X}, t \in \mathbf{R}_+. \tag{6.3.3}$$

Definition The process

$$x(t) := x_{\nu(t)} \tag{6.3.4}$$

is called a *semi-Markov process.* The ergodic theorem for a Markov renewal process and a semi-Markov process, respectively, can be found in [6, 9–11, 14].

6.4 The geometric Markov renewal processes (GMRP)

Let $(x_n, \theta_n)_{n\in \mathbf{Z}_+}$ be a Markov renewal process on the phase space $X \times \mathbf{R}_+$ with the semi-Markov kernel $Q(x,A,t)$ defined in (6.3.3) and $x(t) := x_{\nu(t)}$ be a semi-Markov process where the counting process $\nu(t)$ is defined in (6.3.2). Let $\rho(x)$ be a bounded continuous function on X such that $\rho(x) > -1$. We define the geometric Markov renewal process (GMRP) $\{S_t\}_{t\in \mathbf{R}_+}$ as a stochastic functional S_t defined by

$$S_t := S_0 \prod_{k=1}^{\nu(t)} (1 + \rho(x_k)), \ t \in \mathbf{R}_+, \tag{6.4.1}$$

where $S_0 > 0$ is the initial value of S_t. We call this process $(S_t)_{t\in \mathbf{R}_+}$ a geometric Markov renewal process by analogy with the geometric compound Poisson processes

$$S_t^* = S_0^* \prod_{k=1}^{N(t)} (1 + Y_k), \tag{6.4.2}$$

where $S_0^* > 0$, $N(t)$ is a standard Poisson process, $(Y_k)_{k\in Z_+}$ are i. i. d. random variables. The geometric compound Poisson process $\{S_t^*\}_{t\in \mathbf{R}_+}$ in (6.4.2) is a trading model in many financial applications as a pure jump model [1, 4]. The geometric Markov renewal processes $\{S_t\}_{t\in \mathbf{R}_+}$ in (6.4.1) will be our main trading model in further analysis.

6.4.1 *Jump semi-Markov random evolutions*

Let $C_0(\mathbb{R}_+)$ be the space of continuous functions on $\mathbf{R}_+$ vanishing at infinity and let us define a family of bounded contracting operators $D(x)$ on $C_0(\mathbf{R}_+)$ as follows:

$$D(x)f(s) := f(s(1 + \rho(x)), \ x \in X, s \in \mathbf{R}_+. \tag{6.4.3}$$

With these contraction operators $D(x)$ we define the following jump semi-Markov random evolution (JSMRE) $V(t)$ of the geometric Markov renewal processes $\{S_t\}_{t\in\mathbf{R}_+}$ in (6.4.1):

$$V(t) = \prod_{k=1}^{\nu(t)} D(x_k) := D(x_{\nu(t)}) \circ D(x_{\nu(t)-1}) \circ \ldots \circ D(x_1). \tag{6.4.4}$$

Using (6.4.3) we obtain from (6.4.4)

$$V(t)f(s) = \prod_{k=1}^{\nu(t)} D(x_k)f(s) = f(s\prod_{k=1}^{\nu(t)}(1+\rho(x_k)) = f(S_t), \tag{6.4.5}$$

where S_t is defined in (6.4.1) and $S_0 = s$. Let $Q(x,A,t)$ be a semi-Markov kernel for Markov renewal process $(x_n;\theta_n)_{n\in Z_+}$; that is, $Q(x,A,t) = P(x,A)G_x(t)$ where $P(x,A)$ is the transition probability of the Markov chain $(x_n)_{n\in\mathbf{Z}_+}$ and $G_x(t) := \mathcal{P}(\theta_{n+1} \leq t|x_n = x)$. Let

$$u(t,x) := E_x[V(t)g(x(t))] := E[V(t)g(x(t))/x(0) = x] \tag{6.4.6}$$

be the mean value of the semi-Markov random evolution $V(t)$ in (6.4.5).

The following theorem is proved in [11, 14]:

Theorem 6.1 *The mean value $u(t,x)$ in (6.4.6) of the semi-Markov random evolution $V(t)$ given by the solution of the following Markov renewal equation (MRE):*

$$u(t,x) - \int_0^t \int_X Q(x,dy,ds)D(y)u(t-s,y) = \bar{G}_x(t)g(x), \tag{6.4.7}$$

where $\bar{G}_x(t) = 1 - G_x(t)$, $G_x(t) := \mathcal{P}(\theta_{n+1} \leq t|x_n = x)$, $g(x)$ is a bounded and continuous function on X.

6.4.2 Infinitesimal operators of the GMRP

Let

$$\rho_T(x) := \frac{\rho(x)}{T},\ T > 0$$

and

$$S_t^T := S_0 \prod_{k=1}^{\nu(tT)} (1+\rho_T(x_k)) = S_0 \prod_{k=1}^{\nu(tT)} (1+T^{-1}\rho(x_k)). \tag{6.4.8}$$

In Section 6.5 we have presented detailed information about $\rho_T(x)$ and S_t^T. It can be easily shown that

$$\ln\frac{S_t^T}{S_0} = \sum_{k=1}^{\nu(tT)} \ln(1+\frac{\rho(x_k)}{T}). \tag{6.4.9}$$

To describe martingale properties of the GMRP $(S_t)_{t\in\mathbf{R}_+}$ in (6.4.1) we need to find an infinitesimal operator of the process

$$\eta(t) := \sum_{k=1}^{\nu(t)} \ln(1+\rho(x_k)). \tag{6.4.10}$$

Let $\gamma(t) := t - \tau_{\nu(t)}$ and consider the process $(x(t),\gamma(t))$ on $X\times R_+$. It is a Markov process with infinitesimal operator

$$\hat{Q}f(x,t) := \frac{df}{dt} + \frac{g_x(t)}{\bar{G}_x(t)}\int_X [P(x,dy)f(y,0) - f(x,t)], \tag{6.4.11}$$

where $g_x(t) := \frac{dG_x(t)}{dt}$, $\bar{G}_x(t) = 1 - G_x(t)$, where $f(x,t)\in C(X\times R_+)$. The infinitesimal operator for the process $\ln S(t)$ has the form:

$$\hat{A}f(z,x) = \frac{g_x(t)}{\bar{G}_x(t)}\int_X P(x,dy)[f(z+\ln(1+\rho(y),x) - f(z,x)], \tag{6.4.12}$$

where $z := \ln S_0$. The process $(\ln S(t), x(t), \gamma(t))$ is a Markov process on $R_+\times X\times R_+$ with the infinitesimal operator

$$\hat{L}f(z,x,t) = \hat{A}f(z,x,t) + \hat{Q}f(z,x,t), \tag{6.4.13}$$

where the operators $\hat{A}$ and $\hat{Q}$ are defined in (6.4.12) and (6.4.13), respectively. Thus we obtain that the process

$$\hat{m}(t) := f(\ln S(t), x(t), \gamma(t)) - f(z,x,0) - \int_0^t (\hat{A}+\hat{Q})f(\ln S(u), x(u), \gamma(u))du \tag{6.4.14}$$

is an $\hat{\mathcal{F}}_t$-martingale, where $\hat{\mathcal{F}}_t := \sigma(x(s),\gamma(s); 0\le s\le t)$. If $x(t) := x_{\nu(t)}$ is a Markov process with kernel

$$Q(x,A,t) = P(x,A)(1-e^{-\lambda(x)t}), \tag{6.4.15}$$

namely, $G_x(t) = 1 - e^{-\lambda(x)t}$, then $g_x(t) = \lambda(x)e^{-\lambda(x)t}$, $\hat{G}_x(t) = e^{-\lambda(x)t}$, $\frac{g_x(t)}{\bar{G}_x(t)} = \lambda(x)$ and the operator $\hat{A}$ in (6.4.12) has the form:

$$\hat{A}f(z) = \lambda(x)\int_X P(x,dy)[f(z+\ln(1+\rho(y))) - f(z)]. \tag{6.4.16}$$

The process $(\ln S(t), x(t))$ on $R_+\times X$ is a Markov process with infinitesimal operator

$$\hat{L}f(z,x) = \hat{A}f(z,x) + Qf(z,x), \tag{6.4.17}$$

where

$$Qf(z,x) = \lambda(x)\int_X P(x,dy)(f(y) - f(x)).$$

It follows that the process

$$m(t) := f(\ln S(t), x(t)) - f(z,x) - \int_0^t (\hat{A}+Q)f(\ln S(u), x(u))du \tag{6.4.18}$$

is an $\mathcal{F}_t$-martingale, where $\mathcal{F}_t := \sigma(x(u); 0\le u\le t)$.

6.4.3 Martingale property of the GMRP

Consider the geometric Markov renewal processes $(S_t)_{t\in\mathbb{R}_+}$

$$S_t = S_0 \prod_{k=1}^{\nu(t)} (1+\rho(x_k)). \tag{6.4.19}$$

For $t \in [0,T]$ let us define

$$L_t := L_0 \prod_{k=1}^{\nu(t)} h(x_k), \qquad EL_0 = 1, \tag{6.4.20}$$

where $h(x)$ is a bounded continuous function such that

$$\int_X h(y)P(x,dy) = 1, \qquad \int_X h(y)P(x,dy)\rho(y) = 0. \tag{6.4.21}$$

If $EL_T = 1$, then geometric Markov renewal process S_t in (6.4.19) is an $(\mathcal{F}_t, P^*)$-martingale, where measure P^* is defined as follows:

$$\frac{dP^*}{dP} = L_T, \tag{6.4.22}$$

and

$$\mathcal{F}_t := \sigma(x(s); 0 \le s \le t). \tag{6.4.23}$$

In the discrete case we have

$$S_n = S_0 \prod_{k=1}^{n} (1+\rho(x_k)). \tag{6.4.24}$$

Let $L_n := L_0 \prod_{k=1}^{n} h(x_k)$, $\qquad EL_0 = 1$, where $h(x)$ is defined in (6.4.21). If $EL_N = 1$, then S_n is an $(\mathcal{F}_t, P^*)$-martingale, where $\frac{dP^*}{dP} = L_N$, and $\mathcal{F}_n := \sigma(x_k; 0 \le k \le n)$.

6.5 Averaged geometric Markov renewal processes

In this section we consider the geometric Markov renewal processes $(S_t)_{t\in\mathbf{R}_+}$ in a series scheme. A series scheme means that we consider not only one process S_t, but a series of processes S_t^T which depend on a parameter $T > 0$. The jump in the stock is $T^{-1}\rho(x_k)$ and this jump is small when T is large, and we consider the geometric Markov renewal processes on a larger time interval $[0, tT)$. In this way, a series scheme means that we consider a series of processes S_t^T with small random perturbations $T^{-1}\rho(x_k)$ on a large time interval tT. We are interested in the following question: does S_t^T converge to any S_t of the unperturbed system as $T \to +\infty$? Under different conditions we obtain different results and different kinds of convergence. Under some conditions the averaged geometric Markov renewal processes are obtained as ergodic, merged, and double-merged geometric Markov renewal processes.

6.5.1 Ergodic geometric Markov renewal processes

Let $(x_n)_{n\in\mathbb{Z}_+}$ be a Markov chain on the phase space $(X,\mathcal{X})$. Suppose that the Markov chain $(x_n)_{n\in\mathbf{Z}_+}$ has a stationary distribution $p(A), A\in\mathcal{X}$. The evolution of the geometric Markov renewal process S_t takes place in a stationary regime if the effect of ergodicity of $(x_n)_{n\in\mathbf{Z}_+}$ is sufficiently influenced. This means that S_t should be considered on large intervals of time. This can be done if we consider the counting process $\nu(t)$ in (6.3.2) in new faster time. Let $T>0$ and consider $\nu_T(t):=\nu(tT)$, in new "fast" time. To avoid infinite changes of S_t for finite time under increasing $T\to+\infty$, it is necessary to consider the dependence of the value of jumps of the process S_t on T. It means that the function $\rho(x)$ should depend on T (as $S_{\tau_k}-S_{\tau_k-}=S_{\tau_k-}\rho(x_k)$), i.e., $\rho\equiv\rho_T(x)$ such that $\rho_T(x)\to 0$ uniformly by x. For simplicity we consider

$$\rho_T(x)=\frac{\rho(x)}{T},$$

for all $x\in X$. In this way, S_t in (6.4.1) has the following form:

$$S_t^T=S_0\prod_{k=1}^{\nu(tT)}(1+\rho_T(x_k))=S_0\prod_{k=1}^{\nu(tT)}(1+T^{-1}\rho(x_k)). \tag{6.5.1}$$

In the following we present the main theorem of this chapter.

Theorem 6.2 *Let $m:=\int_X p(dx)\bar{m}(x)$, $\bar{m}(x):=\int_0^\infty(1-G_x(t))dt$, $G_x(t)=\mathbf{P}(\tau_{n+1}-\tau_k<t|x_n=x)$, $\hat{\rho}:=\int_X p(dx)\rho(x)/m$. If $\int_X p(dx)(\rho(x))^2<+\infty$, then the ergodic geometric Markov renewal processes $\{\hat{S}_t\}$ has the following form:*

$$\hat{S}_t=S_0e^{\hat{\rho}t} \tag{6.5.2}$$

for all $t\in R_+$, and $S_0>0$. That is, the dynamics of the ergodic geometric Markov renewal process which describes the dynamics of stock prices is the same as the dynamics of bond prices with interest rate $\hat{\rho}$.

Proof *From (6.5.1) we obtain*

$$\ln\frac{S_t^T}{S_0}=\sum_{k=0}^{\nu(tT)}\ln(1+T^{-1}\rho(x_k)). \tag{6.5.3}$$

Since $\frac{\rho(x)}{T}$ is small for large T, we use the Taylor formula and obtain:

$$\ln(1+\frac{\rho(x)}{T})=\frac{\rho(x)}{T}-1/2(\frac{\rho(x)}{T})^2+r(\frac{\rho(x)}{T})(\frac{\rho(x)}{T})^2, \tag{6.5.4}$$

where function r tends to zero as $T\to+\infty$. Taking into account (6.5.1) and (6.5.4) we obtain:

$$\ln\frac{S_t^T}{S_0} = \frac{1}{T}\sum_{k=0}^{\nu(tT)}\rho(x_k)-\frac{1}{2T^2}\sum_{k=0}^{\nu(tT)}(\rho(x_k))^2 \tag{6.5.5}$$

$$+\frac{1}{T^2}\sum_{k=0}^{\nu(tT)}r(\frac{\rho(x)}{T})(\rho(x_k))^2. \tag{6.5.6}$$

We note that $\nu(tT)$ *has order of growth* tT/m, *where* $m := \int_X p(dx)m(x)$, $m(x) := \int_0^\infty (1 - G_x(t))dt$. *It can be easily shown that the last two terms of the right-hand side in (6.5.5) tend to zero as* $T \to +\infty$. *Using the algorithm of phase averaging for functionals of Markov chain [6, 9, 10, 11, 14] we can show (see the subsections below) that the first term of the right-hand side in (6.5.5) has the following limit:*

$$\lim_{T\to+\infty} \frac{1}{T}\sum_{k=0}^{\nu(Tt)} \rho(x_k) = \hat{\rho}t,$$

where $\hat{\rho} := \int_X p(dx)\rho(x)/m$. *From (6.5.5) we obtain*

$$\lim_{T\to+\infty} \ln\frac{S_t^T}{S_0} = \hat{\rho}t.$$

Thus, $\lim_{T\to+\infty} S_t^T = \hat{S}_t = S_0 e^{\hat{\rho}t}$ *for all* $t \in R_+$.

6.5.1.1 Average scheme

Let

$$G_t^T := T^{-1}\sum_{k=0}^{\nu(tT)} \rho(x_k), \qquad G_0^T = \ln S_0 = \ln s, \tag{6.5.7}$$

and

$$G_n^T := G_{\tau_n T^{-1}}^T, \qquad G_0^T = \ln s.$$

Then

$$G_{n+1}^T - G_n^T = T^{-1}\rho(x_n). \tag{6.5.8}$$

If $f(s) \in C^1(\mathbf{R})$, *then*

$$f(G_{n+1}^T) - f(G_n^T) = T^{-1}\rho(x_n)df(G_n^T)/ds. \tag{6.5.9}$$

Let us define the following functions:

$$\phi^T(s,x) := f(s) + T^{-1}\phi_f^1(s,x), \qquad \forall f \in C^2(\mathbf{R}), \tag{6.5.10}$$

where $\phi_f^1(s,x)$ *is defined by the equation:*

$$(P - I)\phi_f^1(s,x) = (\hat{D} - D(x))f(s), \tag{6.5.11}$$

$$D(x) := \rho(x)d/ds, \qquad \hat{D} := \int_X \pi(dx)D(x) = \int_X \pi(dx)\rho(dx)d/ds. \tag{6.5.12}$$

6.5.1.2 Martingale problem for the limit process $\hat{S}_t$ in average scheme

Let us define the family of functions

$$\psi^T(s,t) := \phi^T(G^T_{[tT]}, x_{[tT]}) - \phi^T(G^T_{[sT]}, x_{[sT]}) - \sum_{j=[sT]}^{[tT]-1} E_\rho[\phi^T(G^T_{j+1}, x_{j+1}) - \phi^T(G^T_J, x_j)|\mathcal{F}_j], \quad (6.5.13)$$

where $\mathcal{F}_j := \sigma(x_k, \tau_k; 0 \le k \le j)$, and ϕ^T are defined in (6.5.10). Functions $\psi^T(0,t)$ are $\mathcal{F}_{[tT]}$-martingale by t, $\psi^T(0,t) := m_T(t)$. Taking into account the expansions (6.5.9) and (6.5.10) we obtain the representation for functions $\psi^T(s,t)$ in (6.5.13):

$$\begin{aligned}\psi^T(s,t) &= f(G^T_{[tT]}) - f(G^T_{[sT]}) + T^{-1}[\phi^1_f(G^T_{[tT]}, x_{[tT]}) - \phi^1_f(G^T_{[sT]}, x_{[sT]}) - \\ &- \sum_{j=[sT]}^{[tT]-1} \{E_\rho[T^{-1}\rho(x_j) df(G^T_j)/ds|\mathcal{F}_j] + T^{-1}(\hat{D} - D(x_j)) f(G^T_j)]\} \\ &= f(G^T_{[tT]}) - f(G^T_{[sT]}) + T^{-1}[\phi^1_f(G^T_{[tT]}, x_{[tT]}) - \phi^1_f(G^T_{[sT]}, x_{[sT]})] \\ &- T^{-1} \sum_{j=[sT]}^{[tT]-1} \hat{D} f(G^T_j).\end{aligned} \quad (6.5.14)$$

The process $m_T(t)$ is a martingale, hence, $E^T[\phi^T(s,t)\eta^s_0]$ is equal to the right-hand side of (6.5.14) for each scalar measurable continuous functional η^s_0, where E^T is an expectation by measure Q_T corresponding to $G^T_{[tT]}$. If the process $G^T_{[tT]}$ converges weakly to some process $\hat{G}_t$ as $T \to +\infty$, then we have

$$0 = E^T[(f(\hat{G}^t) - f(\hat{G}_u) - \int_u^t \hat{D} f(\hat{G}_s) ds)\eta^u_0],$$

namely, the following process

$$f(\hat{G}_t) - f(\hat{G}_u) - \int_0^t \hat{D} f(\hat{G}_v) dv = f(\hat{G}_t) - f(\hat{G}_u) - \int_0^t \bar{\rho}(df(\hat{G}_s)/dg) ds \quad (6.5.15)$$

is a continuous martingale, where $\bar{\rho} := \int_X \pi(dx)\rho(x)$. It means that the process $\hat{G}_t$ satisfies the martingale problem and it is a deterministic process. which follows from (6.5.15):

$$\begin{cases} df(\hat{G}_t)/dt = \bar{\rho} df(\hat{G}_t)/dg \\ f(\hat{G}_0) = f(\ln S_0) = f(\ln s), \end{cases} \quad (6.5.16)$$

and for function $\hat{S}_t$ we have:

$$\hat{S}_t = S_0 e^{\bar{\rho} t}.$$

To obtain the martingale problem for the process G^T_t, we note that

$$G^T_{T^{-1}\tau_{\nu(tT)}} = G^T_{\nu(tT)}, \quad (6.5.17)$$

where G_n^T is defined in (6.5.7). It means that limits for G_t^T and $G_{\nu(tT)}^T$ as $T \to +\infty$ coincide and the limit for $G_{\nu(tT)}^T$ coincides with the limit for $G_{[tT]}^T$ in (6.5.15). Further, taking into account that $T^{-1}\nu(tT) \to t/m$ as $T \to +\infty$, by a probability argument and the renewal theorem, from the previous reasonings, replacing $[tT]$ in the place of $\nu(tT)$, we finally obtain that the process G_t^T converges weakly to the process $\hat{G}_t$ as $T \to +\infty$ which is the solution of the following martingale problem:

$$f(\hat{G}_t) - f(\hat{G}_u) - \int_u^t \hat{\rho}(df(\hat{G}_v)/dg)dv$$

is a continuous martingale, where

$$\hat{\rho} := \int_X \pi(dx)\rho(x)/m.$$

6.5.1.3 Weak convergence of the processes S_t^T in an average scheme

Let Q_T be a measure corresponding to the process G_t^T in (6.5.7). We note that $G_t^T \in D_R[0,+\infty)$, where $D_R[0,+\infty)$ is the Skorokhod space [13]. It is known that for compactness of the family Q_T, it is necessary and sufficient to prove that for any $\gamma > 0$ and for $\alpha > 1/2$

$$Q_T\{|G_{t_2}^T - G_{t_1}^T| \geq \lambda, |G_{t_3}^T - G_{t_2}^T| \geq \lambda\} \leq \frac{1}{\lambda^{2\gamma}}[F(t_3) - F(t_1)]^{2\alpha}, \tag{6.5.18}$$

where F is a continuous nondecreasing function and $t_1 \leq t_2 \leq t_3$. In order to prove (6.5.18) it is enough to prove

$$E\{|G_{t_2}^T - G_{t_1}^T||G_{t_3}^T - G_{t_2}^T| \leq K[t_3 - t_1]^2, \tag{6.5.19}$$

with some constant $K > 0$.

From (6.5.7) we obtain

$$|G_t^T - G_u^T| = |T^{-1} \sum_{k=\nu(uT)+1}^{\nu(tT)} \rho(x_k)| \leq T^{-1} \sup_x \rho(x)|\nu(tT) - \nu(uT) - 1|.$$

We note that

$$\begin{aligned} |G_{t_2}^T - G_{t_1}^T||G_{t_3}^T - G_{t_2}^T| &\leq T^{-2}(\sup_x \rho(x))^2|\nu(t_2T) - \nu(t_1T)||\nu(t_3T) - \nu(t_2)T)| \\ &\leq T^{-2}(\sup_x \rho(x))^2|\nu(t_3T) - \nu(t_1T)|^2. \end{aligned} \tag{6.5.20}$$

6.5.1.4 Characterization of the limiting measure Q for Q_T as $T \to \infty$

It follows from Subsection 6.5.1.3 that there exists a sequence T_n such that the measures Q_{T_n} converge weakly as $T_n \to +\infty$ to some measure Q on $D_R[0,+\infty)$. We want to show that the measure Q is the solution of some martingale problem, namely, the process

$$m(s,t) := f(\hat{G}_t) - f(\hat{G}_s) - \hat{\rho} \int_s^t (df(\hat{G}_u)/dg)du \tag{6.5.21}$$

is a Q-martingale for all $f(g) \in C^1(\mathbf{R})$, and $Em(s,t)\eta_0^s = 0$ for scalar continuous bounded functional η_0^s, where E denotes a mean value by measure Q. But from (6.5.15) it follows that

$$E^T m^T(s,t)\eta_0^s = 0, \tag{6.5.22}$$

and it is only necessary to show that if we take the limit in (6.5.14) as $T \to +\infty$ we obtain (6.5.21). From the equality (6.5.13) we conclude that

$$\lim_{T_n \to +\infty} E^{T_n} m(s,t)\eta_0^s = Em(s,t)\eta_0^s. \tag{6.5.23}$$

Further,

$$\begin{aligned} |E^T m^T(s,t)\eta_0^s - Em(s,t)\eta_0^s| \;\leq\; & |(E^T - E)m(s,t)\eta_0^s| \\ & + E^T|m(s,t) - m^T(s,t)||\eta_0^s| \to_{t \to +\infty} 0. \end{aligned} \tag{6.5.24}$$

That is why there exists the measure Q on $D_R[0,+\infty)$ which solves the martingale problem for the operator $\hat{\rho}\frac{d}{dg}$ (or, equivalently, for the process $\hat{G}_t$ in the form (6.5.15)). Uniqueness of the solution of the martingale problem follows from the fact that the process $\hat{G}_t$ is a deterministic function, since from (6.5.15) it follows that $\hat{G}_t$ satisfies the equation

$$\begin{cases} df(\hat{G}_t)/dt &= \hat{\rho}\frac{df(\hat{G}_t)}{dg} \\ f(\hat{G}_0) &= f(\ln s). \end{cases}$$

This is why $\hat{G}_t = \ln s + \hat{\rho} t$.

6.6 Rates of convergence in ergodic averaging scheme

Let us use the following estimation

$$E[(T^{-1}\sum_{k=1}^{\nu(tT)} \rho(x_k) - t\hat{\rho})] = T^{-1}E[\sum_{k=1}^{\nu(Tt)} \tilde{\rho}(x_k) - tT\hat{\rho}] = T^{-1}O(T) \tag{6.6.1}$$

which follows from the law of large numbers and from $E\rho = EP\rho$, where $\tilde{\rho}(x) := \int_X P(x,dy)\rho(y)$ [6, 9]. The second term in (6.5.5) has the estimation $T^{-1}b(t,\rho)$, where

$$b(t,\rho) := 1/2 \int_X p(dx) \int_X P(x,dy)\rho^2(y)/m,$$

for all $t \in R^+$, which follows from the renewal theorem: $\nu(tT) \equiv tT/m$, for large T. Since $r(T^{-1}\rho(x)) = O_\rho(T)$, where $O_\rho(T)/T \to_{T\to+\infty} 0$, and $T^{-1}\sum_{k=1}^{\nu(tT)}\rho^2(x_k) \to_{T\to+\infty} b(t,\rho)$, where $b(t,\rho)$ is defined above, then the third term in (6.5.5) has the estimation

$$T^{-1}b(t,\rho)O_\rho(T). \tag{6.6.2}$$

Combining (6.6.1)-(6.6.2) we obtain that the right side of (6.5.5) minus $t\hat{\rho}$ has the following estimation:

$$T^{-1}(O(T)+b(t,\rho)+\underline{(}t,\rho)O_\rho(T)) := T^{-1}c(t,\rho,T).$$

Thus, we have from (6.5.5) and (6.6.1)-(6.6.2):

$$\begin{aligned} |S_t^T - \hat{S}_t| &= S_0 e^{t\hat{\rho}}|e^{T^{-1}(O(T)+b(t,\rho)+b(t,\rho)O_\rho(T))} - 1| \\ &= S_0 e^{t\hat{\rho}}|e^{T^{-1}c(t,\rho,T)} - 1|. \end{aligned} \tag{6.6.3}$$

6.7 Merged geometric Markov renewal processes

Theorem 6.3 *The merged GMRP has the form:*

$$\tilde{S}_t = S_0 e^{\int_0^t \hat{\rho}(\hat{x}(s))ds},$$

where $t \in R_+$ and $S_0 > 0$.

Proof *Let us suppose that X consists of r ergodic classes X_i, $i = 1,2,\dots,r$ with stationary distributions $p_i(dx)$, $i = 1,2,\dots,r$ in each class. Then the Markov chain $(x_k)_{k\in\mathbf{Z}_+}$ is merged to the Markov chain $(\hat{x}(s))_{s\in\mathbb{Z}_+}$ in the merged phase space $\hat{X} = \{1,2,...,r\}$ [8, 11, 14]. Taking into account the algorithms of phase merging [9, 11, 12] and expansion (6.5.4) we obtain that $\frac{1}{T}\sum_{k=0}^{\nu(tT)}\rho(x_k)$ is merged to the integral functional*

$$\tilde{\rho}(t) := \int_0^t \hat{\rho}(\hat{x}(s))ds, \tag{6.7.1}$$

where

$$\hat{\rho}(k) := \int_{X_k} p_k(dx)\rho(x)/m(k), \tag{6.7.2}$$

and $m(k) := \int_{X_k} p_k(dx)m(x)$, $m(x) := \int_0^\infty(1-G_x(t))dt$ and $\hat{x}(s)$ is a merged Markov process in the merged phase space $\hat{X}$. In this way, we obtain from (6.5.3)-(6.5.5)

$$\ln\frac{S_t^T}{S_0} \to_{T\to+\infty} \int_0^t \hat{\rho}(\hat{x}(s))ds.$$

Thus, if $S_t^T \to_{T\to+\infty} \tilde{S}_t$, then

$$\tilde{S}_t = S_0 e^{\int_0^t \hat{\rho}(\hat{x}(s))ds}. \tag{6.7.3}$$

It means that the dynamic of merged GMRP is the same as the dynamic of bond prices with various interest rates $\hat{\rho}(k)$, where $k = 1,2,...,r$.

Remark If $k = 1$ in (6.7.3), then $\tilde{S}_t = S_0 e^{\int_0^t \hat{\rho}(\hat{x}(s))ds} = S_0 e^{t\hat{\rho}}$, where $\hat{\rho}$ is defined in (6.7.2). Namely, as $k = 1$, then $\tilde{S}_t$ coincides with $\hat{S}_t$ in (6.5.2).

6.8 Security markets and option prices using generalized binomial models induced by random maps

The special case of (6.2.1) where $N(t) := N(n) = n$ and Y_k are binomial random variables and similarly, the special case of (6.4.1) where $\nu(t) := \nu(n) = n$ and x_k takes two states become classical binomial models which were introduced in [4]. Recall a generalized binomial model induced by position dependent random maps for stock prices and option prices (see Chapter 4). A classical CRR binomial model [4] is a special case of a generalized binomial model induced by random maps where the probabilities of switching from one map to another are constants instead of position dependent. We show that a random map T for a generalized binomial model generates stock prices $S_0, S_1 = T(S_0), \ldots, S_k = T(S_{k-1}), k \geq 1$. If we set $\nu(t) = n, \rho(x_k) = \frac{T(S_{k-1})}{S_{k-1}} - 1, k \geq 1$ in our model (6.4.1), then it can be easily shown that a generalized binomial model is a special case of a geometric Markov renewal process (6.4.1). We have presented some examples in Chapter 4.

6.9 Applications

6.9.1 *Two ergodic classes*

Let $P(x,A) := Prob\{x_{n+1} \in A | x_n = x\}$ be the transition probabilities of supporting embedded reducible Markov chain $\{x_n\}_{n\geq 0}$ in the phase space X. Let us have two ergodic classes X_0 and X_1 of the phase such that:

$$X = X_0 \bigcup X_1, \qquad X_0 \bigcap X_1 = \emptyset. \tag{6.9.1}$$

Let $\{V = 0, 1, \nu\}$ be the measurable merged phase space. A stochastic kernel $P_0(x,A)$ is consistent with the splitting (6.9.1) in the following way:

$$P_0(x, X_k) = 1_k := \begin{cases} 1, & x \in X_k, \\ 0, & x \notin X_k, k = 0, 1. \end{cases}, \tag{6.9.2}$$

Let the supporting embedded Markov chain $(x_n)_{n\in Z_+}$ with the transition probabilities $P_0(x,A)$ be uniformly ergodic in each class $X_k, k = 0, 1$ and have a stationary distribution $\pi_k(dx)$ in the classes $X_k, k = 0, 1$:

$$\pi_k(A) = \int_{X_k} \pi_k(dx) P_0(x, A), \qquad A \subset X_k, \ k = 0, 1. \tag{6.9.3}$$

Let the stationary escape probabilities of the embedded Markov chain $(x_n)_{n\in Z_+}$ with transition probabilities $P(x,A) := Prob\{x_{n+1}\in A|x_n = x\}$ be positive and sufficiently small, that is,

$$q_k(A) = \int_{X_k} \pi_k(dx)P(x,X\setminus X_k) > 0, \quad k = 0,1. \tag{6.9.4}$$

Let the stationary sojourn time in the classes of states be uniformly bounded, namely,

$$0 \leq C \leq m_k := \int_{X_k} \pi_k(dx)m(x) \leq C', \quad k = 0,1, \tag{6.9.5}$$

where

$$m(x) := \int_0^{\infty} \bar{G}(t)dt. \tag{6.9.6}$$

6.9.2 *Algorithms of phase averaging with two ergodic classes*

Merged Markov chain $(\hat{x}_n)_{n\in Z_+}$ in merged phase space $\hat{X} = (e_1,e_0)$ is given by matrix of transition probabilities

$$\hat{P} = (\hat{p}_{kr})_{k,r=0,1}; \tag{6.9.7}$$

$$\hat{p}_{10} = 1-\hat{p}_{11} = \int_{X_1} \pi_1(dx)P(x,X_0) = 1-\int_{X_1} \pi_1(dx)P(x,X_1); \tag{6.9.8}$$

$$\hat{p}_{01} = 1-\hat{p}_{00} = \int_{X_0} \pi_0(dx)P(x,X_1) = 1-\int_{X_0} \pi_0(dx)P(x,X_0); \tag{6.9.9}$$

As $\pi_{kr} \neq 0, k = 0,1$, then $\hat{x}_n$ has virtual transitions. Intensities $\hat{\Lambda}_k$ of sojourn times $\hat{\theta}_k, k = 0,1$, of the merged MRP are calculated by the formula:

$$\hat{\lambda}_k = \frac{1}{m_k}, \qquad m_k = \int_{X_k} \pi_k(dx)m(x), \quad k = 0,1. \tag{6.9.10}$$

And, finally, merged MRP $(\hat{x}_n,\hat{\theta})_{n\in Z_+}$ in merged phase space $\hat{X} = (e_0,e_1)$ is given by stochastic matrix

$$\hat{Q}(t) = (\hat{Q}_{kr})_{k,r=0,1} := \hat{p}_{kr}(1-e^{-\hat{\lambda}_k t}), \quad k,r = 0,1. \tag{6.9.11}$$

Hence, initial semi-Markov system is merged to Markov system with two classes.

6.9.3 *Merging of S_t^T in the case of two ergodic classes*

The merged GMRP in the case of two ergodic classes has the form:

$$\tilde{S}_t = S_0 e^{\int_0^t \hat{\rho}(\hat{x}(s))ds}, \tag{6.9.12}$$

where

$$\begin{aligned}
\hat{\rho}(1) : &= \int_{X_1} \pi_1(dx)\int_{X_1} P(x,dy)\rho(y)/m(1),\\
\hat{\rho}(0) &:= \int_{X_0} \pi_0(dx)\int_{X_0} P(x,dy)\rho(y)/m(0),\\
m(k) &:= \int_{X_k} \pi_k(dx)m(x), \quad k = 0,1.
\end{aligned}$$

Here, $\hat{x}(t)$ is a merged Markov process in $\hat{X} = (e_0, e_1)$ with kernel $\hat{Q}(t)$ in (6.9.11). The dynamic of merged GMRP in (6.9.12) is the dynamic of a bond price with two different interest rates switched by process $\hat{x}(t)$.

6.9.4 Examples for two states ergodic GMRP

The matrix of one-step transition probabilities of phase space is:

$$\hat{P} = \begin{pmatrix} p_{00} & p_{01} \\ p_{10} & p_{11} \end{pmatrix} = \begin{pmatrix} 0.98 & 0.02 \\ 0.02 & 0.98 \end{pmatrix}$$

$S_0 = 10, \quad p_0 = p_1 = 1/2, \qquad \lambda(x_0) = 10, \lambda(x_1) = 12,$

$$\rho(x_k) = \begin{cases} 0.01, & k = 0; \\ 0.02, & k = 1; \end{cases}$$

$G_x(t)$ is the exponential distribution with $\lambda(x)$
therefore $G_x(t) = 1 - e^{-\lambda(x)t}, \qquad \bar{m}(x) = \int_0^\infty (1 - G_x(t))dt = \frac{1}{\lambda(x)}$

$$m = \int_X p(dx)\bar{m}(x) = p_0 \times \frac{1}{\lambda(x_0)} + p_1 \times \frac{1}{\lambda(x_1)}$$

$$\hat{\rho} = \int_X p(dx)\rho(x)/m = \frac{p_0\rho(x_0) + p_1\rho(x_1)}{m}$$

$$\hat{S}_t = S_0 e^{\hat{\rho}t}.$$

Therefore $\hat{\rho} = 0.1636$, $\hat{S}_t = 10e^{0.1636t}$.

6.9.5 Examples for merged GMRP

$$\hat{G}(t) = P(\hat{\theta}_k < t) = 1 - e^{-\hat{\lambda}_k t}$$

$$\hat{Q}(t) = [\hat{p_{kr}} G(t)]_{2\times 2} \quad (k, r = 0, 1)$$

$$\tilde{S}_t = S_0 e^{\int_0^t \hat{\rho}(\hat{x}(s))ds} = S_0 e^{\sum_{i=0}^{v(t)-1} \hat{\rho}(\hat{x}_i)\theta_i + \int_{\tau_{v(t)}}^t \hat{\rho}(\hat{x}_{v(t)})ds}.$$

Therefore, $\tilde{S}_t$ will be split into two cases by the original value of $\hat{x}(0)$:
When $\hat{x}(0) = 0$;

$$\begin{aligned} \tilde{S}_t &= S_0 exp(\Sigma_{i=1}^{\frac{v(t)}{2}} \hat{\rho}(0)\theta_{2i-1} + \Sigma_{i=1}^{\frac{v(t)}{2}} \hat{\rho}(1)\theta_{2i} + (t - \tau_{v(t)})\hat{\rho}(0)). \quad v(t) \textit{ is even} \\ \tilde{S}_t &= S_0 exp(\Sigma_{i=1}^{\frac{v(t)}{2}+\frac{1}{2}} \hat{\rho}(0)\theta_{2i-1} + \Sigma_{i=1}^{\frac{v(t)}{2}-\frac{1}{2}} \hat{\rho}(1)\theta_{2i} + (t - \tau_{v(t)})\hat{\rho}(1)). \\ & \qquad v(t) \textit{ is odd} \end{aligned} \tag{6.9.13}$$

When $\hat{x}(0) = 1$;

$$\tilde{S}_t = S_0 exp(\sum_{i=1}^{\frac{v(t)}{2}} \hat{\rho}(1)\theta_{2i-1} + \sum_{i=1}^{\frac{v(t)}{2}} \hat{\rho}(0)\theta_{2i} + (t - \tau_{v(t)})\hat{\rho}(1)). \quad v(t) \text{ is even}$$

$$\tilde{S}_t = S_0 exp(\sum_{i=1}^{\frac{v(t)}{2}+\frac{1}{2}} \hat{\rho}(1)\theta_{2i-1} + \sum_{i=1}^{\frac{v(t)}{2}-\frac{1}{2}} \hat{\rho}(0)\theta_{2i} + (t - \tau_{v(t)})\hat{\rho}(0)). \quad v(t) \text{ is odd} \tag{6.9.14}$$

$\{x_n\}$ — 4 states Markov chain with transition probability:

$$P = \begin{pmatrix} P_{00} & P_{01} & P_{02} & P_{03} \\ P_{10} & P_{11} & P_{12} & P_{13} \\ P_{20} & P_{21} & P_{22} & P_{23} \\ P_{30} & P_{31} & P_{32} & P_{33} \end{pmatrix} = \begin{pmatrix} 0.49 & 0.49 & 0.01 & 0.01 \\ 0.70 & 0.28 & 0.01 & 0.01 \\ 0.01 & 0.01 & 0.70 & 0.28 \\ 0.01 & 0.01 & 0.49 & 0.49 \end{pmatrix}$$

Merged Markov chain in the merged phase space $\hat{X} = (0,1)$ in each class X_k, $k = 0,1$, since the embedded Markov chain is uniformly ergodic in each class $X_k, k = 0,1$, each of them has a stationary distribution

$$\pi_0(x) = \begin{cases} 0.5868, \\ 0.4132, \end{cases} \quad x \in X_0 \qquad \pi_1(x) = \begin{cases} 0.633, \\ 0.367, \end{cases} \quad x \in X_1.$$

So the matrix of one-step transition probabilities of merged phase space $\hat{X} = (e_1, e_0)$ is:

$$\hat{P} = \begin{pmatrix} \hat{p}_{00} & \hat{p}_{01} \\ \hat{p}_{10} & \hat{p}_{11} \end{pmatrix} = \begin{pmatrix} 0.98 & 0.02 \\ 0.02 & 0.98 \end{pmatrix}$$

$$\rho(x_k) = \begin{cases} 0.01, & k = 0,1 \\ 0.02, & k = 2,3 \end{cases}$$

$\lambda(x_0) = 8, \lambda(x_1) = 10, \lambda(x_2) = 12, \lambda(x_3) = 10$, $S_0 = 10$.

$m_0 = 0.5868 \times \frac{1}{8} + 0.4132 \times \frac{1}{10} = 0.1147$ and $m_1 = 0.633 \times \frac{1}{12} + 0.367 \times \frac{1}{10} = 0.0895$

$$\begin{aligned} \hat{\rho}(1) &:= [0.633 \times (0.70 \times 0.02 + 0.28 \times 0.02) + 0.367 \times (0.49 \times 0.02 + 0.49 \times 0.02)]/0.0895 \\ &= 0.2190 \end{aligned} \tag{6.9.15}$$

$$\begin{aligned} \hat{\rho}(0) &:= [0.5868 \times (0.49 \times 0.01 + 0.49 \times 0.01) + 0.4132 \times (0.70 \times 0.01 + 0.28 \times 0.01)]/0.1147 \\ &= 0.0854 \end{aligned} \tag{6.9.16}$$

Let $\theta_i(i = 1,2,...) = 0.1$; given t, $\exists v(t).s.t.$ $0.1v(t) \leq t < 0.1(v(t)+1)$. Therefore, $\tau_{v(t)} = \theta_i v(t) = 0.1v(t)$.

Therefore, when $\hat{x}(0) = 0$;

$$\begin{aligned} &v(t) \text{ is even}: \\ &\tilde{S}_t = S_0 exp(\Sigma_{i=1}^{\frac{1}{2}\lfloor \frac{t}{0.1} \rfloor} 0.0854 \times 0.1 + \Sigma_{i=1}^{\frac{1}{2}\lfloor \frac{t}{0.1} \rfloor} 0.2190 \times 0.1 + (t - 0.1\lfloor \frac{t}{0.1} \rfloor) \times 0.0854) \\ &v(t) \text{ is odd}: \\ &\tilde{S}_t = S_0 exp(\Sigma_{i=1}^{\frac{1}{2}(\lfloor \frac{t}{0.1} \rfloor + 1)} 0.0854 \times 0.1 + \Sigma_{i=1}^{\frac{1}{2}(\lfloor \frac{t}{0.1} \rfloor - 1)} 0.2190 \times 0.1 + (t - 0.1\lfloor \frac{t}{0.1} \rfloor) \times 0.2190) \end{aligned} \tag{6.9.17}$$

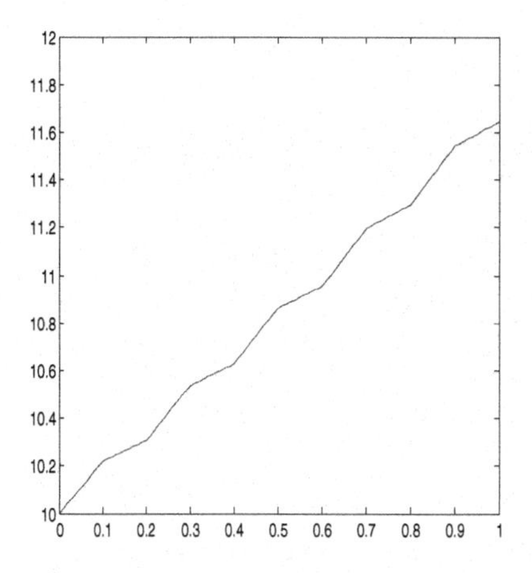

Figure 6.1 *Trend of $\tilde{S}(t)$ w.r.t t in merged GMRP when $\hat{x}(s) = 1,\ S_0 = 10$.*

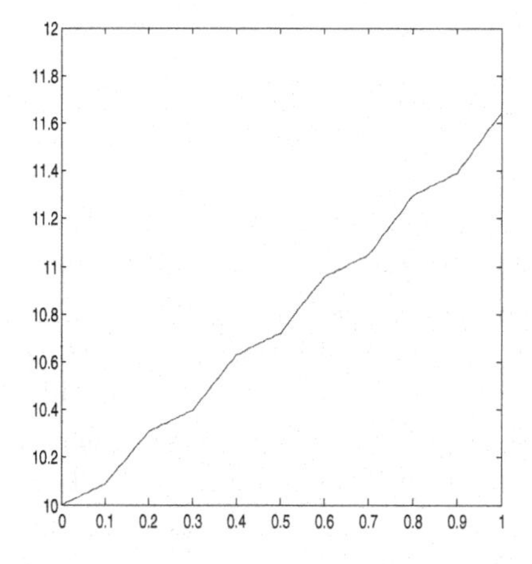

Figure 6.2 *Trend of $\tilde{S}(t)$ w.r.t t in merged GMRP when $\hat{x}(s) = 0,\ S_0 = 10$.*

When $\hat{x}(0) = 1$;

$v(t)$ *is even* :

$$\tilde{S}_t = S_0 exp(\sum_{i=1}^{\frac{1}{2}\lfloor\frac{t}{0.1}\rfloor} 0.2190 \times 0.1 + \sum_{i=1}^{\frac{1}{2}\lfloor\frac{t}{0.1}\rfloor} 0.0854 \times 0.1 + (t - 0.1\lfloor\frac{t}{0.1}\rfloor) \times 0.2190)$$

$v(t)$ *is odd* :

$$\tilde{s}_t = S_0 exp(\sum_{i=1}^{\frac{1}{2}(\lfloor\frac{t}{0.1}\rfloor+1)} 0.2190 \times 0.1 + \sum_{i=1}^{\frac{1}{2}(\lfloor\frac{t}{0.1}\rfloor-1)} 0.0854 \times 0.1 + (t - 0.1\lfloor\frac{t}{0.1}\rfloor) \times 0.0854)$$

Based on this, for double averaged GMRP

$$\check{\rho} = \hat{p}_0\hat{\rho}(0) + \hat{p}_0\hat{\rho}(1)$$

By the ergodicity of the class of merged Markov process,we have $\hat{p}_0 = \hat{p}_1 = 0.500$; therefore

$$\check{\rho} = 0.500\hat{\rho}(0) + 0.500\hat{\rho}(1) = 0.1522, \qquad \check{S}_t = 10e^{0.1522t}$$

References

[1] Aase, K., *Contingent claims valuation when the securities price is a combination of an Ito process and a random point process*, Stoch. Proc. and Their Applic., 28, 185-220, 1988.

[2] Bahsoun, W. and Góra, P., *Position dependent random maps in one and higher dimensions*, Studia Math., 166, 271-286, 2005.

[3] Bahsoun, W., Góra, P., Mayoral, S., and Morales, M., *Random dynamics and finance: constructing implied binomial trees from predetermined stationary density*, Appl. Stochastic Models Bus. Ind., 23, 181-212, 2007.

[4] Cox, J. C., Ross, S.A., and Rubinstein, M., *Option pricing: a simplified approach*, Journal of Financial Economics, 7, 229-264, 1979.

[5] Góra, P. and Boyarsky, A., *Absolutely continuous invariant measures for random maps with position dependent probabilities*, Math. Anal. and Appl. 278, 225-242, 2003.

[6] Islam, M. S. 2010, *Generalized binomial models induced by random dynamical systems and valuation of bonds, annuities and options*, Technical report, Department of Mathematics and Statistics, University of Prince Edward Island, Canada, 2010.

[7] Korolyuk, V.S. and Limnios, N., *Poisson approximation of stochastic systems*, Theory Probab. and Mathem. Statistics, Kiev University, N62, 2002.

[8] Korolyuk, V.S., and Limnios, N., *Stochastic processes in phase merging space*, World Scientific, 2005.

[9] Korolyuk, V.S. and Swishchuk, A.V., *Evolution of systems in random media*, CRC Press, Boca Raton, USA, 1995.

[10] Korolyuk, V.S. and Swishchuk, A.V., *Evolutionary stochastic systems. Algorithms of averaging and diffusion approximation*, Institute of Mathematics Ukrain. Acad. of Sciences, Kiev. (In Russian), 2000.

[11] Pelikan, S., *Invariant densities for random maps of the interval*, Proc. Amer. Math. Soc. 281, 813-825, 1984.

[12] Skorokhod, A., *Studies in the theory of random processes (English translation),*. Ann. Math. Stat., v. 38, n.1, 1967.

[13] Swishchuk, A.V., *Random evolutions and their applications,* Kluwer AP, Dordrecht, 1997.

[14] Swishchuk, A.V., *Random evolutions and their applications: new trends,* Kluwer AP, Dordrecht, 2000.

Chapter 7

Diffusion Approximations of the Geometric Markov Renewal Processes and Option Price Formulas

7.1 Chapter overview

In the previous chapter we introduced the geometric Markov renewal processes (GMRP) as a model for a security market and we considered its approximations in the form of averaged, merged, and double averaged geometric Markov renewal processes. In this chapter we study the geometric Markov renewal processes in a diffusion approximation scheme. Weak convergence analysis and rates of convergence of ergodic geometric Markov renewal processes in diffusion schemes are presented. We present European call option pricing formulas in the case of ergodic, double averaged, and merged diffusion geometric Markov renewal processes.

7.2 Introduction

Let $N(t)$ be a standard Poisson process, $(Y_k)_{k\in Z_+}$ be i. i. d. random variables which are independent of $N(t)$ and $S_0^* > 0$. The geometric compound Poisson process

$$S_t^* = S_0^* \prod_{k=1}^{N(t)} (1+Y_k),\ t>0 \tag{7.2.1}$$

is a trading model in many financial applications with pure jumps [8, p.214]. Motivated by the geometric compound Poisson processes (7.2.1), we have introduced the geometric Markov renewal processes (7.2.2) in the previous chapter defined by

$$S_t := S_0 \prod_{k=1}^{\nu(t)} (1+\rho(x_k)),\ t\in \mathbb{R}_+, \tag{7.2.2}$$

where $S_0 > 0$ is the initial value of S_t. We call this process $(S_t)_{t\in\mathbb{R}_+}$ a geometric Markov renewal process by analogy with the geometric compound Poisson processes

$$S_t^* = S_0^* \prod_{k=1}^{N(t)} (1+Y_k), \tag{7.2.3}$$

where $S_0^* > 0$, $N(t)$ is a standard Poisson process, $(Y_k)_{k \in Z_+}$ are i. i. d. random variables. The geometric compound Poisson process$\{S_t^*\}_{t \in \mathbb{R}_+}$ in (7.2.3) is a trading model in many financial applications as a pure jump model [1, 3, 8]. The geometric Markov renewal processes $\{S_t\}_{t \in \mathbb{R}_+}$ in (7.2.2) will be our main trading model in further analysis. The geometric Markov renewal process (7.2.2) is also known as a switched-switching process. Averaging and diffusion approximation methods are important approximation methods for a switched-switching system. Averaging schemes of the geometric Markov renewal processes (7.2.2) were studied in Chapter 6 (see also [11]).

The singular perturbation technique of a reducible invertible-operator is one of the techniques for the construction of averaging and diffusion schemes for a switched-switching process. Strong ergodicity assumption for the switching process means that the singular perturbation problem has a solution with some additional non-restrictive conditions. Averaging and diffusion approximation schemes for switched-switching processes in the form of random evolutions were studied in [4, p. 157], [8, p. 41]. In this chapter, we introduce diffusion approximation of the geometric Markov renewal processes. We study a discrete Markov-modulated (B,S)-security market described by a geometric Markov renewal process (GMRP). Weak convergence analysis and rates of convergence of ergodic geometric Markov renewal processes in diffusion scheme are presented. We present European call option pricing formulasin the case of ergodic, double averaged, and merged diffusion geometric Markov renewal processes.

7.3 Diffusion approximation of the geometric Markov renewal process (GMRP)

Under an additional balance condition, averaging effect leads to diffusion approximation of the geometric Markov renewal process (GMRP). In fact, we consider the counting process $\nu(t)$ in (7.2.1)(see (6.3.2) for a definition of $\nu(t)$) in the new accelerated scale of time tT^2, that is, $\nu \equiv \nu(tT^2)$. Due to more rapid changes of states of the system under the balance condition, the fluctuations are described by a diffusion processes.

7.3.1 Ergodic diffusion approximation

Let us suppose that balance condition is fulfilled for functional $S_t^T = S_0 \prod_{k=1}^{\nu(tT)}(1 + \rho_T(x_k))$:

$$\hat{\rho} = \int_X p(dx) \int_X P(x,dy)\rho(y)/m = 0, \tag{7.3.1}$$

where $p(x)$ is ergodic distribution of Markov chain $(x_k)_{k \in Z_+}$. Then $\hat{S}(t) = S_0$, for all $t \in R^+$. Consider S_t^T in the new scale of time tT^2 :

$$S_T(t) := S_{tT^2}^T = S_0 \prod_{k=1}^{\nu(tT^2)} (1 + T^{-1}\rho(x_k)). \tag{7.3.2}$$

Due to more rapid jumps of $v(tT^2)$ the process $S_T(t)$ will be fluctuated near the point S_0 as $T \to +\infty$. By similar arguments similar to 6.5.3, 6.5.4, 6.5.5 in (see also (4.3) – (4.5) in [11]), we obtain the following expression:

$$\ln \frac{S_T(t)}{S_0} = T^{-1} \sum_{k=1}^{v(tT^2)} \rho(x_k) - 1/2T^{-2} \sum_{k=1}^{v(tT^2)} \rho^2(x_k) + T^{-2} \sum_{k=1}^{v(tT^2)} r(T^{-1}\rho(x_k))\rho^2(x_k). \tag{7.3.3}$$

Algorithms of ergodic averaging give the limit result for the second term in (7.3.3) (see [8, p. 43], [9, p. 88]):

$$\lim_{T\to+\infty} 1/2T^{-2} \sum_{k=1}^{v(tT^2)} \rho^2(x_k) = 1/2t\hat{\rho}_2, \tag{7.3.4}$$

where $\hat{\rho}_2 := \int_X p(dx) \int_X P(x,dy)\rho^2(y)/m$. Using algorithms of diffusion approximation with respect to the first term in (7.3.3) we obtain [9], p. 88:

$$\lim_{T\to+\infty} T^{-1} \sum_{k=1}^{v(tT^2)} \rho(x_k) = \sigma_\rho w(t), \tag{7.3.5}$$

where $\sigma_\rho^2 := \int_X p(dx)[1/2 \int_X P(x,dy)\rho^2(y) + \int_X P(x,dy)\rho(y)R_0P(x,dy)\rho(y)]/m$, R_0 is a potential [4, p. 68], of $(x_n)_{n\in Z_+}$, $w(t)$ is a standard Wiener process. The last term in (7.3.3) goes to zero as $T \to +\infty$. Let $\hat{S}(t)$ be the limiting process for $S_T(t)$ in (7.3.3) as $T \to +\infty$. Taking limit on both sides of (7.3.3) we obtain

$$\lim_{T\to+\infty} \ln \frac{S_T(t)}{S_0} = \ln \frac{\hat{S}(t)}{S_0} = \sigma_\rho w(t) - 1/2t\hat{\rho}_2, \tag{7.3.6}$$

where σ_ρ^2 and $\hat{\rho}_2$ are defined in (7.3.4) and (7.3.5), respectively. From (7.3.6) we obtain

$$\hat{S}(t) = S_0 e^{\sigma_\rho w(t) - 1/2t\hat{\rho}_2} = S_0 e^{-1/2t\hat{\rho}_2} e^{\sigma_\rho w(t)}. \tag{7.3.7}$$

Thus, $\hat{S}(t)$ satisfies the following stochastic differential equation (SDE):

$$d\hat{S}(t) = \hat{S}(t)[1/2(\sigma_\rho^2 - \hat{\rho}_2)dt + \sigma_\rho dw(t)]. \tag{7.3.8}$$

In this way we have the following corollary:

Corollary 7.1 *The ergodic diffusion GMRP has the form*

$$\hat{S}(t) = S_0 e^{-1/2t\hat{\rho}_2} e^{\sigma_\rho w(t)}, \tag{7.3.9}$$

and it satisfies the following SDE

$$\frac{d\hat{S}(t)}{\hat{S}(t)} = 1/2(\sigma_\rho - \hat{\rho}_2)dt + \sigma_\rho dw(t). \tag{7.3.10}$$

7.3.2 Merged diffusion approximation

Let us suppose that the balance condition satisfies the following

$$\hat{\rho}(k) = \int_{X_k} p_k(dx) \int_{X_k} P(x,dy)\rho(y)/m(k) = 0, \tag{7.3.11}$$

for all $k = 1,2,\dots,r$ where $(x_n)_{n\in Z_+}$ is the supporting embedded Markov chain, p_k is the stationary density for the ergodic component X_k, $m(k)$ is defined in [11] and conditions of reducibility of X are fulfilled. Using the algorithms of merged averaging [4, 8, 9] we obtain from the second part of the right-hand side in (7.3.3):

$$\lim_{T\to+\infty} \frac{1}{2} T^{-1} \sum_{k=1}^{\nu(tT^2)} \rho^2(x_k) = \frac{1}{2}\int_0^t \hat{\rho}_2(\hat{x}(s))ds, \tag{7.3.12}$$

where

$$\hat{\rho}_2(k) := \int_{X_k} p_k(dx) \int_{X_k} P(x,dy)\rho^2(y)/m(k). \tag{7.3.13}$$

using the algorithm of merged diffusion approximation that [4, 8, 9] obtain from the first part of the right-hand side in (7.3.3):

$$\lim_{T\to+\infty} T^{-1} \sum_{k=1}^{\nu(tT^2)} \rho(x_k) = \int_0^t \hat{\sigma}_\rho(\hat{x}(s))dw(s), \tag{7.3.14}$$

where

$$\hat{\sigma}_\rho^2(k) := \int_{X_k} p_k(dx) \int_{X_k} P(x,dy)\rho^2(y) + \int_{X_k} P(x,dy)\rho(y)R_0 \int_{X_k} P(x,dy)\rho(y)/m(k). \tag{7.3.15}$$

The third term in (7.3.3) goes to 0 as $T \to +\infty$. In this way, from (7.3.3) we obtain:

$$\lim_{T\to+\infty} \ln \frac{S_T(t)}{S_0} = \ln \frac{\tilde{S}(t)}{S_0} = \int_0^t \hat{\sigma}_\rho(\hat{x}(s))dw(s) - \frac{1}{2}\int_0^t \hat{\rho}_2(\hat{x}(s))ds, \tag{7.3.16}$$

where $\tilde{S}(t)$ is the limit $S_T(t)$ as $T \to +\infty$. From (7.3.16) we obtain

$$\tilde{S}(t) = S_0 e^{-\frac{1}{2}\int_0^t \hat{\rho}^2(\hat{x}(s))ds + \int_0^t \hat{\sigma}_\rho(\hat{x}(s))dw(s)}. \tag{7.3.17}$$

Stochastic differential equation (SDE) for $\check{S}(t)$ has the following form:

$$\frac{d\tilde{S}(t)}{\tilde{S}(t)} = \frac{1}{2}(\hat{\sigma}_\rho^2(\hat{x}(t)) - \hat{\rho}_2(\hat{x}(t)))dt + \hat{\sigma}_\rho(\hat{x}(t))dw(t), \tag{7.3.18}$$

where $\hat{x}(t)$ is a merged Markov process.

In this way we have the following corollary:

Corollary 7.2 *Merged diffusion GMRP has the form (7.3.17) and satisfies the SDE (7.3.18).*

7.3.3 Diffusion approximation under double averaging

Let us suppose that the phase space $\hat{X} = \{1,2,\dots,r\}$ of the merged Markov process $\hat{x}(t)$ consists of one ergodic class with stationary distributions $(\hat{p}_k; k = \{1,2,\dots r\})$. Let us also suppose that the balance condition is fulfilled:

$$\sum_{k=1}^{r} \hat{p}_k \hat{\rho}(k) = 0. \tag{7.3.19}$$

Then using the algorithms of diffusion approximation under double averaging (see [4], p. 188, [8], p. 49, [9], p. 93) we obtain:

$$\lim_{T\to+\infty} \ln \frac{S_T(t)}{S_0} = \ln \frac{\check{S}(t)}{S_0} = \check{\sigma}_\rho w(t) - \frac{1}{2}\check{\rho}_2 t, \tag{7.3.20}$$

where

$$\check{\sigma}_\rho^2 := \sum_{k=1}^{r} \hat{p}_k \hat{\sigma}_\rho^2(k), \qquad \check{\rho}_2 := \sum_{k=1}^{r} \hat{p}_k \hat{\rho}_2(k), \tag{7.3.21}$$

and $\hat{\rho}_2(k)$ and $\hat{\sigma}_\rho^2(k)$ are defined in (7.3.13) and (7.3.15), respectively. Thus, we obtain from (7.3.20):

$$\check{S}(t) = S_0 e^{-\frac{1}{2}\check{\rho}_2 t + \check{\sigma}_\rho w(t)}. \tag{7.3.22}$$

Corollary 7.3 *The diffusion GMRP under double averaging has the form*

$$\check{S}(t) = S_0 e^{-\frac{1}{2}\check{\rho}_2 t + \check{\sigma}_\rho w(t)}, \tag{7.3.23}$$

and satisfies the SDE

$$\frac{d\check{S}(t)}{\check{S}(t)} = \frac{1}{2}(\check{\sigma}_\rho^2 - \check{\rho}_2)dt + \check{\sigma}_\rho dw(t). \tag{7.3.24}$$

7.4 Proofs

In this section we present proofs of results in Section 7.3. All the above-mentioned results are obtained from the general results for semi-Markov random evolutions [4, 9] in series scheme. The main steps of proof are: 1. Weak convergence of S_t^T in Skorokhod space $D_R[0,+\infty)$ [7, p. 148]; 2). Solution of the martingale problem for the limit process $\hat{S}(t)$; 3). Characterization of the limit measure for the limit process $\hat{S}(t)$; 4). Uniqueness of solution of martingale problem. We also give here the rate of convergence in the diffusion approximation scheme.

7.4.1 Diffusion approximation (DA)

Let

$$G_t^T := T^{-1} \sum_{k=0}^{\nu(tT^2)} \rho(x_k), \quad G_n^T := G_{\tau_n T^{-1}}^T, \quad G_0^T = \ln s, \tag{7.4.1}$$

and the balance condition is satisfied:

$$\hat{\rho} := \int_X p(dx) \int_X P(x,dy)\rho(y) = 0. \tag{7.4.2}$$

Let us define the functions

$$\phi^T(s,x) := f(s) + T^{-1}\phi_f^1(s,x) + T^{-2}\phi_f^2(s,x), \tag{7.4.3}$$

where ϕ_f^1 and ϕ_f^2 are defined as follows:

$$\begin{aligned} (P-I)\phi_f^1(s,x) &= \rho(x)f(s), \\ (P-I)\phi_f^2(s,x) &= [-A(x)+\hat{A}]f(s), \end{aligned} \tag{7.4.4}$$

where

$$\hat{A} := \int_X p(dx)A(x), \tag{7.4.5}$$

and $A(x) := [\rho^2(x)/2 + \rho(x)(R_0 - I)\rho(x)]d^2/ds^2$. From the balance condition (7.4.2) and equality $\Pi(\hat{A} - A(x)) = 0$ it follows that both equations in (7.4.3) simultaneously solvable and the solutions $\phi_f^i(s,x)$ are bounded functions, $i = 1,2$.

We note that

$$f(S_{n+1}^T) - f(G_n^T) = \frac{1}{T}\rho(x_n)\frac{df(x_n)}{ds} \tag{7.4.6}$$

and define

$$\phi^T(s,x) := f(s) + T^{-1}\phi_f^1(s,x) + T^{-2}\phi_f^2(s,x), \tag{7.4.7}$$

where $\phi_f^1(s,x)$ and $\phi_f^2(s,x)$ are defined in (7.4.4) and (7.4.5), respectively. We note, that $G_{n+1}^T - G_n^T = T^{-1}\rho(x_n)$.

7.4.2 Martingale problem for the limiting problem $G_0(t)$ in DA

Let us introduce the family of functions:

$$\begin{aligned} \psi^T(s,t) : &= \phi^T(G_{[tT^2]}^T, x_{[tT^2]}) - \phi^T(G_{[st^2]}^T, x_{[sT^2]}) - \\ & \sum_{j=[sT^2]}^{[tT^2]-1} E[\phi^T(G_j^T, x_{j+1}) - \phi^T(G_j^T, x_j)|\mathcal{F}_j], \end{aligned} \tag{7.4.8}$$

where ϕ^T are defined in (7.4.7) and G_j^T is defined by

$$G_{\frac{\tau_n}{T}}^T = \frac{1}{T}\sum_{k=0}^{n}\rho(x_k). \tag{7.4.9}$$

Functions $\psi^T(s,t)$ are $\mathcal{F}_{[tT^2]}$-martingale by t. Taking into account the expression (7.4.6) and (7.4.7), we find the following expression:

$$\begin{aligned}
\psi^T(s,t) &= f(G^T_{[tT^2]}) - f(G^T_{[sT^2]}) + \varepsilon[\phi^1_f(G^T_{[tT^2]}, x_{[tT^2]}) - \\
&\quad -\phi^1_f(G^T_{[st^2]}, x_{[sT^2]})] + \varepsilon^2[\phi^2_f(G^T_{[tT^2]}, x_{[tT^2]}) - \phi^2_f(G^T_{[sT^2]}, x_{[sT^2]})] - \\
&\quad T^{-1}\sum_{j=[sT^2]}^{[tT^2]-1}\{\rho(x_j)\frac{df(G^T_j)}{dg} + E(\phi^1_f(G^T_j, x_{j+1}) - \phi^2_f(G^T_j, x_j)|\mathcal{F}_j)\} - \\
&\quad -T^{-2}\sum_{j=[sT^2]}^{[tT^2]-1}\{2^{-1}\rho^2(x_j)\frac{df(G^T_j)}{dg} + \rho(x_j)E(\frac{d\phi^1_f(G^T_j, x_{j+1})}{dg}|\mathcal{F}_j) + \\
&\quad +E[\phi^2_f(G^T_j, x_{j+1}) - \phi^2_f(G^T_j, x_j)|\mathcal{F}_j]\} + o(T^{-2}) \\
&= f(G^T_{[tT^2]}) - f(G^T_{[sT^2]}) + [\phi^1_f(G^T_{[tT^2]}, x_{[tT^2]}) - -\phi^1_f(G^T_{[sT^2]}, x_{[sT^2]})] + \\
&\quad T^{-2}[\phi^2_f(G^T_{[tT^2]}, x_{[tT^2]}) - \phi^2_f(G^T_{[sT^2]}, x_{[sT^2]})] \\
&\quad -T^{-2}\sum_{j=[sT^2]}^{[tT^2]-1}\hat{A}f(G^T_j) + O(T^{-2}),
\end{aligned} \tag{7.4.10}$$

where $O(T^{-2})$ is the sum of terms with T^{-2}nd order. Since $\psi^T(0,t)$ is $\mathcal{F}_{[tT^2]}$-martingale with respect to measure Q_T, generated by process $G_T(t)$ in (7.4.1), then for every scalar linear continuous functional η^s_0 we have from (7.4.8)-(7.4.10):

$$\begin{aligned}
0 &= E^T[(\psi^T(s,t)\eta^s_0] \\
&= E^T[(f(G^T_{[tT^2]}) - f(G^T_{[sT^2]}) - T^{-2}\sum_{j=[sT^2]}^{[tT^2]-1}\hat{A}f(G^T_j))\eta^s_0] - \\
&\quad T^{-1}E^T[(\phi^1_f(G^T_{[tT^2]}, x_{[tT^2]}) - \phi^1_f(G^T_{[sT^2]}, x_{[sT^2]}))\eta^s_0] - \\
&\quad -T^{-2}E^T[(\phi^2_f(G^T_{[tT^2]}, x_{[tT^2]}) - \phi^2_f(G^T_{[sT^2]}, x_{[sT^2]}))\eta^s_0] - O(T^{-2}),
\end{aligned} \tag{7.4.11}$$

where E^T is a mean value by measure Q_T. If the process $G^T_{[tT^2]}$ converges weakly to some process $G_0(t)$ as $T \to +\infty$, then from (7.4.11) we obtain

$$0 = E^T[(f(G_0(t)) - f(G_0(s)) - \int_s^t \hat{A}f(G_0(u))du], \tag{7.4.12}$$

i.e., the process

$$f(G_0(t)) - f(G_0(s)) - \int_s^t \hat{A}f(G_0(u))du \tag{7.4.13}$$

is a continuous Q_T-martingale. Since $\hat{A}$ is the second order differential operator and coefficient σ^2_1 is positively defined, where

$$\sigma^2_1 := \int_X \pi(dx)[\rho^2(x)/2 + \rho(x)R_0\rho(x)], \tag{7.4.14}$$

then the process $G_0(t)$ is a Wiener process with variance σ_1^2 in (7.4.14): $G_0(t) = \sigma w(t)$. Taking into account the renewal theorem for $v(t)$, namely, $T^{-1}v(tT^2) \to_{T\to+\infty} t/m$, and the following representation

$$G_t^T = T^{-1}\sum_{k=0}^{v(tT^2)} \rho(x_k) = T^{-1}\sum_{k=0}^{vtT^2)} \rho(x_k) + T^{-1}\sum_{k=[tT^2]+1}^{v(tT^2)} \rho(x_k) \tag{7.4.15}$$

we obtain, replacing $[tT^2]$ by $v(tT^2)$, that process $G_T(t)$ converges weakly to the process $\hat{G}_0(t)$ as $T \to +\infty$, which is the solution of such martingale problem:

$$f(\hat{G}_0(t)) - f(\hat{G}_0(s)) - \int_s^t \hat{A}_0 f(\hat{G}_0(u))du \tag{7.4.16}$$

is a continuous Q_T-martingale, where $\hat{A}_0 := \hat{A}/m$, and $\hat{A}$ is defined in (7.4.5).

7.4.3 Weak convergence of the processes $G_T(t)$ in DA

From the representation of the process $G_T(t)$ it follows that

$$\begin{aligned}\Delta_T(s,t): &= |G_T(t) - G_T(s)| = |T^{-1}\sum_{k=v(sT^2)+1}^{v(tT^2)} \rho(x_k)| \\ &\le T^{-1}\sup_x \rho(x)|v(tT^2) - v(sT^2) - 1|.\end{aligned} \tag{7.4.17}$$

This representation gives the following estimation:

$$|\Delta_T(t_1,t_2)||\Delta_T(t_2,t_3)| \le T^{-2}(\sup_x \rho(x))^2|v(t_3T^2) - v(t_1T^2)|^2. \tag{7.4.18}$$

Taking into account the same reasonings as in [11] we obtain the weak convergence of the processes $G_T(t)$ in DA.

7.4.4 Characterization of the limiting measure Q for Q_T as $T \to +\infty$ in DA

From subsection 7.4.3 (see also Chapter 6 and Subsection 4.1.4 of [11]) it follows that there exists a sequence T_n such that measure Q_{T_n} converges weakly to some measure Q on $D_R[0,+\infty)$ as $T \to +\infty$, where $D_R[0,+\infty)$ is the Skorokhod space [7, p. 148]. This measure is the solution of such a martingale problem: the following process

$$m(s,t) := f(\hat{G}_0(t)) - f(\hat{G}_0(s)) - \int_s^t \hat{A}_0 f(\hat{G}_0(u))du \tag{7.4.19}$$

is a Q-martingale for all $f(g) \in C^2(R)$ and

$$Em(s,t)\eta_0^s = 0, \tag{7.4.20}$$

for scalar continuous bounded functional η_0^s, E is a mean value by measure Q. From (7.4.19) it follows that $E^T m^T(s,t)\eta_0^s = 0$ and it is necessary to show that the limiting passing in (7.4.1) goes to the process in (7.3.12) as $T \to +\infty$. From equality (7.4.11) we find that $\lim_{T_n \to +\infty} E^{T_n} m(s,t)\eta_0^s = Em(s,t)\eta_0^s$. Moreover, from the following expression

$$|E^T m(s,t)\eta_0^s - Em(s,t)\eta_0^s| \le |(E^T - E)m(s,t)\eta_0^s|$$
$$+|E^T|m(s,t) - m^T(s,t)||\eta_0^s| \to_{T\to+\infty} 0, \tag{7.4.21}$$

we obtain that there exists the measure Q on $D_R[0,+\infty)$ which solves the martingale problem for the operator $\hat{A}_0$ (or, equivalently, for the process $\hat{G}_0(t)$ in the form (7.4.12)). Uniqueness of the solution of the martingale problem follows from the fact that operator $\hat{A}_0$ generates the unique semigroup with respect to the Wiener process with variance σ_1^2 in (7.4.14). As long as the semigroup is unique then the limit process $\hat{G}_0(t)$ is unique. See [4, Chapter 1].

7.4.5 Calculation of the quadratic variation for GMRP

If $G_n^T = G_{T^{-1}\tau_n}^T$, the sequence

$$m_n^T := G_n^T - G_0^T - \sum_{k=0}^{n-1} E[G_{k+1}^T - G_k^T|\mathcal{F}_k], \quad G_0^T = g, \tag{7.4.22}$$

is $\mathcal{F}_n$-martingale, where $\mathcal{F}_n := \sigma\{x_k, \theta_k; 0 \le k \le n\}$. From the definition it follows that the characteristic $< m_n^T >$ of the martingale m_n^T has the form

$$< m_n^T >= \sum_{k=0}^{n-1} E[(m_{k+1}^T - m_k^T)^2|\mathcal{F}_k]. \tag{7.4.23}$$

To calculate $< m_n^T >$ let us represent m_n^T in (7.4.22) in the form of martingale-difference:

$$m_n^T = \sum_{k=0}^{n-1} [G_{k+1}^T - E(G_{k+1}^T|\mathcal{F}_k)]. \tag{7.4.24}$$

From representation

$$G_{n+1}^T - G_n^T = \frac{1}{T}\rho(x_n) \tag{7.4.25}$$

it follows that $E(G_{k+1}^T|\mathcal{F}_k) = G_k^T + T^{-1}\rho(x_k)$; that is why

$$G_{k+1}^T - E(G_{k+1}^T|\mathcal{F}_k) = T^{-1}(\rho(x_k) - P\rho(x_k)). \tag{7.4.26}$$

Since from (7.4.22) it follows that

$$m_{k+1}^T - m_k^T = G_{k+1}^T - E(G_{k+1}^T|\mathcal{F}_k) = T^{-1}(\rho(x_k) - P\rho(x_k)), \tag{7.4.27}$$

then substituting (7.4.27) in (7.4.23) we obtain

$$< m_n^T >= T^{-2}\sum_{k=0}^{n-1} [(I-P)\rho)(x_k)]^2. \tag{7.4.28}$$

In an averaging scheme (see Chapter 6 and [11]) for GMRP in the scale of time tT we obtain that $< m^T_{[tT]} >$ goes to zero as $T \to +\infty$ in probability, which follows from (7.4.27):

$$< m^T_{[tT]} >= T^{-2} \sum_{k=0}^{[tT]-1} [(I-P)\rho(x_k)]^2 \to 0 \text{ as } T \to +\infty \tag{7.4.29}$$

for all $t \in R_+$. In the diffusion approximation scheme for GMRP in scale of time tT^2 from (7.4.27) we obtain that characteristic $< m^T_{[tT^2]} >$ does not go to zero as $T \to +\infty$ since

$$< m^T_{[tT^2]} >= T^{-2} \sum_{k=0}^{[tT^2]-1} [(I-P)\rho(x_k)]^2 \to t\sigma_1^2, \tag{7.4.30}$$

where $\sigma_1^2 := \int_X \pi(dx)[(I-P)\rho(x)]^2$.

7.4.6 *Rates of convergence for GMRP*

Consider the representation (7.4.22) for martingale m^T_n. It follows that

$$G^T_n = g + m^T_n + \sum_{k=0}^{n-1} E[G^T_{k+1} - G^T_k | \mathcal{F}_k]. \tag{7.4.31}$$

In a diffusion approximation scheme for GMRP the limit for the process $G^T_{[tT^2]}$ as $T \to +\infty$ will be diffusion process $\hat{S}(t)$ (see (7.3.10)). If $m_0(t)$ is the limiting martingale for $m^T_{[tT^2]}$ in (7.4.22) as $T \to +\infty$, then from (7.4.31) and (7.3.10) we obtain

$$E[G^T_{[tT^2]} - \hat{S}(t)] = E[m^T_{[tT^2]} - m_0(t)] + T^{-1} \sum_{k=0}^{[tT^2]-1} \rho(x_k) - \hat{S}(t). \tag{7.4.32}$$

Since $E[m^T_{[tT^2]} - m_0(t)] = 0$, (because $m^T_{[tT^2]}$ and $m_0(t)$ are zero-mean martingales) then from (7.4.32) we obtain:

$$|E[G^T_{[tT^2]} - \hat{S}(t)]| \leq T^{-1} | \sum_{k=0}^{[tT^2]-1} \rho(x_k) - \hat{S}(t)T |. \tag{7.4.33}$$

Taking into account the balance condition $\int_X \pi(dx)\rho(x) = 0$ and the Central Limit Theorem for a Markov chain [8, p. 98], we obtain

$$| \sum_{k=0}^{[tT^2]-1} \rho(x_k) - \hat{S}(t)T | = C_1(t_0), \tag{7.4.34}$$

where $C_1(t_0)$ is a constant depending on t_0, $t \in [0, t_0]$. From (7.4.33), (7.4.2) and (7.4.32) we obtain:

$$|E[G^T_{[tT^2]} - \hat{S}(t)]| \leq T^{-1} C_1(t_0). \tag{7.4.35}$$

Thus, the rates of convergence in the diffusion scheme has the order T^{-1}.

7.5 Merged diffusion geometric Markov renewal process in the case of two ergodic classes

7.5.1 *Two ergodic classes*

Let $P(x,A) := \mathbb{P}\{x_{n+1} \in A | x_n = x\}$ be the transition probabilities of supporting an embedded reducible Markov chain $\{x_n\}_{n\geq 0}$ in the phase space X. Let us have two ergodic classes X_0 and X_1 of the phase space such that:

$$X = X_0 \cup X_1, \qquad X_0 \cap X_1 = \emptyset. \tag{7.5.1}$$

Let $\{\hat{X} = \{0,1\}, \mathcal{V}\}$ be the measurable merged phase space. A stochastic kernel $P_0(x,A)$ is consistent with the splitting (7.5.1) in the following way:

$$P_0(x,X_k) = 1_k := \left\{ \begin{array}{ll} 1, & x \in X_k, \\ 0, & x \notin X_k, \qquad k = 0,1. \end{array} \right. \tag{7.5.2}$$

Let the supporting embedded Markov chain $(x_n)_{n\in Z_+}$ with the transition probabilities $P_0(x,A)$ is uniformly ergodic in each class $X_k, k = 0,1$ and it has a stationary distribution $\pi_k(dx)$ in the classes X_k, $k = 0,1$:

$$\pi_k(A) = \int_{X_k} \pi_k(dx) P_0(x,A), \qquad A \subset X_k, \quad k = 0,1. \tag{7.5.3}$$

Let the stationary escape probabilities of the embedded Markov chain $(x_n)_{n\in Z_+}$ with transition probabilities $P(x,A) := \mathbb{P}\{x_{n+1} \in A | x_n = x\}$ be positive and sufficiently small; that is,

$$q_k(A) = \int_{X_k} \pi_k(dx) P(x, X \backslash X_k) > 0, \quad k = 0,1. \tag{7.5.4}$$

Let the stationary sojourn time in the classes of states be uniformly bounded, namely,

$$0 \leq C_1 \leq m_k := \int_{X_k} \pi_k(dx) m(x) \leq C_2, \quad k = 0,1, \tag{7.5.5}$$

where

$$m(x) := \int_0^\infty \bar{G}_x(t) dt. \tag{7.5.6}$$

7.5.2 *Algorithms of phase averaging with two ergodic classes*

The merged Markov chain $(\hat{x}_n)_{n\in Z_+}$ in merged phase space $\hat{X}$ is given by matrix of transition probabilities

$$\begin{aligned} \hat{P} &= (\hat{p}_{kr})_{k,r=0,1} \\ \hat{p}_{01} &= 1 - \hat{p}_{11} = \int_{X_1} \pi_1(dx) P(x,X_0) = 1 - \int_{X_1} \pi_1(dx) P(x,X_1) \\ \hat{p}_{01} &= 1 - \hat{p}_{00} = \int_{X_0} \pi_0(dx) P(x,X_1) = 1 - \int_{X_0} \pi_0(dx) P(x,X_0) \end{aligned} \tag{7.5.7}$$

As $\hat{p}_{kr} \neq 0, k = 0,1$, then $\hat{x}_n$ has virtual transitions. Intensities $\hat{\Lambda}_k$ of sojourn times $\hat{\theta}_k, k = 0,1$, of the merged MRP are calculated as follows:

$$\hat{\Lambda}_k = \frac{1}{m_k}, \qquad m_k = \int_{X_k} \pi_k(dx)m(x), \quad k = 0,1. \tag{7.5.8}$$

And, finally, the merged MRP $(\hat{x}_n, \hat{\theta})_{n \in Z_+}$ in the merged phase space $\hat{X}$ is given by the stochastic matrix

$$\hat{Q}(t) = (\hat{Q}_{kr})_{k,r=0,1} := \hat{p}_{kr}(1 - e^{-\hat{\Lambda}_k t}), \quad k,r = 0,1. \tag{7.5.9}$$

Hence, the initial semi-Markov system is merged to a Markov system with two classes.

7.5.3 *Merged diffusion approximation in the case of two ergodic classes*

The merged diffusion GMRP in the case of two ergodic classes has the form:

$$\tilde{S}(t) = S_0 e^{-\frac{1}{2}\int_0^t \hat{\rho}^2(\hat{x}(s))ds + \int_0^t \hat{\sigma}_\rho(\hat{x}(s))dw(s)} \tag{7.5.10}$$

which satisfies the stochastic differential equation (SDE):

$$\frac{d\tilde{S}(t)}{\tilde{S}(t)} = \frac{1}{2}(\hat{\sigma}_\rho^2(\hat{x}(t)) - \hat{\rho}^2(\hat{x}(t)))dt + \hat{\sigma}_\rho(\hat{x}(t))dw(t), \tag{7.5.11}$$

where

$$\begin{aligned}
\hat{\rho}^2(1) &:= \int_{X_1} p_1(dx) \int_{X_1} P(x,dy)\rho^2(y)/m(1), \\
\hat{\rho}^2(0) &:= \int_{X_0} p_0(dx) \int_{X_0} P(x,dy)\rho^2(y)/m(0), \\
\hat{\sigma}_\rho^2(1) &:= \int_{X_1} p_1(dx) \int_{X_1} P(x,dy)\rho^2(y) \\
&\quad + \int_{X_1} P(x,dy)\rho(y)R_0 \int_{X_1} P(x,dy)\rho(y)/m(1), \\
\hat{\sigma}_\rho^2(0) &:= \int_{X_0} p_0(dx) \int_{X_0} P(x,dy)\rho^2(y) \\
&\quad + \int_{X_0} P(x,dy)\rho(y)R_0 \int_{X_0} P(x,dy)\rho(y)/m(0),
\end{aligned} \tag{7.5.12}$$

$\hat{x}(t)$ is a merged Markov process in $\hat{X} = \{0,1\}$ with stochastic matrix $\hat{Q}(t)$ in (7.5.9).

7.6 European call option pricing formulas for diffusion GMRP

7.6.1 *Ergodic geometric Markov renewal process*

As we have seen in Section 7.3, an ergodic diffusion GMRP $\hat{S}(t)$ satisfies the following SDE (see (13.5.2)):

$$\frac{d\hat{S}(t)}{\hat{S}(t)} = 1/2(\sigma_\rho - \hat{\rho}_2)dt + \sigma_\rho dw(t), \tag{7.6.1}$$

where

$$\hat{\rho}_2 = \int_X p(dx) \int_X P(x,dy)\rho^2(y)/m, \tag{7.6.2}$$

$$\sigma_\rho^2 = \int_X p(dx)[1/2 \int_X P(x,dy)\rho^2(y) + \int_X P(x,dy)\rho(y)R_0P(x,dy)\rho(y)/m. \tag{7.6.3}$$

The risk-neutral measure P^* for the process in (7.6.1) is:

$$\frac{dP^*}{P} = \exp\{-\theta t - \frac{1}{2}\theta^2 w(t)\}, \tag{7.6.4}$$

where

$$\theta = \frac{(\frac{1}{2}(\sigma_\rho - \hat{\rho}_2) - r)}{\sigma_\rho}. \tag{7.6.5}$$

Under P^*, the process $e^{-rt}\hat{S}_t$ is a martingale and the process $w^*(t) = w(t) + \theta t$ is a Brownian motion. In this way, in the risk-neutral world, the process $\hat{S}_t$ has the following form

$$\frac{d\hat{S}(t)}{\hat{S}(t)} = rdt + \sigma_\rho dw^*(t), \tag{7.6.6}$$

Using Black-Scholes formula (see [2]) we obtain the European call option pricing formula for our model (7.6.6):

$$C = S_0\Phi(d_+) - Ke^{-rT}\Phi(d_-), \tag{7.6.7}$$

where

$$\begin{aligned} d_+ &= \frac{\ln(S_0/K) + (r + \frac{1}{2}\sigma_\rho t)}{\sigma_\rho\sqrt{t}}, \\ d_- &= \frac{\ln(S_0/K) + (r - \frac{1}{2}\sigma_\rho t)}{\sigma_\rho\sqrt{t}}, \end{aligned} \tag{7.6.8}$$

$\Phi(x)$ is a normal distribution and σ_ρ is defined in (7.6.3).

Remark (Hedging Strategies for GMRP in the Diffusion Approximation (DA) Scheme). The hedging strategies for GMRP in the DA scheme have the following form (see [2] for comparisons):

$$\begin{aligned} \hat{\gamma}_t &= \Phi(\frac{\ln(\hat{S}(t)/K) + (r + \sigma_\rho^2)(T-t)}{\sigma_\rho\sqrt{T-t}}) \\ \hat{\beta}_t &= -\frac{K}{B_0 e^{-rT}}\Phi(\frac{\ln(\hat{S}(t)/K) + (r - \sigma_\rho^2)(T-t)}{\sigma_\rho\sqrt{T-t}}), \end{aligned}$$

where (β_t, γ_t) is an investor's portfolio, $\Phi(x)$ is a standard normal distribution. The capital $\hat{X}_t := \hat{S}(t)\hat{\gamma}_t + B(t)\hat{\beta}_t$ has the form:

$$\hat{X}_t = \hat{S}(t)\hat{\gamma}_t - Ke^{-r(T-t)}\hat{\beta}_t.$$

7.6.2 Double averaged diffusion GMRP

Using the similar arguments as in (7.6.1) - (7.6.7), we can get European call option pricing formula for a double averaged diffusion GMRP in (7.3.24):

$$\frac{d\check{S}(t)}{\check{S}(t)} = 1/2(\check{\sigma}_\rho^2 - \check{\rho}_2)dt + \check{\sigma}_\rho dw(t), \tag{7.6.9}$$

where $\check{\sigma}_\rho^2$ and $\check{\rho}_2$ are defined in (7.3.21),(see also (7.3.13) and (7.3.15)). Namely, the European call option pricing formula for a double averaged diffusion GMRP is :

$$C = S_0\Phi(d_+) - Ke^{-rT}\Phi(d_-), \tag{7.6.10}$$

where

$$\begin{aligned} d_+ &= \frac{\ln(S_0/K) + (r + \frac{1}{2}\check{\sigma}_\rho t)}{\check{\sigma}_\rho\sqrt{t}}, \\ d_- &= \frac{\ln(S_0/K) + (r - \frac{1}{2}\check{\sigma}_\rho t)}{\check{\sigma}_\rho\sqrt{t}}, \end{aligned} \tag{7.6.11}$$

$\Phi(x)$ is a normal distribution and $\check{\sigma}_\rho$ is defined in (7.3.21).

Remark (Hedging Strategies for GMRP in the Double Averaged Diffusion Approximation Scheme). The hedging strategies for GMRP in the double averaged DA scheme has the following form (see [2] for comparisons):

$$\begin{aligned} \check{\gamma}_t &= \Phi(\frac{\ln(\check{S}(t)/K) + (r + \check{\sigma}_\rho^2)(T-t)}{\check{\sigma}_\rho\sqrt{T-t}}) \\ \check{\beta}_t &= -\frac{K}{B_0e^{-rT}}\Phi(\frac{\ln(\check{S}(t)/K) + (r - \check{\sigma}_\rho^2)(T-t)}{\check{\sigma}_\rho\sqrt{T-t}}), \end{aligned}$$

where $(\check{\beta}_t, \check{\gamma}_t)$ is an investor's portfolio, $\Phi(x)$ is a standard normal distribution. The capital $\check{X}_t := \check{S}(t)\check{\gamma}_t + B(t)\check{\beta}_t$ has the form:

$$\check{X}_t = \check{S}(t)\check{\gamma}_t - Ke^{-r(T-t)}\check{\beta}_t.$$

7.6.3 European call option pricing formula for merged diffusion GMRP

From Section 7.3.2, the merged diffusion GMRP has the following form:

$$\frac{d\tilde{S}(t)}{\tilde{S}(t)} = \frac{1}{2}(\hat{\sigma}_\rho^2(\hat{x}(t)) - \hat{\rho}_2(\hat{x}(t)))dt + \hat{\sigma}_\rho(\hat{x}(t))dw(t), \tag{7.6.12}$$

where $\hat{\sigma}_\rho^2$ and $\hat{\rho}_2$ are defined in Section 7.3.2 (see (7.3.18)). Taking into account the result on European call option pricing formula for regime switching geometric Brownian motion (see [9, p.224, corollary]), we obtain the option pricing formula for the merged diffusion GMRP:

$$C = \int C_T^{BS}\left((\frac{z}{T})^{2^{-1}}, T, S_0 \right) F_T^x(dz),$$

where C_T^{BS} is a Black-Scholes value and $F_T^x(dz)$ is a distribution of the random variable

$$z_T^x = \int_0^T \hat{\sigma}_\rho^2(\hat{x}(t))ds,$$

where $\hat{x}(t)$ is a merged Markov process.

7.7 Applications

7.7.1 *Example of two state ergodic diffusion approximation ergodic diffusion approximation*

$$P = \begin{pmatrix} p_{00} & p_{01} \\ p_{10} & p_{11} \end{pmatrix} = \begin{pmatrix} 0.98 & 0.02 \\ 0.02 & 0.98 \end{pmatrix}$$

Due to the ergodicity of the Markov chain:

$$p_0 = p_1 = 0.500, \quad \rho = (0.02, \ -0.02) \quad \lambda(x_0) = 8, \ \ \lambda(x_1) = 10$$

We let $G_x(t)$ be the exponential distribution:

$$G_x(t) = P(\tau_{n+1} - \tau_n < t | x_n = x)$$

$$\bar{m}(x) = \int_0^\infty e^{-\lambda(x)t} dt = \frac{1}{\lambda(x)}, \ m := \int_X p(dx) * \frac{1}{\lambda(x)}.$$

Also the balance condition is fulfilled:

$$\begin{pmatrix} p_0, & p_1 \end{pmatrix} \begin{pmatrix} p_{00} & p_{01} \\ p_{10} & p_{11} \end{pmatrix} \begin{pmatrix} \rho_0 \\ \rho_1 \end{pmatrix} = 0;$$

Therefore, by algorithms of ergodic averaging and diffusion approximation:

$$\begin{aligned} \hat{\rho}_2 &= \int_X p(dx) \int_X P(x, dy) \rho^2(y)/m \\ &= \begin{pmatrix} p_0, & p_1 \end{pmatrix} \begin{pmatrix} p_{00} & p_{01} \\ p_{10} & p_{11} \end{pmatrix} \begin{pmatrix} \rho_0^2 \\ \rho_1^2 \end{pmatrix} / (p_0, \ p_1) \begin{pmatrix} \bar{m}(x_0) \\ \bar{m}(x_1) \end{pmatrix} \end{aligned} \tag{7.7.1}$$

$$\sigma_\rho^2 = \int_X p(dx)[1/2\int_X P(x,dy)\rho^2(y) + \sum_{n=0}^{\infty}(P^n - \Pi)\int_X P(x,dy)\rho(y)\int_X P(x,dy)\rho(y)]/m$$
$$= \int_X p(dx)[1/2\int_X P(x,dy)\rho^2(y) + \sum_{n=0}^{\infty}\int_X P^n(x,dy)\int_X P(y,dz)\rho(z)\int_X P(y,dz)\rho(z)$$
$$-\sum_{n=0}^{\infty}\int_X p(dy)\int_X P(y,dz)\rho(z)\int_X P(y,dz)\rho(z)\mathbf{1}(y)]/m$$
$$= (p_0,\ p_1)[\frac{1}{2}\begin{pmatrix} p_{00} & p_{01} \\ p_{10} & p_{11}\end{pmatrix}\begin{pmatrix}\rho_0^2 \\ \rho_1^2\end{pmatrix} + \sum_{n=0}^{\infty}\begin{pmatrix} p_{00}^n & p_{01}^n \\ p_{10}^n & p_{11}^n\end{pmatrix}\begin{pmatrix}(p_{00}\rho_0 + p_{01}\rho_1)^2 \\ (p_{10}\rho_0 + p_{11}\rho_1)^2\end{pmatrix}$$
$$-\sum_{n=0}^{\infty}\begin{pmatrix} p_0 & p_1 \\ p_0 & p_1\end{pmatrix}\begin{pmatrix}(p_{00}\rho_0 + p_{01}\rho_1)^2 \\ (p_{10}\rho_0 + p_{11}\rho_1)^2\end{pmatrix}]/(p_0,\ p_1)\begin{pmatrix}\bar{m}(x_0) \\ \bar{m}(x_1)\end{pmatrix} \tag{7.7.2}$$

$$\hat{S}(t) = S_0 e^{-1/2t\hat{\rho_2}} e^{\sigma_\rho w(t)} \tag{7.7.3}$$

From those data above, we have

$$\hat{\rho_2} = 3.6 \times 10^{-3}; \tag{7.7.4}$$

$$\sigma_\rho^2 = 1.8 \times 10^{-3}; \tag{7.7.5}$$

$$\therefore\ \hat{S}(t) = 10e^{-1.8\times10^{-3}t + 4.24\times10^{-2}\omega(t)},$$

$$\because\ E(e^{at+b\omega(t)}) = e^{at+\frac{1}{2}b^2t} \qquad Var(e^{at+b\omega(t)}) = e^{2at+2b^2t} - e^{2at+b^2t}$$

$$\therefore\ E(S_t) = 10e^{-9\times10^{-4}t}, \qquad Var(S_t) = 100 - 100e^{-1.8\times10^{-3}t}$$

7.7.2 *Example of merged diffusion approximation*

Since the stationary escape probabilities of the embedded Markov chain are sufficiently small. $\hat{p_{10}}$ and $\hat{p_{01}}$ are sufficiently small.

$\tilde{S}_t$ will be split into two cases by the original value of $\hat{x}(0)$:
When $\hat{x}(0) = 0$

$$\tilde{S}_t = S_0 exp\{-\frac{1}{2}(\sum_{i=1}^{\frac{v(t)}{2}}\hat{\rho}^2(0)\theta_{2i-1} + \sum_{i=1}^{\frac{v(t)}{2}}\hat{\rho}^2(1)\theta_{2i} + (t-\tau_{v(t)})\hat{\rho}^2(0))$$
$$+\sum_{i=1}^{\frac{v(t)}{2}}\hat{\sigma}_\rho(0)(\omega(\sum_{k=1}^{2i-1}\theta_k) - \omega(\sum_{k=1}^{2i-2}\theta_k)) + \sum_{i=1}^{\frac{v(t)}{2}}\hat{\sigma}_\rho(1)(\omega(\sum_{k=1}^{2i}\theta_k) - \omega(\sum_{k=1}^{2i-1}\theta_k))$$
$$+\hat{\sigma}_\rho(0)(\omega(t) - \omega(\tau_{v(t)}))\} \qquad v(t)\ is\ even \tag{7.7.6}$$

$$\tilde{S}_t = S_0 exp\{-\frac{1}{2}(\sum_{i=1}^{\frac{\nu(t)+1}{2}} \hat{\rho}^2(0)\theta_{2i-1} + \sum_{i=1}^{\frac{\nu(t)-1}{2}} \hat{\rho}^2(1)\theta_{2i} + (t-\tau_{\nu(t)})\hat{\rho}^2(1))$$
$$+ \sum_{i=1}^{\frac{\nu(t)+1}{2}} \hat{\sigma}_\rho(0)(\omega(\sum_{k=1}^{2i-1}\theta_k) - \omega(\sum_{k=1}^{2i-2}\theta_k)) + \sum_{i=1}^{\frac{\nu(t)-1}{2}} \hat{\sigma}_\rho(1)(\omega(\sum_{k=1}^{2i}\theta_k) - \omega(\sum_{k=1}^{2i-1}\theta_k))$$
$$+ \hat{\sigma}_\rho(1)(\omega(t) - \omega(\tau_{\nu(t)}))\} \qquad \nu(t)\ is\ odd \tag{7.7.7}$$

When $\hat{x}(0) = 1$

$$\tilde{S}_t = S_0 exp\{-\frac{1}{2}(\sum_{i=1}^{\frac{\nu(t)}{2}} \hat{\rho}^2(1)\theta_{2i-1} + \sum_{i=1}^{\frac{\nu(t)}{2}} \hat{\rho}^2(0)\theta_{2i} + (t-\tau_{\nu(t)})\hat{\rho}^2(1))$$
$$+ \sum_{i=1}^{\frac{\nu(t)}{2}} \hat{\sigma}_\rho(1)(\omega(\sum_{k=1}^{2i-1}\theta_k) - \omega(\sum_{k=1}^{2i-2}\theta_k)) + \sum_{i=1}^{\frac{\nu(t)}{2}} \hat{\sigma}_\rho(0)(\omega(\sum_{k=1}^{2i}\theta_k) - \omega(\sum_{k=1}^{2i-1}\theta_k))$$
$$+ \hat{\sigma}_\rho(1)(\omega(t) - \omega(\tau_{\nu(t)}))\} \qquad \nu(t)\ is\ even \tag{7.7.8}$$

$$\tilde{S}_t = S_0 exp\{-\frac{1}{2}(\sum_{i=1}^{\frac{\nu(t)+1}{2}} \hat{\rho}^2(1)\theta_{2i-1} + \sum_{i=1}^{\frac{\nu(t)-1}{2}} \hat{\rho}^2(0)\theta_{2i} + (t-\tau_{\nu(t)})\hat{\rho}^2(0))$$
$$+ \sum_{i=1}^{\frac{\nu(t)+1}{2}} \hat{\sigma}_\rho(1)(\omega(\sum_{k=1}^{2i-1}\theta_k) - \omega(\sum_{k=1}^{2i-2}\theta_k)) + \sum_{i=1}^{\frac{\nu(t)-1}{2}} \hat{\sigma}_\rho(0)(\omega(\sum_{k=1}^{2i}\theta_k) - \omega(\sum_{k=1}^{2i-1}\theta_k))$$
$$+ \hat{\sigma}_\rho(0)(\omega(t) - \omega(\tau_{\nu(t)}))\} \qquad \nu(t)\ is\ odd \tag{7.7.9}$$

$$P = \begin{pmatrix} 0.49 & 0.49 & 0.01 & 0.01 \\ 0.70 & 0.28 & 0.01 & 0.01 \\ 0.01 & 0.01 & 0.70 & 0.28 \\ 0.01 & 0.01 & 0.49 & 0.49 \end{pmatrix}$$

Merged Markov chain in the merged phase space $\hat{X} = (0,1)$. In each class $X_k\ \ (k=0,1)$, stationary distribution:

$$\begin{cases} \pi_0(x_0) = 0.5868 \\ \pi_0(x_1) = 0.4132 \end{cases} \qquad \begin{cases} \pi_1(x_2) = 0.6330 \\ \pi_1(x_3) = 0.3670 \end{cases}$$

$$\rho(x_k) = \begin{cases} -0.04032, & k=0 \\ 0.05768, & k=1 \\ -0.03571, & k=2 \\ 0.06629, & k=3 \end{cases} \qquad \lambda(x_k) = \begin{cases} 8, & k=0 \\ 10, & k=1 \\ 12, & k=2 \\ 10, & k=3 \end{cases} \qquad S_0 = 10.$$

$$\begin{aligned} m(x_0) &= (0.5868) \times \frac{1}{8} + 0.4132 \times \frac{1}{10} = 0.1147 \\ m(x_1) &= (0.633) \times \frac{1}{12} + 0.367 \times \frac{1}{10} = 0.0895 \end{aligned} \tag{7.7.10}$$

Also, the balance condition is fulfilled:

$$\left(\pi_0(x_0),\ \pi_0(x_1) \right) \begin{pmatrix} p_{00} & p_{01} \\ p_{10} & p_{11} \end{pmatrix} \begin{pmatrix} \rho(x_0) \\ \rho(x_1) \end{pmatrix} = 0;$$

$$\left(\pi_1(x_2),\ \pi_1(x_3) \right) \begin{pmatrix} p_{22} & p_{23} \\ p_{32} & p_{33} \end{pmatrix} \begin{pmatrix} \rho(x_2) \\ \rho(x_3) \end{pmatrix} = 0;$$

$\therefore$ By algorithms of ergodic averaging and diffusion approximation:

$$\begin{aligned} \hat{\rho}^2(0) &= \frac{\int_{X_0} \pi_0(dx) \int_{X_0} P(x,dy)\rho^2(y)}{m(0)}, \\ &= \left(\pi_0^0,\ \pi_0^1 \right) \begin{pmatrix} p_{00} & p_{01} \\ p_{10} & p_{11} \end{pmatrix} \begin{pmatrix} \rho_0^2 \\ \rho_1^2 \end{pmatrix} / \left(\pi_0^0,\ \pi_0^1 \right) \begin{pmatrix} \bar{m}(x_0) \\ \bar{m}(x_1) \end{pmatrix} \\ &= 1.99 \times 10^{-2} \end{aligned} \tag{7.7.11}$$

$$\begin{aligned} \hat{\rho}^2(1) &= \frac{\int_{X_1} \pi_1(dx) \int_{X_1} P(x,dy)\rho^2(y)}{m(1)}, \\ &= \left(\pi_1^2,\ \pi_1^3 \right) \begin{pmatrix} p_{22} & p_{23} \\ p_{32} & p_{33} \end{pmatrix} \begin{pmatrix} \rho_2^2 \\ \rho_3^2 \end{pmatrix} / \left(\pi_1^2,\ \pi_1^3 \right) \begin{pmatrix} \bar{m}(x_2) \\ \bar{m}(x_3) \end{pmatrix} \\ &= 2.64 \times 10^{-2} \end{aligned} \tag{7.7.12}$$

$$\begin{aligned} \hat{\sigma_\rho}^2(1) &= \int_{X_1} \pi_1(dx)[1/2 \int_{X_1} P(x,dy)\rho^2(y) \\ &\quad + \sum_{n=0}^{\infty} (P^n - \Pi) \int_{X_1} P(x,dy)\rho(y) \int_{X_1} P(x,dy)\rho(y)]/m_1 \\ &= \int_{X_1} \pi_1(dx)[1/2 \int_{X_1} P(x,dy)\rho^2(y) \\ &\quad + \sum_{n=0}^{\infty} \int_{X_1} P^n(x,dy) \int_{X_1} P(y,dz)\rho(z) \int_{X_1} P(y,dz)\rho(z) \\ &- \sum_{n=0}^{\infty} \int_{X_1} p(dy) \int_{X_1} P(y,dz)\rho(z) \int_{X_1} P(y,dz)\rho(z)\mathbf{1}(y)]/m_1 \\ &= (\pi_1^2,\ \pi_1^3)[\frac{1}{2} \begin{pmatrix} p_{22} & p_{23} \\ p_{32} & p_{33} \end{pmatrix} \begin{pmatrix} \rho_2^2 \\ \rho_3^2 \end{pmatrix} + \sum_{n=0}^{\infty} \begin{pmatrix} p_{22}^n & p_{23}^n \\ p_{32}^n & p_{33}^n \end{pmatrix} \begin{pmatrix} (p_{22}\rho_2 + p_{23}\rho_3)^2 \\ (p_{32}\rho_2 + p_{33}\rho_3)^2 \end{pmatrix} \\ &- \sum_{n=0}^{\infty} \begin{pmatrix} p_2 & p_3 \\ p_2 & p_3 \end{pmatrix} \begin{pmatrix} (p_{22}\rho_2 + p_{23}\rho_3)^2 \\ (p_{32}\rho_2 + p_{33}\rho_3)^2 \end{pmatrix}]/(\pi_1^2,\ \pi_1^3) \begin{pmatrix} m(x_2) \\ m(x_3) \end{pmatrix} \\ &= 2.83 \times 10^{-2} \end{aligned} \tag{7.7.13}$$

$$\begin{aligned}
\hat{\sigma_\rho}^2(0) &= \int_{X_0} \pi_0(dx)[1/2 \int_{X_0} P(x,dy)\rho^2(y) \\
&\quad + \sum_{n=0}^{\infty} (P^n - \Pi) \int_{X_0} P(x,dy)\rho(y) \int_{X_0} P(x,dy)\rho(y)]/m_0 \\
&= \int_{X_0} \pi_0(dx)[1/2 \int_{X_0} P(x,dy)\rho^2(y) \\
&\quad + \sum_{n=0}^{\infty} \int_{X_0} P^n(x,dy) \int_{X_0} P(y,dz)\rho(z) \int_{X_0} P(y,dz)\rho(z) \\
&\quad - \sum_{n=0}^{\infty} \int_{X_0} p(dy) \int_{X_0} P(y,dz)\rho(z) \int_{X_0} P(y,dz)\rho(z)\mathbf{1}(y)]/m_0 \\
&= (\pi_0^0,\ \pi_0^1)[\frac{1}{2}\begin{pmatrix} p_{00} & p_{01} \\ p_{10} & p_{11} \end{pmatrix}\begin{pmatrix} \rho_0^2 \\ \rho_1^2 \end{pmatrix} + \sum_{n=0}^{\infty}\begin{pmatrix} p_{00}^n & p_{01}^n \\ p_{10}^n & p_{11}^n \end{pmatrix}\begin{pmatrix} (p_{00}\rho_0 + p_{01}\rho_1)^2 \\ (p_{10}\rho_0 + p_{11}\rho_1)^2 \end{pmatrix} \\
&\quad - \sum_{n=0}^{\infty}\begin{pmatrix} p_0 & p_1 \\ p_0 & p_1 \end{pmatrix}\begin{pmatrix} (p_{00}\rho_0 + p_{01}\rho_1)^2 \\ (p_{10}\rho_0 + p_{11}\rho_1)^2 \end{pmatrix}]/(\pi_0^0,\ \pi_0^1)\begin{pmatrix} m(x_0) \\ m(x_1) \end{pmatrix} \\
&= 2.11 \times 10^{-2}
\end{aligned} \tag{7.7.14}$$

Let $\theta_i(i = 0,1,2,...) = 0.1$; given t, $\exists v(t).s.t.\ 0.1v(t) \le t < 0.1(v(t)+1)\ \therefore \tau_{v(t)} = \theta_i v(t) = 0.1v(t)$.
$\therefore$ When $\hat{x}(0) = 0$;

$$\begin{aligned}
\tilde{S}(t) =& 10 \times exp\{-\frac{1}{2}(\sum_{i=1}^{\frac{1}{2}\lfloor\frac{t}{0.1}\rfloor} 1.99 \times 10^{-2} \times 0.1 + \sum_{i=1}^{\frac{1}{2}\lfloor\frac{t}{0.1}\rfloor} 2.64 \times 10^{-2} \times 0.1 + (t - 0.1\lfloor\frac{t}{0.1}\rfloor) \times 1.99 \times 10^{-2}) \\
&+ \sum_{i=1}^{\frac{1}{2}\lfloor\frac{t}{0.1}\rfloor} 0.1453(\omega(0.1(2i-1)) - \omega(0.1(2i-2))) + \sum_{i=1}^{\frac{1}{2}\lfloor\frac{t}{0.1}\rfloor} 0.1681(\omega(0.1 \times 2i) \\
&- \omega(0.1(2i-1))) + 0.1453(\omega(t) - \omega(0.1\lfloor\frac{t}{0.1}\rfloor))\} \qquad v(t)\ is\ even
\end{aligned} \tag{7.7.15}$$

$$\begin{aligned}
\tilde{S}(t) =& 10 \times exp\{-\frac{1}{2}(\sum_{i=1}^{\frac{1}{2}(\lfloor\frac{t}{0.1}\rfloor+1)} 1.99 \times 10^{-2} \times 0.1 + \sum_{i=1}^{\frac{1}{2}(\lfloor\frac{t}{0.1}\rfloor-1)} 2.64 \times 10^{-2} \times 0.1 + (t - 0.1\lfloor\frac{t}{0.1}\rfloor) \\
&\times 2.64 \times 10^{-2}) + \sum_{i=1}^{\frac{1}{2}(\lfloor\frac{t}{0.1}\rfloor+1)} 0.1453(\omega(0.1(2i-1)) - \omega(0.1(2i-2))) + \sum_{i=1}^{\frac{1}{2}(\lfloor\frac{t}{0.1}\rfloor-1)} 0.1681 \\
&(\omega(0.1 \times 2i) - \omega(0.1(2i-1))) + 0.1681(\omega(t) - \omega(0.1\lfloor\frac{t}{0.1}\rfloor))\} \qquad v(t)\ is\ odd
\end{aligned} \tag{7.7.16}$$

When $\hat{x}(0) = 1$;

$$\begin{aligned}
\tilde{S}(t) =& 10 \times exp\{-\frac{1}{2}(\sum_{i=1}^{\frac{1}{2}\lfloor\frac{t}{0.1}\rfloor} 2.64 \times 10^{-2} \times 0.1 + \sum_{i=1}^{\frac{1}{2}\lfloor\frac{t}{0.1}\rfloor} 1.99 \times 10^{-2} \times 0.1 + (t - 0.1\lfloor\frac{t}{0.1}\rfloor) \times 2.64 \times 10^{-2}) \\
&+ \sum_{i=1}^{\frac{1}{2}\lfloor\frac{t}{0.1}\rfloor} 0.1681(\omega(0.1(2i-1)) - \omega(0.1(2i-2))) + \sum_{i=1}^{\frac{1}{2}\lfloor\frac{t}{0.1}\rfloor} 0.1453(\omega(0.1 \times 2i) \\
&- \omega(0.1(2i-1))) + 0.1681(\omega(t) - \omega(0.1\lfloor\frac{t}{0.1}\rfloor))\} \qquad v(t)\ is\ even
\end{aligned} \tag{7.7.17}$$

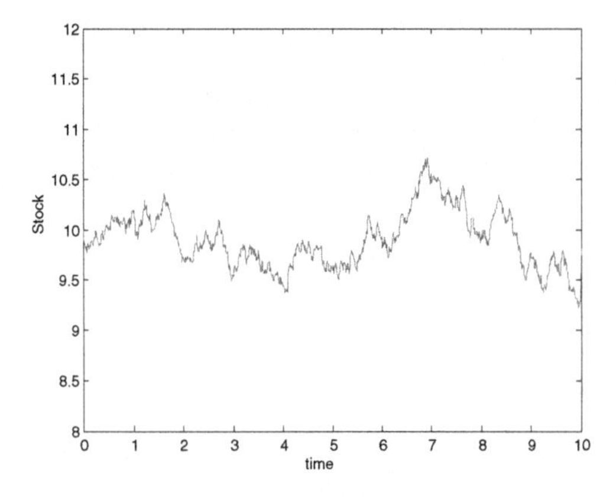

Figure 7.1 *Sample path of $\tilde{S}(t)$ w.r.t t in ergodic diffusion approximation.*

$$\tilde{S}(t) = 10 \times exp\{-\frac{1}{2}(\sum_{i=1}^{\frac{1}{2}(\lfloor\frac{t}{0.1}\rfloor+1)} 2.64 \times 10^{-2} \times 0.1 + \sum_{i=1}^{\frac{1}{2}(\lfloor\frac{t}{0.1}\rfloor-1)} 1.99 \times 10^{-2} \times 0.1 + (t - 0.1\lfloor\frac{t}{0.1}\rfloor) \times 1.99 \times 10^{-2}) + \sum_{i=1}^{\frac{1}{2}(\lfloor\frac{t}{0.1}\rfloor+1)} 0.1681(\omega(0.1(2i-1)) - \omega(0.1(2i-2))) + \sum_{i=1}^{\frac{1}{2}(\lfloor\frac{t}{0.1}\rfloor-1)} 0.1453 (\omega(0.1 \times 2i) - \omega(0.1(2i-1))) + 0.1453(\omega(t) - \omega(0.1\lfloor\frac{t}{0.1}\rfloor))\} \quad \nu(t) \text{ is odd} \tag{7.7.18}$$

$\therefore$

$$E(\tilde{S}(t)) = \begin{cases} 10e^{-1.614\times10^{-5}\lfloor\frac{t}{0.1}\rfloor+9.288\times10^{-4}t} & when\ \hat{x}(t) = 1, \\ 10e^{1.614\times10^{-5}\lfloor\frac{t}{0.1}\rfloor+6.061\times10^{-4}t} & when\ \hat{x}(t) = 0, \end{cases} \tag{7.7.19}$$

$$Var(\tilde{S}(t)) = \begin{cases} 100e^{-3.896\times10^{-4}\lfloor\frac{t}{0.1}\rfloor+3.010\times10^{-2}t} - 100e^{-3.228\times10^{-5}\lfloor\frac{t}{0.1}\rfloor+1.858\times10^{-3}t} \\ when\ \hat{x}(t) = 1; \\ 100e^{3.896\times10^{-4}\lfloor\frac{t}{0.1}\rfloor+2.230\times10^{-2}t} - 100e^{3.228\times10^{-5}\lfloor\frac{t}{0.1}\rfloor+1.212\times10^{-3}t} \\ when\ \hat{x}(t) = 0; \end{cases} \tag{7.7.20}$$

Under double averaging with stationary distribution:
$\hat{p}_k = \frac{1}{2},\ k = 0, 1;\quad S_0 = 10$
$\breve{\rho}_2 = 0.5 \times 1.99 \times 10^{-2} + 0.5 \times 2.64 \times 10^{-2} = 2.32 \times 10^{-2}$ $\quad \breve{\sigma}_{\rho}^2 = 0.5 \times 2.83 \times 10^{-2} + 0.5 \times 2.11 \times 10^{-2} = 2.47 \times 10^{-2}$

$$\therefore S_T(t) \simeq 10e^{-1.16\times10^{-2}t+0.1572\omega(t)}$$

$$E(S_T(t)) = 10e^{7.5\times10^{-4}t} \quad Var(S_T(t)) = 100e^{2.62\times10^{-2}t} - 100e^{1.5\times10^{-3}t}$$

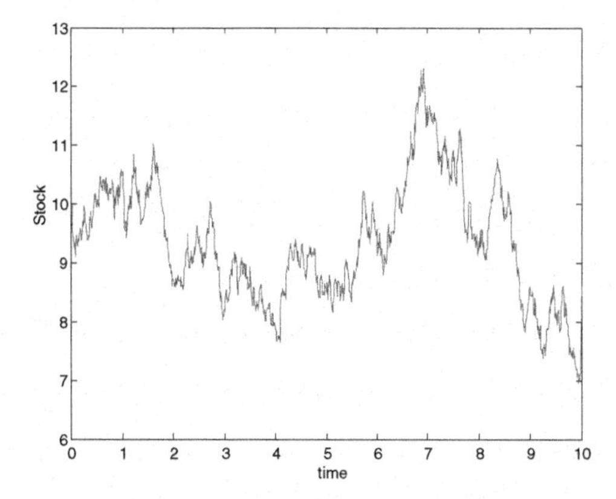

Figure 7.2 *Sample path of $\tilde{S}(t)$ w.r.t t in merged diffusion approximation when $\hat{x}(s) = 1$.*

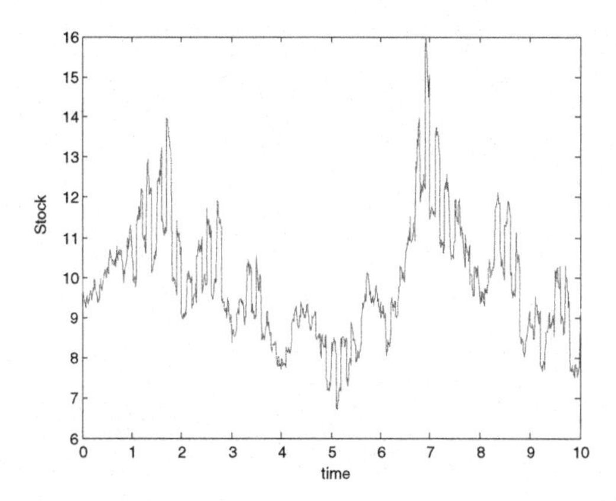

Figure 7.3 *Sample path of $\tilde{S}(t)$ w.r.t t in merged diffusion approximation when $\hat{x}(s) = 0$.*

7.7.3 Call option pricing for ergodic GMRP

Using data from the example of ergodic diffusion GMRP:

$S_0 = 10, \quad K = 10, \qquad \sigma_\rho^2 = 1.8 \times 10^{-3}, \qquad r = 0.01$

$$
\begin{aligned}
d_+ &= \frac{ln(S_0/K) + (r + 0.5\sigma_\rho^2)T}{\sigma_\rho \sqrt{T}} = \frac{(0.01 + 0.5 \times 1.8 \times 10^{-3})T}{0.0424\sqrt{T}} \\
d_- &= \frac{ln(S_0/K) + (r - 0.5\sigma_\rho^2)T}{\sigma_\rho \sqrt{T}} = \frac{(0.01 - 0.5 \times 1.8 \times 10^{-3})T}{0.0424\sqrt{T}}
\end{aligned}
\tag{7.7.21}
$$

$$\begin{aligned} C &= 10\Phi(\frac{(0.01+0.5\times 1.8\times 10^{-3})T}{0.0424\sqrt{T}}) \\ &\quad -10e^{-0.01T}\Phi(\frac{(0.01-0.5\times 1.8\times 10^{-3})T}{0.0424\sqrt{T}}) \end{aligned} \tag{7.7.22}$$

When T=1, $C = 10\times 0.6014 - 10e^{-0.01}\times 0.5850 = 0.2222$

7.7.4 Call option pricing formulas for double averaged GMRP

Let $\hat{p}_0 = \hat{p}_1 = \frac{1}{2}$ be the stationary distributions, $S_0 = 10,\ K = 10,\ T = 0.5,\ r = 0.01$
Since

$$\breve{\sigma}_\rho^2 = \sum_{k=0}^{r} \hat{p}_k \hat{\sigma}_\rho^2(k) \qquad \breve{p}_2 = \sum_{k=0}^{r} \hat{p}_k \hat{\rho}_2(k)$$

$$\therefore\ \breve{\rho}_2 == 0.5\times 1.99\times 10^{-2} + 0.5\times 2.64\times 10^{-2} = 2.32\times 10^{-2}$$

$$\breve{\sigma}_\rho^2 = 0.5\times 2.83\times 10^{-2} + 0.5\times 2.11\times 10^{-2} = 2.47\times 10^{-2}$$

$$\begin{aligned} d_+ &= \frac{ln(S_0/K)+(r+0.5\breve{\sigma}_\rho^2)T}{\breve{\sigma}_\rho\sqrt{t}} = \frac{(0.01+0.5\times 2.47\times 10^{-2})T}{0.1572\sqrt{T}} \\ d_- &= \frac{ln(S_0/K)+(r-0.5\breve{\sigma}_\rho^2)T}{\breve{\sigma}_\rho\sqrt{t}} = \frac{(0.01-0.5\times 2.47\times 10^{-2})T}{0.1572\sqrt{T}} \end{aligned} \tag{7.7.23}$$

$$\begin{aligned} C &= 10\Phi(\frac{(0.01+0.5\times 2.47\times 10^{-2})T}{0.1572\sqrt{T}}) \\ &\quad -10e^{-0.01T}\Phi(\frac{(0.01-0.5\times 2.47\times 10^{-2})T}{0.1572\sqrt{T}}). \end{aligned} \tag{7.7.24}$$

When T=1, $C = 10\times 0.5565 - 10e^{-0.01}\times 0.4940 = 0.6742$

References

[1] Aase, K., *Contingent claims valuation when the securities price is a combination of an Ito process and a random point process*, Stoch. Proc. & Their Applic., 28 (1988), 185-220.

[2] Black, F. and Scholes, M., *The pricing of options and corporate liabilities*, Journal of Political Economy, 81 (1973), 635-654.

[3] Cox, J. C., Ross, S.A., and Rubinstein, M., *Option pricing: a simplified approach*, Journal of Financial Economics, 7 (1979) 229-264.

[4] Korolyuk, V.S. and Swishchuk, A.V., *Evolution of systems in random media*, CRC Press, Boca Raton, USA, 1995.

[5] Korolyuk, V.S. and Limnios, N., *Poisson approximation of stochastic systems*, Theory Probab. and Mathem. Statistics, Kiev University, N62, 2002.

[6] Korolyuk, V.S., and Limnios, N., *Stochastic processes in phase merging space*, World Scientific, 2005.

[7] Skorokhod, A. *Studies in the theory of random processes (English translation),* Ann. Math. Stat., Vol. 38, No.1, 1967.

[8] Swishchuk, A.V., *Random evolutions and their applications: new trends,* Kluwer AP, Dordrecht, 2000.

[9] Swishchuk, A.V., *Random evolutions and their applications,* Kluwer AP, Dordrecht, 1997.

[10] Swishchuk, A.V., (B,S,X)*-securities markets*, Ukrainian Mathem. Congress, Kiev, Ukraine, 21-23 August 2001. Abstract of Communications.

[11] Swishchuk, A.V. and Islam, M. S., *The geometric Markov renewal processes with application to finance*, Stochastic Analysis and Application, 10, 4, 2011.

Chapter 8

Normal Deviation of a Security Market by the Geometric Markov Renewal Processes

8.1 Chapter overview

In this chapter, we consider the geometric Markov renewal process as a model for a security market. Normal deviations of the geometric Markov renewal processes (GMRP) for ergodic averaging and double averaging schemes are derived. Some applications in finance are presented.

8.2 Normal deviations of the geometric Markov renewal processes

Algorithms of averaging define the averaged systems (or models) which may be considered as the first approximation. Algorithms of diffusion under balance condition define diffusion models which may be considered as the second approximation. In this section we consider the algorithms of construction of the first and the second approximation in the case when the balance condition is not fulfilled.

8.2.1 Ergodic normal deviations

Let us consider the normal deviated process

$$w_T(t) := \sqrt{T}(\alpha_T(t) - \hat{\rho}t), \tag{8.2.1}$$

where $\alpha_T(t) := T^{-1}\sum_{k=1}^{\nu(tT)} \rho(x_k)$, and $\hat{\rho}$ is defined in [9] (see Chapter 6, Theorem 6.2, see also Theorem 4.1 in [9]) and $\hat{\rho} \neq 0$. The above process $w_T(t)$ defines deviations of the initial model $\alpha_T(t)$ in scale of time tT from the averaged model $\hat{\rho}t$ (see [9], proof of Theorem 4.1). It is known (see [4]) that under large T the model $w_T(t)$ has the properties of the Wiener process. From (8.2.1) we obtain

$$\alpha_T(t) = \hat{\rho}t + \frac{1}{\sqrt{T}} w_T(t). \tag{8.2.2}$$

Rewrite $w_T(t)$ in (8.2.1) in the following form:

$$w_T(t) = T^{-1/2} \sum_{k=1}^{\nu(tT)} [\rho(x_k) - \hat{\rho}] - \hat{\rho} T^{-1/2}(tT - \nu(tT)). \qquad (8.2.3)$$

The second term in (8.2.3) goes to zero as $T \to +\infty$. For the first term in (8.2.3) we note that function $P\rho - \hat{\rho}$ satisfies the balance condition with respect to the measure $p(dx)$, as $\int_X p(dx)[P\rho - \hat{\rho}] = 0$. This is why we can apply the algorithms of diffusion approximation (see [10]) to the first term in (8.2.3), where instead of function $P\rho$ it needs the function $P\rho - \hat{\rho}$. Taking limit as $T \to \infty$ we obtain from (8.2.3) :

$$\lim_{T \to \infty} w_T(t) = \hat{\sigma} \hat{w}(t), \qquad (8.2.4)$$

where $\hat{w}(t)$ is a standard Wiener process with diffusion coefficient

$$\hat{\sigma} := \int_X p(dx)(P\rho - \hat{\rho})\mathbf{R}_0(P\rho - \hat{\rho}) + 2^{-1}(P\rho - \hat{\rho})^2]/m, \qquad (8.2.5)$$

which follows from the algorithms of normal deviations [4]. In (8.2.5) R_0 is a potential [8] of $(x_n)_{n \in Z_+}$. Thus, we obtain the double approximation for $\alpha_T(t)$ in an ergodic normal deviations scheme:

$$\alpha_T(t) \simeq \hat{\rho} t + T^{-1/2} \hat{\sigma} \hat{w}(t), \qquad (8.2.6)$$

where $\hat{\sigma}$ is defined in (8.2.5). Hence,

$$\ln \frac{S_T(t)}{S_0} \simeq \hat{\rho} t + T^{-1/2} \hat{\sigma} \hat{w}(t). \qquad (8.2.7)$$

Corollary 8.1 *Ergodic normal deviated GMRP has the form:*

$$S_T(t) \simeq S_0 e^{\hat{\rho} t + T^{-1/2} \hat{\sigma} \hat{w}(t)}, \qquad (8.2.8)$$

or, in stochastic differential equation (SDE) form

$$\frac{dS_T(t)}{S_T(t)} \simeq (\hat{\rho} + \frac{1}{2} + T^{-1} \hat{\sigma}^2) dt + T^{-1/2} \hat{\sigma} d\hat{w}(t). \qquad (8.2.9)$$

8.2.2 *Reducible (merged) normal deviations*

Let us suppose that the balance condition is not fulfilled :

$$\hat{\rho}(k) = \int_{X_k} p_k(dx) \int_{X_k} P(x, dy)\rho(y)/m(k) \neq 0, \qquad (8.2.10)$$

for all $k = 1, 2, \ldots, r$ where $(x_n)_{n \in Z_+}$ is the supporting embedded Markov chain, p_k is the stationary density for the ergodic component X_k, $m(k) := \int_{X_k} p_k(dx) m(x)$,

$m(x) := \int_0^\infty (1 - G_x(t))dt$ and conditions of reducibility of X are fulfilled. Let us consider the normal deviated process

$$\tilde{w}_T(t) := \sqrt{T}(\alpha_T(t) - \tilde{\rho}(t)), \tag{8.2.11}$$

where $\alpha_T(t) := T^{-1}\sum_{k=1}^{\nu(tT)} \rho(x_k)$, and $\tilde{\rho}(t) = \int_0^t \hat{\rho}(\hat{x}(s))ds$, and $\hat{\rho}(k)$ is defined in (8.2.10). In this case the construction of the normal deviated process for $\alpha_T(t)$ in reducible case consists of the fact see [3, 4, 7, 8], that $\tilde{w}_T(t)$ is a stochastic Ito integral under large T :

$$\tilde{w}_T(t) \simeq \int_0^t \tilde{\sigma}(\hat{x}(s))dw(s), \tag{8.2.12}$$

where

$$\tilde{\sigma}(k) := \int_{X_k} p_k(dx)[(P\rho - \hat{\rho}(k))\mathbf{R}_o(P\rho - \hat{\rho}(k)) + 2^{-1}(P\rho - \hat{\rho}(k))^2]/m(k), \tag{8.2.13}$$

for all $k = 1, 2, \ldots, r$. Thus, double approximation of GMRP has the following form:

$$\alpha_T(t) \simeq \tilde{\rho}(t) + T^{-1/2}\int_0^t \tilde{\sigma}(\hat{x}(s))dw(s). \tag{8.2.14}$$

From (8.2.11) and (8.2.14) it follows that

$$\ln\frac{S_T(t)}{S_0} \simeq \tilde{\rho}(t) + T^{-1/2}\int_0^t \tilde{\sigma}(\hat{x}(s))dw(s), \tag{8.2.15}$$

or,

$$S_T(t) \simeq e^{\tilde{\rho}(t) + T^{-1/2}\int_0^t \tilde{\sigma}(\hat{x}(s))dw(s)}. \tag{8.2.16}$$

Corollary 8.2 *The reducible normal deviated GMRP has the form:*

$$S_T(t) \simeq S_0 e^{\tilde{\rho}(t) + T^{-1/2}\int_0^t \tilde{\sigma}(\hat{x}(s))dw(s)},$$

or, in the form of SDE

$$\frac{dS_T(t)}{S_T(t)} \simeq (\hat{\rho}(\hat{x}(t)) + \frac{1}{2}T^{-1}\tilde{\sigma}^2(\hat{x}(t)))dt + T^{-1/2}\tilde{\sigma}(\hat{x}(t))dw(t), \tag{8.2.17}$$

where $\hat{\rho}(k)$ and $\tilde{\sigma}(k)$ are defined in (8.2.10) and (8.2.13), respectively.

8.2.3 Normal deviations under double averaging

Let us suppose that merged phase space $\hat{X}$ of the merged Markov process $\hat{x}(t)$ consists of one ergodic class with stationary distribution $(\hat{p}_k)_{k=1,2,\ldots,N}$. Then using algorithms of double averaging [4] it can easily be shown that

$$\alpha_T(t) \simeq \check{\rho}t, \tag{8.2.18}$$

where $\check{\rho} := \sum_{i=1}^{\nu(tT)} \hat{p}_k \hat{\rho}(k)$, and $\hat{\rho}(k)$ are defined in (8.2.10). Thus,

$$\ln \frac{S_t^T}{S_0} \simeq \ln \frac{\check{S}_t}{S_0} = t\check{\rho}.$$

Let $S_t^T :\simeq \check{S}_t$. Then $\check{S}_t = S_0 e^{t\check{\rho}}$.

Let us suppose that $\check{\rho} \neq 0$, and let us consider the normal deviated process

$$\check{w}_T(t) := \sqrt{T}(\alpha_T(t) - \check{\rho}t). \tag{8.2.19}$$

Normal deviations of the initial process under double averaging consists of the fact that process $\check{w}_T(t)$ in (8.2.19) under large T is a Wiener process with diffusion coefficient $\check{\sigma}^2$:

$$\check{\sigma}^2 := \sum_{k=1}^{r} \hat{p}_k \tilde{\sigma}^2(k), \tag{8.2.20}$$

where

$$\tilde{\sigma}^2(k) := \int_{X_k} p_k(dx)[(P\rho - \hat{\rho}(k))\mathbf{R}_o(P\rho - \hat{\rho}(k)) + 2^{-1}(P\rho - \hat{\rho}(k))^2]/m(k), \tag{8.2.21}$$

for all $k = 1, 2, \ldots, r$. That is

$$\check{w}_T(t) \simeq \check{\sigma}^2 w(t),$$

where $w(t)$ is a standard Wiener process. In this way, double approximation of $\alpha_T(t)$ in (8.2.18) is expressed in the form:

$$\alpha_T(t) \simeq \check{\rho}t + T^{-1/2}\check{\sigma}^2 w(t). \tag{8.2.22}$$

From (8.2.19) and (8.2.22) it follows that

$$\ln \frac{S_T(t)}{S_0} \simeq \check{\rho}t + T^{-1/2}\check{\sigma}^2 w(t),$$

or, equivalently,

$$S_T(t) \simeq S_0 e^{\check{\rho}t + T^{-1/2}\check{\sigma}^2 w(t)}.$$

Corollary 8.3 *Normal deviated GMRP under double averaging has the form:*

$$S_T(t) \simeq S_0 e^{\check{\rho}t + T^{-1/2}\check{\sigma}^2 w(t)}, \tag{8.2.23}$$

or, in the form of SDE,

$$\frac{dS_T(t)}{S_T(t)} \simeq (\check{\rho} + \frac{1}{2}T^{-1}\check{\sigma}^2)dt + T^{-1/2}\check{\sigma}^2 dw(t). \tag{8.2.24}$$

8.3 Applications

8.3.1 Example of two state ergodic normal deviated GMRP

$$P = \begin{pmatrix} p_{00} & p_{01} \\ p_{10} & p_{11} \end{pmatrix} = \begin{pmatrix} 0.98 & 0.02 \\ 0.02 & 0.98 \end{pmatrix}$$

Due to the ergodicity of the Markov chain: $p(x_k) = 0.500$, $k = 0,1$. $S_0 = 10$, $T = 10$,

$$\rho(x_k) = \begin{cases} 0.05, & k=0 \\ -0.03, & k=1 \end{cases}, \quad \lambda(x_k) = \begin{cases} 8, & k=0 \\ 10, & k=1 \end{cases},$$

$$P\rho(x) = \int_X \rho(y)P(x,dy), \qquad m = \int_X p(dx) * \frac{1}{\lambda(x)}$$

$\therefore$

$$\begin{aligned}
\hat{\rho} &= \frac{\int_X p(dx) \int_X P(x,dy)\rho(y)}{m} \\
&= \begin{pmatrix} p_0, & p_1 \end{pmatrix} \begin{pmatrix} p_{00} & p_{01} \\ p_{10} & p_{11} \end{pmatrix} \begin{pmatrix} \rho_0 \\ \rho_1 \end{pmatrix} / \begin{pmatrix} p_0, & p_1 \end{pmatrix} \begin{pmatrix} \bar{m}(x_0) \\ \bar{m}(x_1) \end{pmatrix} \qquad (8.3.1) \\
&= 8.89 \times 10^{-2}
\end{aligned}$$

$$\begin{aligned}
\hat{\sigma}^2 &= \int_X p(dx)[(P\rho - m\hat{\rho})R_0(P\rho - m\hat{\rho}) + \frac{1}{2}(P\rho - m\hat{\rho})^2]/m \\
&= \int_X p(dx)[\sum_{n=0}^{\infty}(P^n - \Pi)(\int_X P(x,dy)\rho(y) - \int_X p(dx)\int_X P(x,dy)\rho(y))^2 \\
&+ \frac{1}{2}(\int_X P(x,dy)\rho(y) - \int_X p(dx)\int_X P(x,dy)\rho(y))^2]/m \\
&= \int_X p(dx)[\sum_{n=0}^{\infty}\int_X P^n(x,dy)(\int_X P(y,dz)\rho(z) - \int_X p(dy)\int_X p(y,dz)\rho(z))^2 \\
&- \sum_{n=0}^{\infty}\int_X p(dy)(\int_X P(y,dz)\rho(z) - \int_X p(dy)\int_X P(y,dz)\rho(z))^2\mathbf{1}(y) \\
&+ \frac{1}{2}(\int_X P(x,dy)\rho(y) - \int_X p(dx)\int_X P(x.dy)\rho(y))^2]/m \\
&= (p_0, \ p_1)[\sum_{n=0}^{\infty}\begin{pmatrix} p_{00}^n & p_{01}^n \\ p_{10}^n & p_{11}^n \end{pmatrix}\begin{pmatrix} (p_{00}\rho_0 + p_{01}\rho_1 - m\hat{\rho})^2 \\ (p_{10}\rho_0 + p_{11}\rho_1 - m\hat{\rho})^2 \end{pmatrix} \\
&- \sum_{n=0}^{\infty}\begin{pmatrix} p_0 & p_1 \\ p_0 & p_1 \end{pmatrix}\begin{pmatrix} (p_{00}\rho_0 + p_{01}\rho_1 - m\hat{\rho})^2 \\ (p_{10}\rho_0 + p_{11}\rho_1 - m\hat{\rho})^2 \end{pmatrix} \\
&+ \frac{1}{2}\begin{pmatrix} (p_{00}\rho_0 + p_{01}\rho_1 - m\hat{\rho})^2 \\ (p_{10}\rho_0 + p_{11}\rho_1 - m\hat{\rho})^2 \end{pmatrix}]/m \\
&= 6.6 \times 10^{-3}
\end{aligned} \qquad (8.3.2)$$

$$\therefore\ S_T(t) \approx S_0 e^{\hat{\rho}t+T^{-1/2}\hat{\sigma}\hat{\omega}(t)} = 10e^{8.89\times10^{-2}t+T^{-1/2}8.12\times10^{-2}\hat{\omega}(t)} \tag{8.3.3}$$

$$\because\ E(e^{at+b\omega(t)}) = e^{at+\frac{1}{2}b^2 t} \qquad Var(e^{at+b\omega(t)}) = e^{2at+2b^2 t} - e^{2at+b^2 t}$$

$$\therefore\ E(S_T(t)) = 10e^{(8.89\times10^{-2}+\frac{1}{2T}6.6\times10^{-3})t}$$

$$Var(S_T(t)) = 100e^{(0.1778+\frac{1}{T}1.32\times10^{-2})t} - 100e^{(0.1778+\frac{1}{T}6.6\times10^{-3})t}$$

8.3.2 Example of merged normal deviations in 2 classes

Since the stationary escape probabilities of the embedded Markov chain are sufficiently small, $\hat{p_{10}}$ and $\hat{p_{01}}$ are sufficiently small.

$S_T(t)$ will be split into two cases by the original value of $\hat{x}(0)$:

$\therefore$,when $\hat{x}(0) = 0$;

$$\begin{aligned} S_T(t) = S_0 exp\{&\sum_{i=1}^{\frac{v(t)}{2}} \hat{\rho}(0)\theta_{2i-1} + \sum_{i=1}^{\frac{v(t)}{2}} \hat{\rho}(1)\theta_{2i} + (t-\tau_{v(t)})\hat{\rho}(0) \\ &+T^{-1/2}(\sum_{i=1}^{\frac{v(t)}{2}} \hat{\sigma}_\rho(0)(\omega(\sum_{k=1}^{2i-1}\theta_k) - \omega(\sum_{k=1}^{2i-2}\theta_k)) \\ &+\sum_{i=1}^{\frac{v(t)}{2}} \hat{\sigma}_\rho(1)(\omega(\sum_{k=1}^{2i}\theta_k) - \omega(\sum_{k=1}^{2i-1}\theta_k)) \\ &+\hat{\sigma}_\rho(0)(\omega(t)-\omega(\tau_{v(t)})))\} \qquad v(t)\ is\ even \end{aligned} \tag{8.3.4}$$

$$\begin{aligned} \tilde{S}_t = S_0 exp\{&\sum_{i=1}^{\frac{v(t)+1}{2}} \hat{\rho}(0)\theta_{2i-1} + \sum_{i=1}^{\frac{v(t)-1}{2}} \hat{\rho}(1)\theta_{2i} + (t-\tau_{v(t)})\hat{\rho}(1) \\ &+T^{-1/2}(\sum_{i=1}^{\frac{v(t)+1}{2}} \hat{\sigma}_\rho(0)(\omega(\sum_{k=1}^{2i-1}\theta_k) - \omega(\sum_{k=1}^{2i-2}\theta_k)) \\ &+\sum_{i=1}^{\frac{v(t)-1}{2}} \hat{\sigma}_\rho(1)(\omega(\sum_{k=1}^{2i}\theta_k) - \omega(\sum_{k=1}^{2i-1}\theta_k)) \\ &+\hat{\sigma}_\rho(1)(\omega(t)-\omega(\tau_{v(t)})))\} \qquad v(t)\ is\ odd \end{aligned} \tag{8.3.5}$$

when $\hat{x}(0) = 1$;

$$
\begin{aligned}
S_T(t) = S_0 exp\{&\sum_{i=1}^{\frac{\nu(t)}{2}} \hat{\rho}(1)\theta_{2i-1} + \sum_{i=1}^{\frac{\nu(t)}{2}} \hat{\rho}(0)\theta_{2i} + (t-\tau_{\nu(t)})\hat{\rho}(1) \\
&+ T^{-1/2}(\sum_{i=1}^{\frac{\nu(t)}{2}} \hat{\sigma}_\rho(1)(\omega(\sum_{k=1}^{2i-1}\theta_k) - \omega(\sum_{k=1}^{2i-2}\theta_k)) \\
&+ \sum_{i=1}^{\frac{\nu(t)}{2}} \hat{\sigma}_\rho(0)(\omega(\sum_{k=1}^{2i}\theta_k) - \omega(\sum_{k=1}^{2i-1}\theta_k)) \\
&+ \hat{\sigma}_\rho(1)(\omega(t) - \omega(\tau_{\nu(t)})))\} \qquad \nu(t) \text{ is even}
\end{aligned}
\tag{8.3.6}
$$

$$
\begin{aligned}
\tilde{S}_t = S_0 exp\{&\sum_{i=1}^{\frac{\nu(t)+1}{2}} \hat{\rho}(1)\theta_{2i-1} + \sum_{i=1}^{\frac{\nu(t)-1}{2}} \hat{\rho}(0)\theta_{2i} + (t-\tau_{\nu(t)})\hat{\rho}(0) \\
&+ T^{-1/2}(\sum_{i=1}^{\frac{\nu(t)+1}{2}} \hat{\sigma}_\rho(1)(\omega(\sum_{k=1}^{2i-1}\theta_k) - \omega(\sum_{k=1}^{2i-2}\theta_k)) \\
&+ \sum_{i=1}^{\frac{\nu(t)-1}{2}} \hat{\sigma}_\rho(0)(\omega(\sum_{k=1}^{2i}\theta_k) - \omega(\sum_{k=1}^{2i-1}\theta_k)) \\
&+ \hat{\sigma}_\rho(0)(\omega(t) - \omega(\tau_{\nu(t)})))\} \qquad \nu(t) \text{ is odd}
\end{aligned}
\tag{8.3.7}
$$

$$
P = \begin{pmatrix} p_{00} & p_{01} & p_{02} & p_{03} \\ p_{10} & p_{11} & p_{12} & p_{13} \\ p_{20} & p_{21} & p_{22} & p_{23} \\ p_{30} & p_{31} & p_{32} & p_{33} \end{pmatrix} = \begin{pmatrix} 0.49 & 0.49 & 0.01 & 0.01 \\ 0.70 & 0.28 & 0.01 & 0.01 \\ 0.01 & 0.01 & 0.70 & 0.28 \\ 0.01 & 0.01 & 0.49 & 0.49 \end{pmatrix}
$$

Due to the ergodic properties of (x_n), stationary distribution in each class X_0, X_1:

$$
\begin{cases} \pi_0(x_0) = 0.5868, & x_0 \in X_0 \\ \pi_0(x_1) = 0.4132, & x_1 \in X_0 \end{cases} \qquad \begin{cases} \pi_1(x_2) = 0.6330, & x_2 \in X_1 \\ \pi_1(x_3) = 0.3670, & x_3 \in X_1 \end{cases}
$$

$$
\lambda(x_0) = 8,\ \lambda(x_1) = 10,\ \lambda(x_2) = 12,\ \lambda(x_3) = 10,\ S_0 = 10,\quad T = 10,
$$

$$
\rho(x_k) = \begin{cases} 0.03, & k = 0 \\ -0.02, & k = 1 \end{cases}, \qquad \rho(x_k) = \begin{cases} 0.02, & k = 2 \\ -0.02, & k = 3 \end{cases}
$$

$$
m(k) := \int_{X_k} \pi_k(dx) m(x) \tag{8.3.8}
$$

$$\begin{aligned}\hat{\rho}(1) &= \frac{\int_{X_1} p(dx)\int_{X_1} P(x,dy)\rho(y)}{m(1)} \\ &= \begin{pmatrix} p(x_2), & p(x_3) \end{pmatrix}\begin{pmatrix} p_{22} & p_{23} \\ p_{32} & p_{33} \end{pmatrix}\begin{pmatrix} \rho_2 \\ \rho_3 \end{pmatrix} / \begin{pmatrix} p(x_2), & p(x_3) \end{pmatrix}\begin{pmatrix} \bar{m}(x_2) \\ \bar{m}(x_3) \end{pmatrix} \\ &= 0.0594\end{aligned} \tag{8.3.9}$$

$$\begin{aligned}\hat{\rho}(0) &= \frac{\int_{X_0} p(dx)\int_{X_0} P(x,dy)\rho(y)}{m(0)} \\ &= \begin{pmatrix} p(x_0), & p(x_1) \end{pmatrix}\begin{pmatrix} p_{00} & p_{01} \\ p_{10} & p_{11} \end{pmatrix}\begin{pmatrix} \rho_0 \\ \rho_1 \end{pmatrix} / \begin{pmatrix} p(x_0), & p(x_1) \end{pmatrix}\begin{pmatrix} \bar{m}(x_0) \\ \bar{m}(x_1) \end{pmatrix} \\ &= 0.0806\end{aligned} \tag{8.3.10}$$

$$\begin{aligned}\tilde{\sigma}^2(1) &= \int_{X_1} \pi_1(dx)[(P\rho - m_1\hat{\rho}(1))R_0(P\rho - m_1\hat{\rho}(1)) + \frac{1}{2}(P\rho - m_1\hat{\rho}(1))^2]/m_1 \\ &= \int_{X_1} \pi_1(dx)[\sum_{n=0}^{\infty}(P^n - \Pi)(\int_{X_1} P(x,dy)\rho(y) - \int_{X_1}\pi_1(dx)\int_{X_1} P(x,dy)\rho(y))^2 \\ &+ \frac{1}{2}(\int_{X_1} P(x,dy)\rho(y) - \int_{X_1}\pi_1(dx)\int_{X_1} P(x,dy)\rho(y))^2]/m_1 \\ &= \int_{X_1} \pi_1(dx)[\sum_{n=0}^{\infty}\int_{X_1} P^n(x,dy)(\int_{X_1} P(y,dz)\rho(z) - \int_{X_1}\pi_1(dy)\int_{X_1} p(y,dz)\rho(z))^2 \\ &- \sum_{n=0}^{\infty}\int_{X_1} p(dy)(\int_{X_1} P(y,dz)\rho(z) - \int_{X_1}\pi_1(dy)\int_{X_1} P(y,dz)\rho(z))^2\mathbf{1}(y) \\ &+ \frac{1}{2}(\int_{X_1} P(x,dy)\rho(y) - \int_{X_1}\pi_1(dx)\int_{X_1} P(x.dy)\rho(y))^2]/m_1 \\ &= (\pi_1^2, \ \pi_1^3)[\sum_{n=0}^{\infty}\begin{pmatrix} p_{22}^n & p_{23}^n \\ p_{32}^n & p_{33}^n \end{pmatrix}\begin{pmatrix} (p_{22}\rho_2 + p_{23}\rho_3 - m_1\hat{\rho}(1))^2 \\ (p_{32}\rho_2 + p_{33}\rho_3 - m_1\hat{\rho}(1))^2 \end{pmatrix} \\ &- \sum_{n=0}^{\infty}\begin{pmatrix} p_2 & p_3 \\ p_2 & p_3 \end{pmatrix}\begin{pmatrix} (p_{22}\rho_2 + p_{23}\rho_3 - m_1\hat{\rho}(1))^2 \\ (p_{32}\rho_2 + p_{33}\rho_3 - m_1\hat{\rho}(1))^2 \end{pmatrix} \\ &+ \frac{1}{2}\begin{pmatrix} (p_{22}\rho_2 + p_{23}\rho_3 - m_1\hat{\rho}(1))^2 \\ (p_{32}\rho_2 + p_{33}\rho_3 - m_1\hat{\rho}(1))^2 \end{pmatrix}]/m_1 \\ &= 2.4 \times 10^{-3}\end{aligned} \tag{8.3.11}$$

$$
\begin{aligned}
\tilde{\sigma}^2(0) &= \int_{X_0} \pi_0(dx)[(P\rho - m_0\hat{\rho}(0))R_0(P\rho - m_0\hat{\rho}(0)) + \frac{1}{2}(P\rho - m_0\hat{\rho}(0))^2]/m_0 \\
&= \int_{X_0} \pi_0(dx)[\sum_{n=0}^{\infty}(P^n - \Pi)(\int_{X_0} P(x,dy)\rho(y) - \int_{X_0}\pi_0(dx)\int_{X_0} P(x,dy)\rho(y))^2 \\
&+ \frac{1}{2}(\int_{X_0} P(x,dy)\rho(y) - \int_{X_0}\pi_0(dx)\int_{X_0} P(x,dy)\rho(y))^2]/m_0 \\
&= \int_{X_0} \pi_0(dx)[\sum_{n=0}^{\infty}\int_{X_0} P^n(x,dy)(\int_{X_0} P(y,dz)\rho(z) - \int_{X_0}\pi_0(dy)\int_{X_0} p(y,dz)\rho(z))^2 \\
&- \sum_{n=0}^{\infty}\int_{X_0} p(dy)(\int_{X_0} P(y,dz)\rho(z) - \int_{X_0}\pi_0(dy)\int_{X_0} P(y,dz)\rho(z))^2\mathbf{1}(y) \\
&+ \frac{1}{2}(\int_{X_0} P(x,dy)\rho(y) - \int_{X_0}\pi_0(dx)\int_{X_0} P(x.dy)\rho(y))^2]/m_0 \\
&= (\pi_0^0,\ \pi_0^1)[\sum_{n=0}^{\infty}\begin{pmatrix} p_{00}^n & p_{01}^n \\ p_{10}^n & p_{11}^n \end{pmatrix}\begin{pmatrix} (p_{00}\rho_0 + p_{01}\rho_1 - m_0\hat{\rho}(0))^2 \\ (p_{10}\rho_0 + p_{11}\rho_1 - m_0\hat{\rho}(0))^2 \end{pmatrix} \\
&- \sum_{n=0}^{\infty}\begin{pmatrix} p_0 & p_1 \\ p_0 & p_1 \end{pmatrix}\begin{pmatrix} (p_{00}\rho_0 + p_{01}\rho_1 - m_0\hat{\rho}(0))^2 \\ (p_{10}\rho_0 + p_{11}\rho_1 - m_0\hat{\rho}(0))^2 \end{pmatrix} \\
&+ \frac{1}{2}\begin{pmatrix} (p_{00}\rho_0 + p_{01}\rho_1 - m_0\hat{\rho}(0))^2 \\ (p_{10}\rho_0 + p_{11}\rho_1 - m_0\hat{\rho}(0))^2 \end{pmatrix}]/m_0 \\
&= 3 \times 10^{-3}
\end{aligned} \tag{8.3.12}
$$

Let $\theta_i(i = 1,2,...) = 0.1$; given t, $\exists \nu(t).s.t.\ 0.1\nu(t) \leq t < 0.1(\nu(t)+1)\ \therefore \tau_{\nu(t)} = \theta_i \nu(t) = 0.1\nu(t)$
$\therefore$, when $\hat{x}(0) = 0$;

$$
\begin{aligned}
S_T(t) = 10 * exp\{ &\sum_{i=1}^{\frac{1}{2}\lfloor\frac{t}{0.1}\rfloor} 8.06 \times 10^{-2} * 0.1 \\
&+ \sum_{i=1}^{\frac{1}{2}\lfloor\frac{t}{0.1}\rfloor} 5.94 \times 10^{-2} * 0.1 + (t - 0.1\lfloor\frac{t}{0.1}\rfloor) * 8.06 \times 10^{-2} \\
&+ T^{-1/2}(\sum_{i=1}^{\frac{1}{2}\lfloor\frac{t}{0.1}\rfloor} 5.48 \times 10^{-2}(\omega(0.1(2i-1)) - \omega(0.1(2i-2))) \\
&+ \sum_{i=1}^{\frac{1}{2}\lfloor\frac{t}{0.1}\rfloor} 4.90 \times 10^{-2}(\omega(0.2i) \\
&- \omega(0.1(2i-1))) + 5.48 \times 10^{-2}(\omega(t) - \omega(0.1\lfloor\frac{t}{0.1}\rfloor)))\} \qquad \nu(t)\ is\ even
\end{aligned} \tag{8.3.13}
$$

$$S_T(t) = 10 * exp\{\sum_{i=1}^{\frac{1}{2}(\lfloor \frac{t}{0.1} \rfloor+1)} 8.06 \times 10^{-2} * 0.1$$
$$+ \sum_{i=1}^{\frac{1}{2}(\lfloor \frac{t}{0.1} \rfloor-1)} 5.94 \times 10^{-2} * 0.1 + (t - 0.1\lfloor \frac{t}{0.1} \rfloor) * 5.94$$
$$\times 10^{-3} + T^{-1/2}(\sum_{i=1}^{\frac{1}{2}(\lfloor \frac{t}{0.1} \rfloor+1)} 5.48 \times 10^{-2}(\omega(0.1(2i-1)) - \omega(0.1(2i-2)))$$
$$+ \sum_{i=1}^{\frac{1}{2}(\lfloor \frac{t}{0.1} \rfloor-1)} 4.90 \times 10^{-2}$$
$$(\omega(0.2i) - \omega(0.1(2i-1))) + 4.90 \times 10^{-2}(\omega(t) - \omega(0.1\lfloor \frac{t}{0.1} \rfloor)))\} \quad v(t) \text{ is odd} \tag{8.3.14}$$

When $\hat{x}(0) = 1$;

$$S_T(t) = 10 * exp\{\sum_{i=1}^{\frac{1}{2}\lfloor \frac{t}{0.1} \rfloor} 5.94 \times 10^{-2} * 0.1$$
$$+ \sum_{i=1}^{\frac{1}{2}\lfloor \frac{t}{0.1} \rfloor} 8.06 \times 10^{-2} * 0.1 + (t - 0.1\lfloor \frac{t}{0.1} \rfloor) * 5.94 \times 10^{-2}$$
$$+ T^{-1/2}(\sum_{i=1}^{\frac{1}{2}\lfloor \frac{t}{0.1} \rfloor} 4.90 \times 10^{-2}(\omega(0.1(2i-1)) - \omega(0.1(2i-2)))$$
$$+ \sum_{i=1}^{\frac{1}{2}\lfloor \frac{t}{0.1} \rfloor} 5.48 \times 10^{-2}(\omega(0.2i)$$
$$- \omega(0.1(2i-1))) + 4.90 \times 10^{-2}(\omega(t) - \omega(0.1\lfloor \frac{t}{0.1} \rfloor)))\} \quad v(t) \text{ is even} \tag{8.3.15}$$

$$S_T(t) = 10 * exp\{\sum_{i=1}^{\frac{1}{2}(\lfloor \frac{t}{0.1} \rfloor+1)} 5.94 \times 10^{-2} * 0.1 + \sum_{i=1}^{\frac{1}{2}(\lfloor \frac{t}{0.1} \rfloor+1)} 8.06 \times 10^{-2} * 0.1 + (t - 0.1\lfloor \frac{t}{0.1} \rfloor)$$
$$* 8.06 \times 10^{-2} + T^{-1/2}(\sum_{i=1}^{\frac{1}{2}(\lfloor \frac{t}{0.1} \rfloor+1)} 4.9 \times 10^{-2}(\omega(0.1(2i-1)) - \omega(0.1(2i-2)))$$
$$+ \sum_{i=1}^{\frac{1}{2}(\lfloor \frac{t}{0.1} \rfloor-1)} 5.48 \times 10^{-2}(\omega(0.2i) - \omega(0.1(2i-1)))$$
$$+ 5.48 \times 10^{-2}(\omega(t) - \omega(0.1\lfloor \frac{t}{0.1} \rfloor)))\} \quad v(t) \text{ is odd} \tag{8.3.16}$$

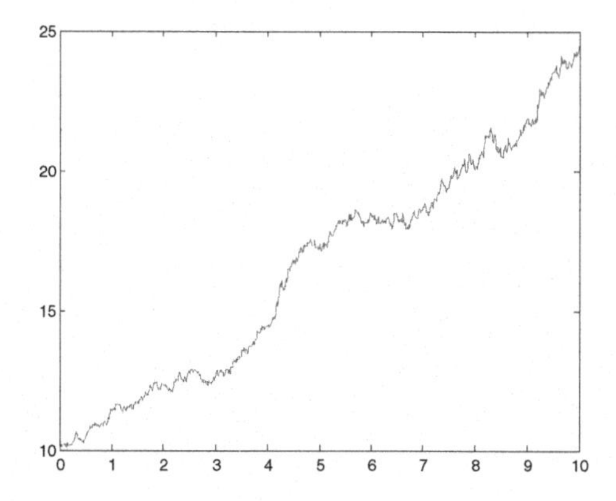

Figure 8.1 *Sample path of $\tilde{S}(t)$ w.r.t t in ergodic normal deviation.*

$\therefore$

$$E(\tilde{S}(t)) = \begin{cases} 10e^{(-1.1\times10^{-3}-\frac{1}{T}1.5\times10^{-5})\lfloor\frac{t}{0.1}\rfloor+(8.06\times10^{-2}+\frac{1}{T}1.5\times10^{-3})t} & when\ \hat{x}(t)=0, \\ 10e^{(1.1\times10^{-3}+\frac{1}{T}1.5\times10^{-5})\lfloor\frac{t}{0.1}\rfloor+(5.94\times10^{-2}+\frac{1}{T}1.2\times10^{-3})t} & when\ \hat{x}(t)=1, \end{cases} \tag{8.3.17}$$

$$Var(\tilde{S}(t)) = \begin{cases} 100e^{(2.2\times10^{-3}-\frac{1}{T}6\times10^{-5})\lfloor\frac{t}{0.1}\rfloor+(1.61\times10^{-1}+\frac{1}{T}6\times10^{-3})t} & \\ -100e^{(-2.2\times10^{-3}-\frac{1}{T}3\times10^{-5})\lfloor\frac{t}{0.1}\rfloor+(1.61\times10^{-1}+\frac{1}{T}3\times10^{-3})t} & when\ \hat{x}(t)=0; \\ 100e^{(2.2\times10^{-3}+\frac{1}{T}6\times10^{-5})\lfloor\frac{t}{0.1}\rfloor+(1.19\times10^{-1}+\frac{1}{T}4.8\times10^{-3})t} & \\ -100e^{(2.2\times10^{-3}+\frac{1}{T}3\times10^{-5})\lfloor\frac{t}{0.1}\rfloor+(1.19\times10^{-1}+\frac{1}{T}2.4\times10^{-3})t} & when\ \hat{x}(t)=1; \end{cases} \tag{8.3.18}$$

Under double averaging with stationary distribution $\hat{p}_k = \frac{1}{2},\ k=0,1$

$$\breve{\rho} = 0.5*(0.0806)+0.5*0.0594 = 0.07$$

$$\breve{\sigma}^2 = 0.5*2.4\times10^{-3}+0.5*3\times10^{-3} = 2.7\times10^{-3}$$

$$\therefore S_T(t) \simeq 10e^{0.07t+T^{-1/2}0.0520\omega(t)}$$

8.4 European call option pricing formula for normal deviated GMRP

8.4.1 Ergodic GMRP

As we have seen in previous sections, an ergodic normal deviated GMRP $S_T(t)$ satisfies the following SDE:

$$\frac{dS_T(t)}{S_T(t)} \simeq (\hat{\rho}+\frac{1}{2}T^{-1}\hat{\sigma}^2)dt+T^{-1/2}\hat{\sigma}d\hat{\omega}(t) \tag{8.4.1}$$

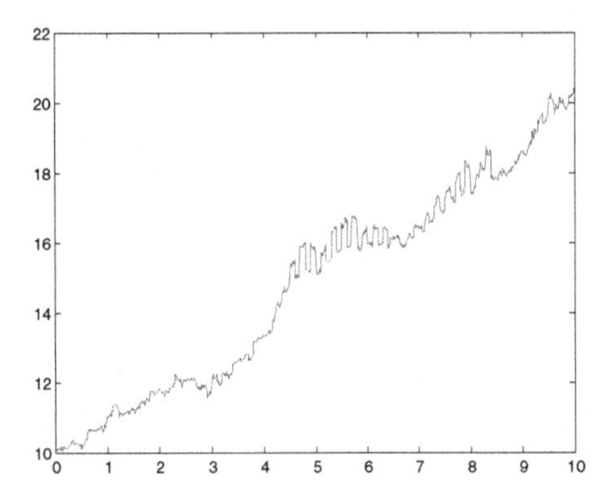

Figure 8.2 *Sample path of $S_T(t)$ w.r.t t in merged normal deviation when $\hat{x}(s) = 1$.*

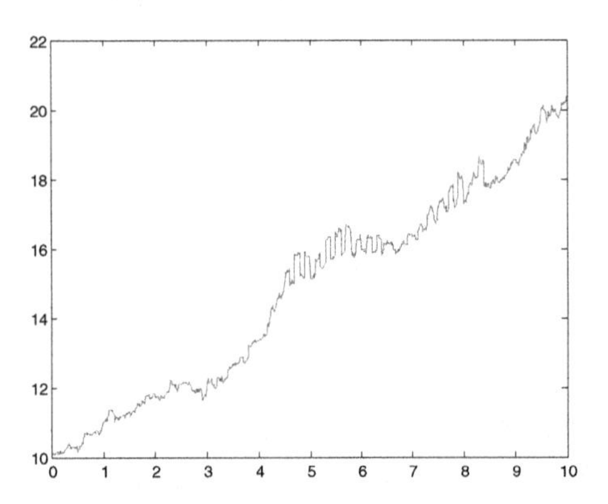

Figure 8.3 *Sample path of $S_T(t)$ w.r.t t in merged normal deviation when $\hat{x}(s) = 0$.*

where

$$\hat{\rho} = \frac{\int_X p(dx) \int_X P(x,dy)\rho(y)}{m}, \hat{\sigma}^2 := \int_X p(dx)[(P\rho - \hat{\rho})R_0(P\rho - \hat{\rho}) + 2^{-1}(P\rho - \hat{\rho})^2]/m. \tag{8.4.2}$$

The risk-neutral measure P^* for the process in (8.4.1) is:

$$\frac{dP^*}{dP} = exp\{-\theta\omega(t) - \frac{1}{2}\theta^2 t\}, \tag{8.4.3}$$

where

$$\theta = \frac{\hat{\rho} + \frac{1}{2}T^{-1}\hat{\sigma}^2 - r}{T^{-1/2}\sigma_\rho}. \tag{8.4.4}$$

is the market price of risk.

Under P^*, the process $e^{-rt}S_T(t)$ is a martingale and the process $\hat{\omega}^*(t) = \hat{\omega}(t) +$

θt is a Brownian motion. In this way, in the risk-neutral world, the process $S_T(t)$ has the following form

$$\frac{dS_T(t)}{S_T(t)} = rdt + T^{-1/2}\hat{\sigma}d\hat{\omega}^*(t). \tag{8.4.5}$$

Using Black-Scholes formula we obtain the European call option pricing formula for our model (8.4.5):

$$C = S_0\Phi(d_+) - Ke^{-rT}\Phi(d_-), \tag{8.4.6}$$

where

$$d_+ = \frac{ln(S_0/K) + (r + (1/2)T^{-1}\hat{\sigma}^2)T}{\hat{\sigma}}, d_- = \frac{ln(S_0/K) + (r - (1/2)T^{-1}\hat{\sigma}^2)T}{\hat{\sigma}},$$

$\Phi(x)$ is a normal distribution and $\hat{\sigma}$ is defined above, T is the time scale interval.

Remark (Hedging Strategies for GMRP in the Normal Deviation (ND) Scheme). The hedging strategies for GMRP in the ND scheme has the following form (see [2] for comparisons):

$$\begin{aligned}
\hat{\gamma}_t^T &= \Phi\left(\frac{\ln(S_T(t)/K) + (r + \hat{\sigma}^2 T^{-1})(T-t)}{\hat{\sigma}T^{-1/2}\sqrt{T-t}}\right) \\
\hat{\beta}_t^T &= -\frac{K}{B_0 e^{-rT}}\Phi\left(\frac{\ln(S_T(t)/K) + (r - \hat{\sigma}^2 T^{-1})(T-t)}{\hat{\sigma}T^{-1/2}\sqrt{T-t}}\right),
\end{aligned}$$

where $(\hat{\beta}_t^T, \hat{\gamma}_t^T)$ is an investor's portfolio, $\Phi(x)$ is a standard normal distribution. The capital $\hat{X}_t^T := S_T(t)\hat{\gamma}_t^T + B(t)\hat{\beta}_t^T$ has the form:

$$\hat{X}_t^T = S_T(t)\hat{\gamma}_t^T - Ke^{-r(T-t)}\hat{\beta}_t^T.$$

8.4.2 *Double averaged normal deviated GMRP*

Using arguments similar to those in (8.2.1)–(8.2.5), we can get European call option pricing formula for double averaged normal deviated GMRP in (8.2.23):

$$\frac{dS_T(t)}{S_T(t)} \simeq (\breve{\rho} + \frac{1}{2}T^{-1}\breve{\sigma}^2)dt + T^{-1/2}\breve{\sigma}d\omega(t). \tag{8.4.7}$$

where $\breve{\sigma}^2$ and $\breve{\rho}$ are defined in (8.2.18), and(8.2.21). Namely, the European call option pricing formula for a double averaged normal deviated GMRP is:

$$C = S_0\Phi(d_+) - Ke^{-rT}\Phi(d_-), \tag{8.4.8}$$

where

$$\begin{aligned}
d_+ &= \frac{ln(S_0/K) + (r + (1/2)T^{-1}\breve{\sigma}^2)T}{\breve{\sigma}}, \\
d_- &= \frac{ln(S_0/K) + (r - (1/2)T^{-1}\breve{\sigma}^2)T}{\breve{\sigma}},
\end{aligned} \tag{8.4.9}$$

$\Phi(x)$ is a normal distribution and $\breve{\sigma}$ is defined above; T is the time scale interval.

Remark (Hedging Strategies for GMRP in the Double Averaged Normal Deviation (ND) Scheme). The hedging strategies for GMRP in the double averaged ND scheme has the following form (see [2] for comparisons):

$$\begin{aligned}\check{\gamma}_t &= \Phi(\frac{\ln(S_T(t)/K)+(r+\check{\sigma}^2T^{-1})(T-t)}{\check{\sigma}T^{-1/2}\sqrt{T-t}})\\ \check{\beta}_t &= -\frac{K}{B_0e^{-rT}}\Phi(\frac{\ln(\check{S}(t)/K)+(r-\check{\sigma}^2T^{-1})(T-t)}{\check{\sigma}T^{-1/2}\sqrt{T-t}}),\end{aligned}$$

where $(\check{\beta}_t, \check{\gamma}_t)$ is an investor's portfolio, $\Phi(x)$ is a standard normal distribution. The capital $\check{X}_t := \check{S}(t)\check{\gamma}_t + B(t)\check{\beta}_t$ has the form:

$$\check{X}_t = \check{S}(t)\check{\gamma}_t - Ke^{-r(T-t)}\check{\beta}_t.$$

8.4.3 *Call option pricing for ergodic GMRP*

Using data from the example of ergodic Normal Deviated GMRP:

$$S_0 = 10, \quad K = 10, \qquad \hat{\sigma}^2 = 6.6\times 10^{-3}, \qquad r = 0.01, \quad T = 2$$

$$\begin{aligned}d_+ &= \frac{ln(S_0/K)+(r+0.5T^{-1}\hat{\sigma}^2)T_0}{T^{-1/2}\hat{\sigma}T_0^{1/2}} = \frac{(0.01+0.5*0.5*6.6\times 10^{-3})T_0}{5.74\times 10^{-2}T_0^{1/2}}\\ d_- &= \frac{ln(S_0/K)+(r-0.5T^{-1}\hat{\sigma}^2)T_0}{T^{-1/2}\hat{\sigma}T_0^{1/2}} = \frac{(0.01-0.5*0.5*6.6\times 10^{-3})T_0}{5.74\times 10^{-2}T_0^{1/2}}\end{aligned} \tag{8.4.10}$$

$$\begin{aligned}C &= 10\Phi(\frac{(0.01+0.5*0.5*6.6\times 10^{-3})T_0}{5.74\times 10^{-2}T_0^{1/2}})\\ &\quad -10e^{-0.01T_0}\Phi(\frac{(0.01-0.5*0.5*6.6\times 10^{-3})T_0}{5.74\times 10^{-2}T_0^{1/2}})\end{aligned} \tag{8.4.11}$$

When $T_0 = 1, \quad C = 0.2814$

8.4.4 *Call option pricing formulas for double averaged GMRP*

Let $\hat{p}_0 = \hat{p}_1 = \frac{1}{2}$ be the stationary distributions, $S_0 = 10,\ K = 10,\ T = 2,\ r = 0.01$
Since

$$\check{\sigma}_\rho^2 = \sum_{k=0}^{r} \hat{p}_k\hat{\sigma}_\rho^2(k) \qquad \check{\rho}_2 = \sum_{k=0}^{r} \hat{p}_k\hat{\rho}_2(k)$$

$$\begin{aligned}&\therefore \check{\sigma}^2 = 0.5*2.4\times 10^{-3} + 0.5*3\times 10^{-3} = 2.7\times 10^{-3}\\ &\check{\rho} = 0.5*(0.0806) + 0.5*0.0594 = 0.07\end{aligned} \tag{8.4.12}$$

$$\begin{aligned}d_+ &= \frac{ln(S_0/K)+(r+0.5T^{-1}\check{\sigma}^2)T_0}{T^{-1/2}\check{\sigma}T_0^{1/2}} = \frac{(0.01+6.75\times 10^{-4})T_0}{3.67\times 10^{-2}T_0^{1/2}}\\ d_- &= \frac{ln(S_0/K)+(r-0.5T^{-1}\check{\sigma}^2)T_0}{T^{-1/2}\check{\sigma}T_0^{1/2}} = \frac{(0.01-6.75\times 10^{-4})T_0}{3.67\times 10^{-2}T_0^{1/2}}\end{aligned} \tag{8.4.13}$$

When $T_0 = 0.5$,

$$C = 10\Phi(\frac{(0.01+6.75\times10^{-4})T_0}{3.67\times10^{-2}T_0^{1/2}}) - 10e^{-0.01}\Phi(\frac{(0.01-6.75\times10^{-4})T_0}{3.67\times10^{-2}T_0^{1/2}})$$
$$= 0.2011 \tag{8.4.14}$$

8.5 Martingale property of GMRP

Consider the geometric Markov renewal processes $(S_t)_{t\in\mathbb{R}_+}$

$$S_t = S_0 \prod_{k=1}^{\nu(t)} (1+\rho(x_k)). \tag{8.5.1}$$

For $t \in [0,T]$ let us define

$$L_t = L_0 \prod_{k=1}^{\nu(t)} h(x_k), \quad EL_0 = 1, \tag{8.5.2}$$

where h(x) is a bounded continuous function such that

$$\int_X h(y)P(x,dy) = 1; \quad \int_X h(y)P(x,dy)\rho(y) = 0. \tag{8.5.3}$$

If $EL_T = 1$, then geometric Markov renewal process S_t in (8.5.1) is an (F_t, P^*)-martingale, where measure P^* is defined as follows:

$$\frac{dP^*}{dP} = L_T; \tag{8.5.4}$$

and

$$F_t := \sigma(x(s); 0 \le s \le t). \tag{8.5.5}$$

In the discrete case we have

$$S_n = S_0 \prod_{k=1}^{n} (1+\rho(x_k)). \tag{8.5.6}$$

Let $L_n := L_0 \prod_{k=1}^{n} h(x_k)$, $EL_0 = 1$, where $h(x)$ is defined in (8.5.3). If $EL_N = 1$, then S_n is an (F_t, P^*)-martingale, where $\frac{dP^*}{dP} = L_N$, and $F_n := \sigma(x_k; 0 \le k \le n)$.

8.6 Option pricing formulas for stock price modelled by GMRP

In this section we consider the option pricing formula for European call options, which describes the dynamic of stock prices by GMRP in discrete and continuous time cases. Let $f(S_N) = (S_N - K)^+ := max(S_N - K, 0)$, and $S_N = S_0 \prod_{k=1}^{N}(1+\rho(x_k))$, where S_N is the stock price at time N, K is the strike price, and N is the maturity of option. Then the price $C_N(y)$ of a European call option in discrete time is:

$$C_N(y) = E^*[(1+r)^{-N} f(S_T)|F_0] = (1+r)^{-N} \int_X \cdots \int_X f(S_0 \prod_{i=1}^{N} (1+\rho(y_i))) P^*(y_{i-1}, dy_i), \tag{8.6.1}$$

where $y_0 = y$,and $P^*(x,A) = hP$ is the distribution of x_n with respect to P^*.

Let $f(S_T) = (S_T - K)^+$, and $S_T = S_0 \prod_{k=1}^{\nu(tT)} (1+\rho(x_k))$, where S_T is the stock price at time T and T is the maturity of the option. Then the price $C_T(y)$ of a European call option in continuous time is

$$\begin{aligned} C_T(y) &= E^*[e^{-rT} f(S_T)|F_0] = E^*[E^*[e^{-rT} f(S_T)|F_0]] \\ &= e^{-rT} \sum_{k=0}^{+infty} P(\nu(T) = k) \int_X \cdots \int_X f(S_0 \prod_{i=1}^{N} (1+\rho(y_i))) P^*(y_{i-1}, dy_i), \end{aligned} \tag{8.6.2}$$

where $P^*(x,A) = hP$ is a distribution of x_n with respect to P^*, $y_0 = y$.

Remark. In the case of Poisson process $\nu(t) \equiv N(t)$ we obtain from (8.6.2):

$$C_T(y) = e^{-rT} \sum_{k=0}^{+infty} \frac{e^{-\lambda T} (\lambda T)^k}{k!} \int_X \cdots \int_X f(S_0 \prod_{i=1}^{N} (1+\rho(y_i))) P^*(y_{i-1}, dy_i). \tag{8.6.3}$$

8.7 Examples of option pricing formulas modelled by GMRP

8.7.1 Example of two states in discrete time

The price $C_N(y)$ of a European call option in discrete time is:

$$\begin{aligned} C_N(y) &= E^*[(1+r)^{-N} f(S_T)|F_0] \\ &= (1+r)^{-N} [\sum_{i=1}^{M} \sum_{j=1}^{M} \cdots \sum_{k=1}^{M} (S_0 \prod_{i=1}^{N} (1+\rho(y_i)) - K)_+ p_{li} p_{ij} \cdots p_{jk}] \end{aligned} \tag{8.7.1}$$

Here M is the number of states in the Markov chain, p_{ij} is entry of transition probability matrix with respect to P^*.

$$P^* = \begin{pmatrix} p_{11} & p_{12} & p_{13} & p_{14} \\ p_{21} & p_{22} & p_{23} & p_{24} \\ p_{31} & p_{32} & p_{33} & p_{34} \\ p_{41} & p_{42} & p_{43} & p_{44} \end{pmatrix} = \begin{pmatrix} 0.25 & 0.25 & 0.25 & 0.25 \\ 0.1 & 0.1 & 0.5 & 0.3 \\ 0.3 & 0.1 & 0.2 & 0.4 \\ 0.15 & 0.25 & 0.4 & 0.2 \end{pmatrix} \quad \rho = \begin{pmatrix} 0.05 \\ -0.03 \\ 0.02 \\ -0.04 \end{pmatrix}$$

$S_0 = 100, \quad K = 100, \quad r = 0.01, \quad N = 3.$
Call Option in discrete time is $C_N(x_1) = 2.4804, C_N(x_2) = 2.3720, C_N(x_3) = 2.5712, C_N(x_4) = 2.4014$.

8.7.2 Generalized example in continuous time in Poisson case

In the case of Poisson process $v(t) \equiv N(t)$, the price $C_T(y)$ of European call option in continuous time is:

$$\begin{aligned} C_T(y) &= E^*[e^{-rT} f(S_T)|F_0] \\ &= e^{-rT} \sum_{k=0}^{+\infty} \frac{e^{-\lambda T}(\lambda T)^k}{k!} \sum_{i=1}^{k} \sum_{j=1}^{k} \cdots \sum_{q=1}^{k} (S_0 \prod_{i=1}^{k}(1+\rho(y_i)) - K)_+ p_{ij} p_{jk} \cdots p_{pq}. \end{aligned} \tag{8.7.2}$$

Let

$$P^* = \begin{pmatrix} p_{11} & p_{12} & p_{13} & p_{14} \\ p_{21} & p_{22} & p_{23} & p_{24} \\ p_{31} & p_{32} & p_{33} & p_{34} \\ p_{41} & p_{42} & p_{43} & p_{44} \end{pmatrix} = \begin{pmatrix} 0.25 & 0.25 & 0.25 & 0.25 \\ 0.1 & 0.1 & 0.5 & 0.3 \\ 0.3 & 0.1 & 0.2 & 0.4 \\ 0.15 & 0.25 & 0.4 & 0.2 \end{pmatrix} \quad \rho = \begin{pmatrix} 0.05 \\ -0.03 \\ 0.02 \\ -0.04 \end{pmatrix}$$

p_{ij} is entry of transition probability matrix with respect to P^*. $\lambda = 2$, $T = 1$, $r = 0.01$, $S_0 = K = 100$.

Calculating $C_T(y)$ in averaging scheme,$\pi(i)$ is the stationary distribution of embedded Markov chain x_i.

If at the maturity time T, counting process $N(t) \leq 5$:

$$C_T(x_1) = 1.8563$$

If at the maturity time T, counting process $N(t) \leq 10$:

$$C_T(x_1) = 1.9160$$

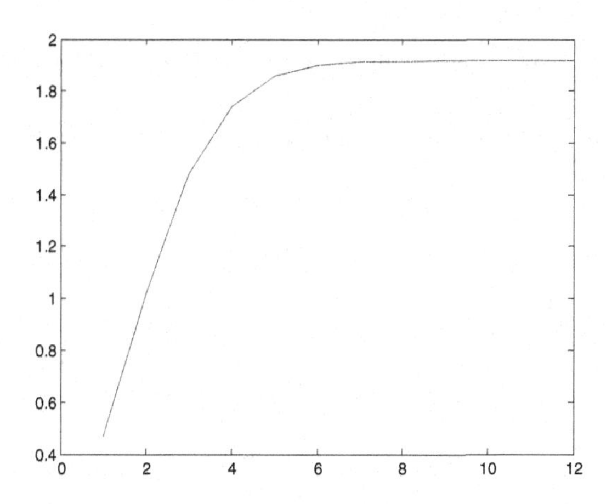

Figure 8.4 *Curves approaching option price when $N(t) \to \infty$.*

If at the maturity time T, counting process $N(t) \leq 11$:

$$C_T(x_1) = 1.9161$$

If at the maturity time T, counting process $N(t) \leq 12$:

$$C_T(x_1) = 1.9161$$

$\therefore$, call option when maturity is T=1: $C_T(x_1) = 1.9161$.

References

[1] Aase, K., *Contingent claims valuation when the securities price is a combination of an Ito process and a random point process*, Stoch. Proc. & Their Applic., 28, 185-220, 1988.

[2] Cox, J.C., Ross, S.A., and Rubinstein, M., *Option pricing: a simplified approach*, Journal of Financial Economics, 7 229-264, 1979.

[3] Korolyuk, V.S. and Swishchuk, A.V., *Evolution of systems in random media*, CRC Press, Boca Raton, USA, 1995

[4] Korolyuk, V.S. and Swishchuk, A.V., *Evolutionary stochastic systems. Algorithms of averaging and diffusion approximation,* Institute of Mathematics Ukrain. Acad. of Sciences, Kiev (In Russian), 2000.

[5] Korolyuk, V.S. and Limnios, N., *Poisson approximation of stochastic systems*, Theory Probab. and Mathem. Statistics, Kiev University, N62, 2002.

[6] Korolyuk, V.S., and Limnios, N., *Stochastic processes in phase merging space*, World Scientific, 2005.

[7] Skorokhod, A., *Studies in the theory of random processes (English translation),*. Ann. Math. Stat., 38, 1, 1967.

[8] Swishchuk, A.V., *Random evolutions and their applications: new trends,* Kluwer AP, Dordrecht, 2000.

[9] Swishchuk, A.V., *Random evolutions and their applications,* Kluwer AP, Dordrecht, 1997.

[10] Swishchuk, A.V. and Islam, M.S., *The geometric Markov renewal processes with application to finance*, Stochastic Analysis and Application, 29, 2, 2010.

[11] Swishchuk, A.V. and Islam, M.S., *Diffusion Approximations of the geometric Markov renewal processes and option price formulas*, International Journal of Stochastic Analysis, 2010, Article ID 347105, 21 pages, doi:10.1155/2010/347105.

Chapter 9

Poisson Approximation of a Security Market by the Geometric Markov Renewal Processes

9.1 Chapter overview

In this chapter, we consider the geometric Markov renewal processes as a model for a security market. We introduce a Poisson averaging scheme for the geometric Markov renewal processes. Financial applications in a Poisson approximation scheme of the geometric Markov renewal processes are presented.

9.2 Averaging in Poisson scheme

In this section we consider averaging of GMRP in Poisson scheme. In the limit we obtain compound Poisson process with deterministic drift.

Let $\rho_k^T(x,\omega) \equiv \rho_k^T(x)$ be a sequence of random variables for all $x \in X$ and for all $T > 0$. Let us consider the process S_t^T in a series scheme:

$$S_t^T = S_0 \prod_{k=1}^{\nu(tT)} (1+\rho_k^T(x_k,\omega)) = S_0 \prod_{k=1}^{\nu(tT)} (1+T^{-1}\rho_k(x_k)). \tag{9.2.1}$$

We note that this scheme is more general than previous, because of the random variables $\rho_k^T(x,\omega)$. From (9.2.1) it follows that

$$\ln \frac{S_t^T}{S_0} = \sum_{k=1}^{\nu(tT)} \ln(1+\rho_k^T(x_k)) \equiv T^{-1} \sum_{k=1}^{\nu(tT)} \rho_k(x_k), \tag{9.2.2}$$

for large T. Here, $(x_k;\tau_k)_{k\in Z_+}$ is a Markov renewal process, $(x_k)_{k\in Z_+}$ is an embedded Markov chain, $(\tau_k)_{k\in Z_+}$ are moments of jumps, $\mathbb{P}(\theta_{k+1} \le t | x_k = x) = 1 - e^{-q(x)t}$, $\theta_k := \tau_{k+1} - \tau_k$, and $P(x,A) := \mathbb{P}(x_{k+1} \in A | x_k = x)$. We note that $x(t) := x_{\nu(t)}$ is a Markov process with transition kernel

$$Q(x,A,t) = P(x,A)(1-e^{-q(x)t}), \tag{9.2.3}$$

where $\nu(t)$ is a counting process. We suppose that $x(t)$ is an ergodic Markov process with a stationary distribution $\pi(dx)$. Then $(x_n)_{n\in Z_+}$ is an ergodic Markov chain with stationary distribution $p(dx)$:

$$\pi(dx)q(x) = qp(dx), \qquad q := \int_X \pi(dx)q(x). \tag{9.2.4}$$

Let

$$F_x^T(z) := \mathcal{P}(\rho_k^T(x,\omega) \leq z),$$

for all $z \in R$. We also suppose that for a fixed sequence x_k, the sequence $\rho_k^T(x_k)$ is an independent random variable for $k \in Z_+$. Using the result by Korolyuk and Limnios [5] we obtain that right-hand side of (9.2.2) converges weakly to compound Poisson process $P(t) := \sum_{k=1}^{N_0(t)} \alpha_k^0 + a_0 qt$ with deterministic drift:

$$\sum_{k=1}^{\nu(tT)} \rho_k^T(x_k) \simeq P(t) := \sum_{k=1}^{N_0(t)} \alpha_k^0 + a_0 qt, \tag{9.2.5}$$

where α_k^0 are i.i.d. random variables with distribution function $F^0(z)$:

$$Eg(\alpha_k^0) = \int_R g(z)F^0(dz) = \hat{F}(g)/\hat{F}(1), \tag{9.2.6}$$

$$\hat{F}(q) := \int_X p(dx)F_x(g), \qquad \hat{F}(1) := \int_X p(dx)F_x(R).$$

Here $F_x(g)$ is such that $\int_X g(z)F_x(dz) = T^{-1}[F_x(g) + o_T(x,g)]$, where $\sup_x |F_x(g)| \leq +\infty$, $g(z)$ is such that $g(z)/z^2 \to_{|z|\to 0} 0$, and $F_x(g) = \int_R g(z)F_x(dz)$. The above compound Poisson process $N_0(t)$ is defined by intensity q_0:

$$q_0 := q\hat{F}(1).$$

The above value a_0 is defined as

$$a_0 := \hat{a} - \hat{F}(1)E\alpha_1^0,$$

where $\hat{a} := \int_X p(dx)a(x)$ and $a(x)$ is defined asymptotically:

$$\int_R zF_x^T(dz) = T^{-1}[a(x) + o_T(x)]$$

with $\sup_x |a(x)| \leq a < +\infty$ and $0_T(x) \to_{T\to+\infty} o$. Let $S_P(t) := \lim_{T\to+\infty} S_t^T$. Then,

$$\ln \frac{S_t^T}{S_0} \to_{T\to+\infty} \ln \frac{S_P(t)}{S_0} = P(t) = \sum_{k=1}^{N_0(t)} \alpha_k^0 + a_0 qt,$$

or,

$$S_P(t) = S_0 e^{\sum_{k=1}^{N_0(t)} \alpha_k^0 + a_0 qt}.$$

We note that

$$\sum_{k=1}^{N_0(t)} \alpha_k^0 = \int_0^t \int_0^{+\infty} y\mu(dy;ds),$$

where μ is a measure of jumps of the process $N_0(t)$. In the above we have proved the following corollary.

Corollary 9.1 *The Poisson GMRP has the form:*

$$\ln \frac{S_P(t)}{S_0} = \int_0^t \int_0^{+\infty} y\mu(dy;ds) + a_0 qt,$$

or, in the form of SDE

$$\frac{dS_P(t)}{S_P(t)} = \int_0^{+\infty} y\mu(dy;ds) + a_0 q dt. \tag{9.2.7}$$

Poisson GMRP is the solution of the SDE in (9.2.7). It means that the dynamic of stock price in Poisson scheme is described by Poisson GMRP and it is the Merton model.

9.3 Option pricing formula under Poisson scheme

In this section we consider option pricing formula for European call options, which is described by dynamic of stock prices as GMRP in discrete and continuous time cases.

Let $f(S_N) = (S_N - K)^*$, and $S_N = S_0 \prod_{k=1}^{N}(1+\rho(x_k))$.
Then

$$\begin{aligned} C_N(y) &= E^*[(1+r)^{-N} f(S_N)|\mathcal{F}_0] \\ &= (1+r)^{-N} \int_X \dots \int_X f(S_0 \prod_{i=1}^{N}(1+\rho(y_i)))P^*(y_{i-1}, dy_i), \end{aligned} \tag{9.3.1}$$

where $y_0 = y$, and $P^*(x,A) = hP$ is the distribution of x_n with respect to P^*.

Let $f(S_T) = (S_T - K)^+$, and $S_T = S_0 \prod_{k=1}^{\nu(tT)}(1+\rho(x_k))$.
Then

$$C_T(y) = E^*[(1+r)^{-N} f(S_T)|\mathcal{F}_0] = E^*[E^*[(1+r)^{-N} f(S_T)|\nu(T)|\mathcal{F}_0]] =$$

$$= (1+r)^{-N} \sum_{k=0}^{+\infty} \mathcal{P}(\nu(T)=k) \int_X \dots \int_X f(S_0 \prod_{i=1}^{k}(1+\rho(y_i))P^*(y_{i-1}, dy_i)), \tag{9.3.2}$$

where $P^*(x,A) = hP$ is a distribution of x_n with respect to P^*, h is defined in (6.4.21), $y_0 = y$.

In the case of Poisson process $\nu(t) \equiv N(t)$ we obtain from (9.3.2):

$$C_T(y) = (1+r)^{-N} \sum_{k=0}^{+\infty} \frac{e^{-\lambda T}(\lambda T)^k}{k!} \int_X \cdots \int_X f(S_0 \prod_{i=1}^{k}(1+\rho(y_i)))P^*(y_{i-1}, dy_i).$$

9.4 Application of Poisson approximation with a finite number of jump values

Corollary 1

Let the increment process $\sum_{k=1}^{\nu(tT)} \rho_k^T(x_k)$ with a finite number of jump values

$$P(\rho_k^{1/T}(x) = \frac{1}{T}a_m) = q_m - \frac{1}{T}p_m(x), \ 1 \le m \le M, P(\rho_k^x = b_m) = \frac{1}{T}p_m(x), 1 \le m \le M, \tag{9.4.1}$$

with the additional obvious relation

$$\sum_{m=1}^{M} q_m = 1 \tag{9.4.2}$$

Then the increment process $\sum_{k=1}^{\nu(tT)} \rho_k^T(x_k)$ and its stochastic exponential converges weakly to the Poisson process $P(t) := \sum_{k=1}^{N_0(t)} \alpha_k^0 + a_0 q t$ determined by the distribution function of the big jumps

$$\begin{aligned} &P(\alpha_k^0 = b_m) = p_m^0, \quad 1 \le m \le M, \qquad p_m^0 = \frac{\hat{p}_m}{\hat{p}}, \\ &\hat{p}_m = \int_E \rho(dx) p_m(x), \quad 1 \le m \le M, \qquad \hat{p} = \sum_{m=1}^{M} \hat{p}_m. \end{aligned} \tag{9.4.3}$$

The intensity of the counting Poisson process $\nu_0(t), t \ge 0$, is defined by

$$q_0 := q\hat{p}, \tag{9.4.4}$$

and the drift parameter a_0 is given by

$$a_0 = \sum_{m=1}^{M} a_m q_m$$

The intensity of jumps q_0 is determined by the averaged initial probability $\hat{p}$ and the averaged intensity q of the switched Markov process.

9.4.1 Applications in finance

Assume that the process $A = (A_t)_{0 \le t \le T}$ is a special case of Levy jump-diffusion, i.e., a Brownian motion plus a compensated compound Poisson process. The paths of this process can be described by

$$L_t = bt + \sigma W_t + (\sum_{k=1}^{N_t} J_k - t\lambda\kappa) \tag{9.4.5}$$

where $b \in R$, $\sigma \in R_{\geq 0}$, $W = (W_t)_{0\leq t\leq T}$ is a standard Brownian motion, $N = (N_t)_{0\leq t\leq T}$ is a Poisson process with parameter λ (i.e., $E[N_t] = \lambda t$) and $J = (J_k)_{k\geq 1}$ is an i.i.d sequence of random variables with probability distribution F and $E[J] = \kappa < \infty$. Hence, F describes the distribution of the jumps, which arrive according to the Poisson process. All sources of randomness are mutually independent.

The triplet (b, c, ν) is called the Lévy or Characteristic triplet and the exponent in

$$E[e^{iux}] = exp[ibu - \frac{u^2 c}{2} + \int_R (e^{iux} - 1 - iux1_{\{|x|<1\}})\nu(dx)].$$

here the random variable L_t of the Levy jump-diffusion is infinitely divisible with Levy triplet $b = b\cdot t, c = \sigma^2 \cdot t$ and $\nu = (\lambda F)\cdot t$.

9.4.1.1 Risk neutral measure

Under the risk neutral measure, denoted by $\bar{P}$, we model the asset price process as an exponential Levy process

$$S_t = S_0 \exp L_t$$

where the Levy process L has the triplet $(\bar{b}, \bar{c}, \bar{\nu})$.

The process L has the canonical decomposition

$$L_t = \bar{b}t + \sqrt{\bar{c}}\bar{W}_t + \int_0^t \int_R x(\nu^L - \bar{\nu}^L)(ds, dx) \tag{9.4.6}$$

where $\bar{W}$ is a $\bar{P}$-Brownian motion and $\bar{\nu}^L$ is the $\bar{P}$-compensator of the jump measure μ^L.

Because we have assumed that $\bar{P}$ is a risk neutral measure, the asset price has a mean rate of return $\mu \triangleq r - \delta$ and the discounted and re-invested process $(e^{(r-\delta)t} S_t)_{0\leq t\leq T}$, is a martingale under $\bar{P}$. Here $r \geq 0$ is the risk-free interest rate, $\delta \geq 0$ the continuous dividend yield of the asset. Therefore, the drift term $\bar{b}$ takes the form

$$\bar{b} = r - \delta - \frac{\bar{c}}{2} - \int_R (e^x - 1 - x)\bar{\nu}(dx). \tag{9.4.7}$$

In the example $L_t = \sum_{k=1}^{N_0(t)} \alpha_k^0 + a_0 qt$, which is a compound Poisson process with deterministic drift, the canonical decomposition

$$\begin{aligned} L_t &= \bar{b}t + \int_0^t \int_R x * u^L(ds, dx) - E(\sum_{k=1}^{N_0(t)} \alpha_k^0) \\ &= \bar{b}t + \int_0^t \int_R x * u^L(ds, dx) - \int_0^t \int_0^\infty x * \lambda\bar{F}(dx)(ds) \end{aligned} \tag{9.4.8}$$

here $\bar{F}(dx)$ is the distribution of jumps under risk neutral measure, so $\bar{\nu}^L(dt,dx) = \lambda * \bar{F}(dx)(dt)$. And the drift term $\bar{b}$ takes the form

$$\bar{b} = r - \int_R (e^x - 1 - x)\bar{\nu}(dx); \tag{9.4.9}$$

here $\bar{\nu}(dx) = \lambda \bar{F}(dx)$.

9.4.1.2 *On market incompleteness*

Assume that the price process of a financial asset is modeled as an exponential Levy process under both the real and the risk neutral measure. Assume that these measures, denoted P and $\bar{P}$, are equivalent and denote the triplet of the Levy process under P and $\bar{P}$ by (b,c,ν) and $(\bar{b},\bar{c},\bar{\nu})$, respectively.

Now, applying Girsanov's theorem we get that these triplets are related via $\bar{c} = c, \bar{\nu} = Y \cdot \nu$ and

$$\bar{b} = b + c\beta + x(Y-1) * \nu, \tag{9.4.10}$$

where (β, Y) is the tuple of functions related to the density process. On the other hand, from the martingale condition we get that

$$\bar{b} = r - \frac{\bar{c}}{2} - (e^x - 1 - x) * \bar{\nu}. \tag{9.4.11}$$

Equating and using $\bar{c} = c, \bar{\nu} = Y \cdot \nu$, we have that

$$\begin{aligned} &0 = b + c\beta + x(Y-1) * \nu - r + \frac{\bar{c}}{2} + (e^x - 1 - x) * \bar{\nu} \\ \Leftrightarrow &0 = b - r + c(\beta + \frac{1}{2}) + ((e^x - 1)Y - x) * \nu. \end{aligned} \tag{9.4.12}$$

Therefore, we have one equation but two unknown parameters, β and Y stemming from the change of measure. Every solution tuple (β, Y) of the above equation corresponds to a different *equivalent martingale measure*, which explains why the market is not complete. The tuple (β, Y) could also be termed the tuple of "*market price of risk.*"

Example (Compound Poisson model). Let us consider the compounded Poisson model, where the driving motion is a compound Poisson process with intensity $\lambda > 0$ and jump size α_k, i.e. $L_t = bt + \sum_{k=1}^{N_t} \alpha_k$ and $\nu(dx) = \lambda F(dx)$. Then, the equation (9.4.10) has a unique solution for Y, which is

$$\begin{aligned} &0 = b - r + ((e^x - 1)Y - x) * \lambda F(dx) \\ \Leftrightarrow &0 = b - r + ((e^{\alpha_1} - 1)Y - \alpha_1)\lambda p_1 + \dots ((e^{\alpha_N} - N)Y - \alpha_N)\lambda p_N \\ \Leftrightarrow &Y = \frac{r - b + (\alpha_1 p_1 + \dots \alpha_N p_N)\lambda}{((e^{\alpha_1} - 1)p_1 + \dots (e^{\alpha_N} - 1)p_N)\lambda}. \end{aligned} \tag{9.4.13}$$

Therefore, the martingale measure is unique and the market is complete. We could call the quantity Y in (9.4.13) the market price of jump risk.

Moreover, we can also check that plugging (9.4.13) into (9.4.12), we recover the martingale condition (9.4.11); indeed, we have that

$$\begin{aligned}\bar{b} &= b + \lambda(Y-1)\sum_{i=1}^{N} p_i\alpha_i \\ &= b + \lambda(Y-1)\sum_{i=1}^{N} p_i\alpha_i + Y\lambda\sum_{i=1}^{N}(e^{\alpha_i}-1)p_i \\ &= r - \bar{\lambda}\sum_{i=1}^{N}(e^{\alpha_i}-1-\alpha_i)p_i,\end{aligned} \tag{9.4.14}$$

where we have used (9.4.13) and that $\bar{\nu} = Y\cdot\nu$, which is the current framework translates to $\bar{\lambda} = Y\lambda$.

9.4.2 Example

For the given increment process

$$A_\varepsilon(t) := \sum_{k=1}^{\nu(t/\varepsilon)} \alpha_k^\varepsilon(x_k), \quad t \geq 0,$$

with n states embedded Markov chain $(x_k, k \geq 0)$.

$M = 5, n = 5$

we have matrix

$$\{\alpha_k^\varepsilon(x)\}_{n\times M}\{a_m\}_{1\times M} = [0.0002, 0.0001, 0.0005, 0.001, 0.0005],$$

$$\{b_m\}_{1\times M} = [0.001, 0.0015, -0.001, 0.0005, -0.0005],$$

$$\{q_m\}_{1\times M} = [0.1, 0.3, 0.2, 0.3, 0.1],$$

and $\{p_m(x)\}_{n\times M}$ with random entries from 0 to 1.

$\pi(B)$, stationary distribution of Markov process $x(t)$, is given by [1/6,1/6,2/9,2/9,2/9]; and $\rho(B)$, stationary distribution of the embedded Markov chain $x_k, k \geq 0$ is given by the relation:.$\pi(dx)q(x) = qp(dx), \quad q := \int_X \pi(dx)q(x)$.

The intensity of sojourn times the point process of jump times is given by $\{q(x)\}_{1\times n} = [50, 50, 55, 55, 55]$.

$\therefore$

$$A_0 = \sum_{k=1}^{N_0(t)} \alpha_k^0 + a_0 qt$$

.

Applying the canonical decomposition under risk-neutral measure, this compound Poisson model is

$$
\begin{aligned}
L_t &= \bar{b}t + \int_0^t \int_R x(\mu^L - \bar{\nu}^L)(ds,dx) \\
&= \bar{b}t + \int_0^t \int_R x\mu^L(ds,dx) - \bar{\lambda}tE[\alpha_k^0]
\end{aligned}
\tag{9.4.15}
$$

$$
\bar{b} = r - \int_R (e^x - 1 - x)\bar{\nu}(dx) \tag{9.4.16}
$$

Here,$\bar{\nu}^L = Y * \nu^L = Y * \lambda tF(dx), \quad \bar{\nu} = Y * \nu = Y * \lambda * F(dx).$

Plugging parameters λ, r, b and the distribution of α_k^0 into (9.4.13),when the distribution of jumps are given, we have that a sample for Y: $Y = 1.0772$

Therefore, the drift term $\bar{b} = r - Y * \int_R \lambda(e^x - 1 - x)F(dx) = 0.0229,$ the compensator $\int_0^t \int_R x\bar{\nu}^L(ds,dx) = Y * \lambda t \int_R xF(dx) = 0.0455t$
under risk-neutral measure,

$$
L_t = 0.0229t - 0.0455t + \sum_{i=1}^{N_t} \alpha_i^0 = -0.0156t + \sum_{i=1}^{N_t} \alpha_i^0 \tag{9.4.17}
$$

Simulation Assume that we want to simulate the Levy jump-diffusion

$$
L_t = bt + \sum_{k=1}^{N_t} J_k
$$

where $N_t \sim \text{Poisson}(\lambda t)$ and $J \sim F(dx)$.

We can simulate a discretized trajectory of the Levy jump L at fixed points $t_1, \ldots, t_n$ as follows:

- generate a Poisson random variate N with parameter λT;
- generate N random variates τ_k uniformly distributed in [0,T]; these variates correspond to the jump times;
- simulate the law of jump size J, i.e., simulate random variates J_k with law $F(dx)$.

Method for pricing options is to use a Monte Carlo simulation. The main advantage of this method is that complex and exotic derivatives can be treated easily — which is very important in applications, since little is known about functionals of Levy processes. Moreover, options on several assets can also be handled easily using Monte Carlo simulations. The main drawback of Monte Carlo methods is the slow computational speed.

We briefly sketch the pricing of a European call option on a Levy driven asset. The payoff of the call option with strike K at the time of maturity T is $g(S_T) =$

$(S_T - K)^+$ and the price is provided by the discounted expected payoff under a risk-neutral measure, i.e.,

$$C_T(S,K) = e^{-rT} E[(S_T - K)^+].$$

The crux of pricing European options with Monte Carlo methods is to simulate the terminal value of asset price $S_T = S_0 \exp L_T$. Let S_{T_k} for $k = 1, \ldots, N$ denote the simulated values; then, the option price $C_T(S,K)$ is estimated by the average of the prices for the simulated asset values, that is

$$\hat{C}_T(S,K) = e^{-rT} \frac{1}{N} \sum_{k=1}^{N} (S_{T_k} - K)^+,$$

and by the Law of Large Numbers we have that

$$\hat{C}_T(S,K) \to C_T(S,K) \quad as \quad N \to \infty$$

When the sample $N = 5000, S_0 = K = 100$, interest rate $r = 0.01$, the European call option price for this Levy-jump is

$$\hat{C}_T(S,K) = 2.6022$$

References

[1] Aase, K., *Contingent claims valuation when the securities price is a combination of an Ito process and a random point process*, Stoch. Proc. & Their Applic., 28, 185-220, 1988.

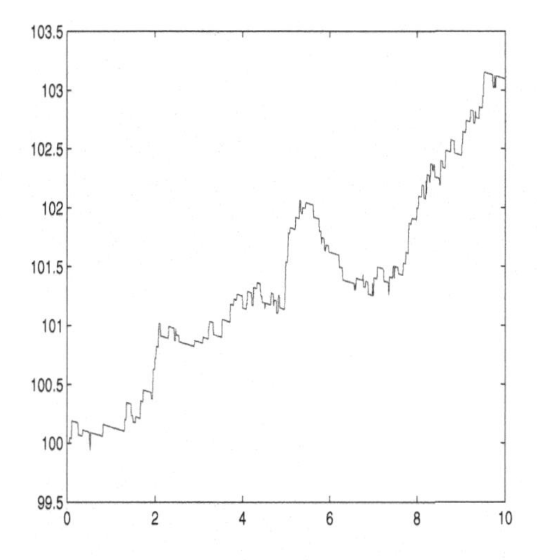

Figure 9.1 *Sample path of S_t w.r.t t in Poisson scheme under risk-neutral measure.*

[2] Cox, J.C., Ross, S.A., and Rubinstein, M., *Option pricing: a simplified approach*, Journal of Financial Economics, 7 229-264, 1979.

[3] Korolyuk, V.S. and Limnios, N., *Poisson approximation of stochastic systems*, Theory Probab. and Mathem. Statistics, Kiev University, N62, 2002.

[4] Korolyuk, V.S., and Limnios, N., *Stochastic processes in phase merging space*, World Scientific, 2005.

[5] Korolyuk, V.S. and Swishchuk, A.V., *Evolution of systems in random media*, CRC Press, Boca Raton, USA, 1995.

[6] Korolyuk, V.S. and Swishchuk, A.V., *Evolutionary stochastic systems. Algorithms of averaging and diffusion approximation,* Institute of Mathematics Ukrain. Acad. of Sciences, Kiev (In Russian), 2000.

[7] Swishchuk, A.V., *Random evolutions and their applications,* Kluwer AP, Dordrecht, 1997.

[8] Swishchuk, A.V., *Random evolutions and their applications: new trends,* Kluwer AP, Dordrecht, 2000.

[9] Swishchuk, A.V. and Islam, M.S., *The geometric Markov renewal processes with application to finance*, Stochastic Analysis and Application, 2010.

[10] Swishchuk, A.V. and Islam, M.S., *Diffusion approximations of the geometric Markov renewal processes and option price formulas*, International Journal of Stochastic Analysis, Volume 2010, Article ID 347105, 21 pages, doi:10.1155/2010/347105.

Chapter 10

Stochastic Stability of Fractional RDS in Finance

10.1 Chapter overview

The stability of the zero state of the stock price or the capital is useful from a financial point of view. If we know the conditions under which the stock price $S(t)$ (or the capital) tends to zero ($S(t) \to 0$), when $t \to +\infty$, (these are stability conditions), then we shall also know the conditions under which the stock price (or the capital) tends to infinity ($S(t) \to +\infty$), when $t \to +\infty$ (these are instability conditions). This is a desirable situation for an investor.

From the other side, some counterexamples were presented in [6, 7] to show that an uncritical application of the usual methods of continuous-time portfolio optimization can be misleading in the case of a stochastic opportunity set. This remark is valid for both the classical stochastic control approach (Merton, for example) and so-called martingale approach. They consider portfolio problems with stochastic inrerest rates, stochastic volatility, and stochastic market price of risk, and show that these portfolio problems *can be unstable*. For example, rates explode and expected rollover returns are infinite even if the rollover period is arbitrarily short (see References [13, 14]) (giving rise to infinite expectations under the risk-adjusted probability measure). As a sequence, such models cannot price, for example, one of the most widely used hedging instruments on the Euromoney market, namely the Eurodollar futures contract. If instead one models the effective annual rate these problems disappear.

We consider in this chapter the stochastic stability of fractional (B,S)-markets, that is, financial markets with a stochastic behavior that is caused by a random process with long-range dependence, fractional Brownian motion. Three financial models are considered. They arose as a result of different approaches to the definition of the stochastic integral with respect to fractional Brownian motion. The stochastic stability of fractional Brownian markets with jumps is also considered. In Section 10.6, Appendix, we give some definitions of stability, Lyapunov indices, and some results on rates of convergence of fractional Brownian motion, which we use in our development of stochastic stability.

As we already mentioned, we study here the stochastic stability of stochastic models in finance driven by fractional Brownian motion (fBm), not pricing results. Of course, the model with only fractional Brownian motion is not arbitrage-free. However, we could add a Brownian motion to the fBm noise and make it arbitrage-free (see, e.g., [10]; this book contains applications of fBm in finance for so-called mixed Brownian fractional Brownian models, Chapter 5) and study pricing results as well. The results on stochastic stability of models with Brownian models are well-known, but there are no results on stochastic stability of models with fBm.

10.2 Fractional Brownian motion as an integrator

Fractional Brownian motion, $B_t^H, t \geq 0)$, with a Hurst index $H \in (0,1)$ is a Gaussian process with zero mean and covariance function

$$EB_t^H B_s^H = \frac{1}{2}(t^{2H} + s^{2H} - |t-s|^{2H}), \quad s,t \geq 0.$$

We consider a modification of B_t^H that has continuous trajectories and $B_0^H = 0$. B_t^H has stationary increments and is self-similar; that is $B_{\alpha t}^H = \alpha^H B_t^H, \quad \alpha > 0$. If $H = \frac{1}{2}$ then $B_t^H = W_t$, standard Brownian motion. If $H > \frac{1}{2}$ then B_t^H is a process with long-range dependence, that is $\sum_{n=1}^{\infty} r(n) = \infty$, where $r(n) := E[B_1^H(B_{n+1}^H - B_n^H)]$. Due to its properties, the process B_t^H with index $H \in (\frac{1}{2}, 1)$ is a useful tool for various applications, including finance [9]. Nevertheless the financial market constructed with respect to fBm admits arbitrage if we use stochastic integration theory based on ordinary products (see [1, 6]). We denote the corresponding integral as

$$I_t^1 := \int_0^t f(s,\omega)\delta B_s^H = \lim \sum_k \frac{1}{2}[f(u_k,\omega) + f(u_{k+1},\omega)][B_{u_{k+1}}^H - B_{u_k}^H],$$

where the limit in probability is taken as the mesh of the finite partitions $\{u_k\}$ of the interval $[0,t]$ goes to zero. The integral I_t^1 has nonzero mean; it is called the fractional Stratonovich (or pathwise) integral [4], similar to the integral with respect to standard Brownian motion.

The following example of A. N. Shiryaev ([15]), see also [5, 12]), demonstrates there are arbitrage possibilities for a market with a self-financing strategy based on the Stratonovich integral.

Example 1. Let $r > 0$, and $B_t = e^{rt}$ and $S_t = e^{rt + B_t^H}, \quad H \in [\frac{1}{2}, 1)$ be the price of a bond and a stock, respectively, at time $t \geq 0$. Then the (B,S)-market can be described by the following system of differential equations

$$\begin{aligned} dB_t &= rB_t dt, \\ dS_t &= S_t(rdt + \delta B_t^H), \end{aligned}$$

where δB_t^H denotes the Stratonovich differential. Consider a portfolio of the form

$\pi_t = (\beta_t, \gamma_t)$, where $\beta_t = 1 - e^{2B_t^H}$, $\gamma_t = 2(e^{B_t^H} - 1)$. Then

$$X_t^\pi = \beta_t B_t + \gamma_t S_t = e^{rt}(e^{B_t^H} - 1)^2 = \int_0^t \beta_s dB_s^H + \int_0^t \gamma_t \delta S_s.$$

This implies that the strategy π_t is self-financing. Further, $X_0^\pi = 0$. However, $X_t^\pi \geq 0$, and $EX_t^\pi > 0, \quad t > 0$. Therefore, there is arbitrage. Such an approach to stochastic integration, using the Stratonovich integral, will be called the Stratonovich scheme (see [1]).

The second approach to stochastic integration with respect to fBm is developed in [1, 2, 4]. Such an integral will be denoted by

$$I_t^2 = \int_0^t f(s,\omega)dB_t^H.$$

It is based on Wick product and defined by

$$\int_0^t f(s,\omega)dB_t^H = \lim \sum_k f(u_k,\omega) \diamond [B_{u_{k+1}}^H - B_{u_k}^H],$$

where $\diamond$ denotes the Wick product, and the integral has zero mean. The connection between I_t^1 and I_t^2 and their properties, in particular, their Ito formula, are discussed in [1, 2, 4]. Moreover, References [2] contains a "white noise calculus" for a fBm defined on the probability space $(\Omega, F, P) = (S'(R), F, P_H)$, where $S'(R)$ is the space of rapidly decreasing smooth functions, F is the Borel σ-field and P_H is the measure introduced according to the Bochner–Minlos theorem by the formula

$$\int_{S'(R)} exp\{i < \omega, f >\}dP_H(\omega) = exp\{-\frac{1}{2} \parallel f \parallel_H^2\},$$

where $\parallel f \parallel_H^2 = H(2H-1)\int_{R^2} f(s)f(t)|s-t|^{2H-2}dsdt$. The process is then defined by

$$B_t^h := < I[0,t], \omega > .$$

Such approach will be called the Hu–Oksendal scheme.

Reference [2] describes another construction: $(\Omega, F, P) = (S'(R), F, P_H)$, where P is the measure introduced according to the Bochner–Minlos theorem by the formula

$$\int_{S'(R)} exp\{i < \omega, f >\}dP(\omega) = exp\{-\frac{1}{2} \parallel f \parallel^2\},$$

where $\parallel f \parallel^2 = \int_{R^2} f^2(t)dt$. The process B_H is defined by $B_H(t) := < M_H I[0,t] >$. Here $M_H f(x) = \int_R |t-x|^{3/2-H} f(t)dt$. This approach permits one to consider different fBm on the same probability space and we can define a linear combination:

$$B_m(t) := \sigma_1 B_1^{H_1} + \ldots + \sigma_m B_1^{H_m}, \quad \sigma_i > 0, \quad H_i \in (0,1).$$

The construction of the stochastic integral with respect to fBm is similar in [4] and [2], but in the second approach we can also consider

$$I_t^3 := \int_0^t f(s,\omega)dB_M(s) = \sum_{i=1}^{m} \int_0^t f(s,\omega)\sigma_i dB_s^{H_i}, \quad i=1,2,...,m.$$

This construction will be called the Elliott–van der Hoek scheme.

10.3 Stochastic stability of a fractional (B,S)-security market in Stratonovich scheme

In this section, we consider stability almost sure, in mean, in mean square for fractional (B,S)-security markets, later fractional Brownian markets, without and with jumps.

10.3.1 Definition of fractional Brownian market in Stratonovich scheme

A fractional (B,S)-market in the Stratonovich scheme without jumps is defined by the system of stochastic differential equations

$$\begin{cases} dB_t &= rB_t dt, \\ dS_t &= S_t(\mu dt + \sigma \delta B_t^H), \quad H \in (0,1). \end{cases} \tag{10.1}$$

Here B_t and S_t correspond to bonds and stocks, δB_t^H is the Stratonovich differential with respect to fBm.

We define the corresponding to the stochastic term in (10.1) integral as

$$\int_0^t f(s,\omega)\delta B_s^H = \lim \sum_k \frac{1}{2}[f(u_k,\omega)+f(u_{k+1},\omega)][B_{u_{k+1}}^H - B_{u_k}^H],$$

where the limit in probability is taken as the mesh of the finite partitions $\{u_k\}$ of the interval $[0,t]$ goes to zero. This integral has nonzero mean. It is called the fractional Stratonovich (pathwise) integral [4], similar to the integral with respect to standard Brownian motion.

The system (10.1) is equivalent to

$$\begin{cases} B_t &= B_0 e^{rt}, \\ S_t &= S_0 e^{\mu dt + \sigma B_t^H}, \quad H \in (0,1), \quad S_0, B_0 > 0. \end{cases} \tag{10.2}$$

Here $r > 0$ is the risk free interest rate, $\mu \in R$ is the appreciation rate of S_t, and $\sigma > 0$ is the volatility coefficient of S_t.

10.3.2 Stability almost sure, in mean and mean square of fractional Brownian markets without jumps in Stratonovich scheme

Almost Sure Stability.

From Lemma and Corollary (see Appendix, Section 10.6) we obtain the following result.

Lemma 1. Let $(B_t^H, t \geq 0)$ be a fBm with $H \in (0,1)$. Then $B_t^H t^{-H-\varepsilon} \to 0$ as $t \to +\infty$ almost sure for any $\varepsilon > 0$.

Corollary 1.

$$\lim_{t\to+\infty} t^{-1} B_t^H = 0, \quad H \in (0,1)$$

almost sure.

Consider almost sure stability of (B,S)-market (10.2) with respect to Lyapunov index (see Appendix).

In this case, the Lyapunov index is:

$$\lambda_{a.s.} := \lim_{t\to+\infty} \frac{\ln S_t}{t} = \mu + \lim_{t\to+\infty} \frac{\sigma B_t^H}{t} = \mu,$$

almost sure, according to Corollary 1.

Thus, we obtain the following result.

Lemma 2.

1. If $\mu < 0$, then the process S_t is stable almost sure; if $\mu > 0$ then S_t is unstable almost sure.

2. If $\mu < r$, then the discount process S_t/B_t is stable almost sure; if $\mu > r$ then S_t/B_t is unstable almost sure.

Stability in Mean and Mean Square.

For investigation of stability in the mean we calculate

$$E\exp\{\sigma B_t^H\} = \frac{1}{t^H\sqrt{2\pi}} \int_R e^{\sigma y - \frac{y^2}{2t^{2H}}} dy = e^{\frac{\sigma^2 t^{2H}}{2}}.$$

Similarly,

$$E(\exp\{\sigma B_t^H\})^m = \exp\{\frac{\sigma^2 t^{2H} m^2}{2}\}, \quad m \geq 1. \tag{10.3}$$

In particular,

$$E(\exp\{\sigma B_t^H\})^2 = \exp\{2\sigma^2 t^{2H}\}.$$

It follows from (3) that

$$ES_t = S_0 e^{\mu t + \frac{\sigma^2 t^{2H}}{2}}.$$

Therefore, the Lyapunov index (see Appendix) equals

$$\lambda_{mean} = \mu + \lim_{t\to+\infty} \frac{\sigma^2 t_{2H-1}}{2} = \begin{cases} +\infty, & H > 1/2, \\ \mu, & H < 1/2, \\ \mu + \frac{\sigma^2}{2}, & H = 1/2. \end{cases} \tag{10.4}$$

For $m = 2$ we obtain:

$$\lambda_{square} = 2\mu + \lim_{t\to+\infty} \frac{\sigma^2 m^2 t_{2H-1}}{2} = \begin{cases} +\infty, & H > 1/2, \\ 2\mu, & H < 1/2, \\ 2\mu + 2\sigma^2, & H = 1/2. \end{cases} \tag{10.5}$$

In this way, we have the following result.

Lemma 3.

I. The dynamics of stock price is *stable in mean* if:
1) $\mu < 0$ for $H < 1/2$;
2) $\mu + \frac{\sigma^2}{2} < 0$ for $H = 1/2$.
It is *unstable in mean* for $H > 1/2$.
II. The dynamics of stock price is *stable in mean-square* if:
1) $\mu < 0$ for $H < 1/2$;
2) $\mu + \sigma^2 < 0$ for $H = 1/2$.
It is *unstable in mean-square* for $H > 1/2$.
III. Discount stock price $\frac{S_t}{B_t}$ is *stable in mean* if:
1) $\mu < r$ for $H < 1/2$;
2) $2\mu - 2r + \sigma^2 < 0$ for $H = 1/2$.
It is *unstable in mean* for $H = 1/2$.
IV. Discount stock price is *stable in mean-square* if:
1) $\mu < r$ for $H < 1/2$;
2) $\mu - r + \sigma^2 < 0$ for $H = 1/2$.
It is *unstable in mean-square* for $H > 1/2$.

From this Lemma 3 we can see why the case $H > 1/2$ was convenient for finance applications (see Hu and Oksendal): stock prices and discount stock prices are unstable and hence increase over time.

10.3.3 Stability almost sure, in mean and mean-square of fractional Brownian markets with jumps in Stratonovich scheme

Consider fractional Brownian market with jumps in a Stratonovich scheme. Let N_t be Poisson process with intensity $\alpha > 0$ and moments of jumps $\tau_k, \quad k \geq 1$. Let also $(U_k, \quad k \geq 0)$ be sequence of independent identically distributed random variables taking their values from $(-1, +\infty)$. We suppose that on the intervals $[\tau_i, \tau_{i+1}), \quad i \geq 1$ the stock price S_t follows (1) and in the moments τ_i S_{τ_i} has jumps, more precisely,

$$S_{\tau_i} - S_{\tau_i-} = S_{\tau_i-} U_i$$

or

$$S_{\tau_i} = (1 + U_i) S_{\tau_i-}.$$

The number of jumps will equal N_t on the interval $[0, t]$; therefore the value of the process S_t equals to

$$S_t = S_0 e^{\mu t + \sigma B_t^H} \prod_{k=1}^{N_t} (1 + U_k), \tag{10.6}$$

which is easy to verify by induction (see [17], Proposition 3). Such process S_t is a solution of stochastic integral equation

$$S_t = S_0 + \int_0^t S_u(\mu du + \sigma \delta B_u^H) + \sum_{k=1}^{N_t} S_{\tau_k -} U_k$$

with Stratonovich stochastic integral.

In this way, fractional Brownian market with jumps is defined by the system of two equations:

$$\begin{cases} B_t &= B_0 e^{rt}, \quad B_0 > 0, \\ S_t &= S_0 e^{\mu t + \sigma B_t^H} \prod_{k=1}^{N_t} (1+U_k), \quad H \in (0,1), \quad S_0 > 0. \end{cases}$$

Stability Almost Sure with Jumps.

From (6) we obtain that Lyapunov index for S_t equals to

$$\lambda_{a.s.} := \lim_{t\to\infty} \frac{S_t}{t} = \mu + \lim_{t\to\infty} \frac{\sigma B_t^H}{t} + \lim_{t\to\infty} \frac{1}{t} \sum_{k=1}^{N_t} \ln(1+U_k) = \mu + \alpha E \ln(1+U_k),$$

almost sure, where we used Corollary from Appendix, renewal theorem [3] for N_t, which states that

$$t^{-1} N_t \to \alpha$$

almost sure when $t \to \infty$, and the strong law of large numbers [3] for $y_y := \frac{1}{t} \sum_{k=1}^{t} \ln(1+U_k)$, which states that

$$y_t \to E \ln(1+U_1)$$

almost sure when $t \to \infty$.

By analogical reasonings, we have that discount stock price $\frac{S_t}{B_t}$, where $B_t = B_0 e^{rt}, \quad r > 0$, is stable almost sure if $\mu + \alpha E \ln(1+U_1) < r$. Otherwise, it is unstable.

From here, we obtain the following result.

Lemma 4.

I. The process S_t is *stable almost sure* for $\mu + \alpha E \ln(1+U_1) < 0$ and *unstable* for $\mu + \alpha E \ln(1+U_1) > 0$, $H \in (0,1)$.

II. The *discount stock price* $\frac{S_t}{B_t}$, where $B_t = B_0 e^{rt}, \quad r > 0$, is *stable almost sure* if $\mu + \alpha E \ln(1+U_1) < r$. Otherwise, the *discount stock price* is *unstable*.

Stability in Mean with Jumps.

Study of stability of fBm with jumps in mean will be based on the following relation from [17]

$$E \prod_{k=1}^{N_t} (1+U_k) = e^{\alpha E U_1 t}. \tag{10.7}$$

From (10.4), (10.6) and (10.7) we obtain

$$ES_t = S_0 e^{\mu t + \frac{\sigma^2 t^{2H}}{2} + \alpha EU_1 t}.$$

Lyapunov index for ES_t equals to

$$\begin{aligned}\lambda_{mean} &= \lim_{t\to\infty} \frac{\ln ES_t}{t} \\ &= \mu + \lim_{t\to\infty} \frac{\sigma^2}{2} t^{2H-1} + \alpha EU_1 \\ &= \begin{cases} +\infty, & H > 1/2, \\ \mu + \alpha EU_1, & H < 1/2, \\ \mu + \frac{\sigma^2}{2} + \alpha EU_1, & H = 1/2. \end{cases}\end{aligned}$$

Stability of fBm with jumps in mean for discount stock price is based on the following relation

$$E\frac{S_t}{B_t} = S_0 e^{(\mu - r)t + \frac{\sigma^2 t^{2H}}{2} + \alpha EU_1 t}.$$

which follows from (9.7).

Lyapunov index for $E\frac{S_t}{B_t}$ equals to

$$\begin{aligned}\lambda_{mean} &= \lim_{t\to\infty} \frac{\ln E\frac{S_t}{B_t}}{t} \\ &= (\mu - r) + \lim_{t\to\infty} \frac{\sigma^2}{2} t^{2H-1} + \alpha EU_1 \\ &= \begin{cases} +\infty, & H > 1/2, \\ \mu - r + \alpha EU_1, & H < 1/2, \\ \mu - r + \frac{\sigma^2}{2} + \alpha EU_1, & H = 1/2. \end{cases}\end{aligned}$$

From here, we obtain the following result.

Lemma 5.

I. The process S_t in (6) is *stable in mean* if $\mu + \alpha EU_1 < 0$ for $H < 1/2$, and if $\mu + \frac{\sigma^2}{2} + \alpha EU_1 < 0$, for $H = 1/2$. And it is *unstable* for $H > 1/2$.

II. The *discount stock price* $\frac{S_t}{B_t}$ is *stable in mean* if $\mu + \alpha EU_1 < r$ for $H < 1/2$, and if $\mu + \frac{\sigma^2}{2} + \alpha EU_1 < r$, for $H = 1/2$. And the *discount stock price* is *unstable* for $H > 1/2$.

Stability in Mean Square with Jumps.

For mean-square stability we use the following relation from [17]:

$$E(\prod_{k=1}^{N_t}(1+U_k))^2 = e^{\alpha t(EU_1^2 + 2EU_1)}. \tag{10.8}$$

Taking into account (10.3) and (10.8) we have

$$E(S_t)^2 = S_0^2 e^{2\mu t} e^{2\sigma^2 t^{2H}} e^{\alpha t(EU_1^2+2EU_1)}.$$

In this case, the Lyapunov index equals to

$$\begin{aligned}\lambda_{square} &= \lim_{t\to\infty} \frac{\ln E(S_t)^2}{t} \\ &= 2\mu + 2\lim_{t\to\infty} \sigma^2 t^{2H-1} + \alpha(EU_1^2 + 2EU_1) =\end{aligned}$$

$$\begin{cases} +\infty, & H > 1/2, \\ 2\mu + \alpha(EU_1^2 + 2EU_1), & H < 1/2, \\ 2\mu + 2\sigma^2 + \alpha(EU_1^2 + 2U_1), & H = 1/2. \end{cases}$$

For discount stock price we have the following Lyapunov index:

$$\begin{aligned}\lambda_{square} &= \lim_{t\to\infty} \frac{\ln E(\frac{S_t}{B_t})^2}{t} \\ &= 2(\mu - r) + 2\lim_{t\to\infty} \sigma^2 t^{2H-1} + \alpha(EU_1^2 + 2EU_1) =\end{aligned}$$

$$\begin{cases} +\infty, & H > 1/2, \\ 2(\mu - r) + \alpha(EU_1^2 + 2EU_1), & H < 1/2, \\ 2(\mu - r) + 2\sigma^2 + \alpha(EU_1^2 + 2U_1), & H = 1/2. \end{cases}$$

From here we obtain the following result.

Lemma 6.

I. The process S_t in (6) is *stable in mean square* if $2\mu + \alpha(EU_1^2 + 2EU_1) < 0$ for $H < 1/2$, and if $2\mu + 2\sigma^2 + \alpha(EU_1^2 + 2EU_1) < 0$, for $H = 1/2$. And it is *unstable in mean square* for $H > 1/2$.

II. The *discount stock price* $\frac{S_t}{B_t}$ is *stable in mean* if $2\mu + \alpha(EU_1^2 + 2EU_1) < 2r$ for $H < 1/2$, and if $2\mu + 2\sigma^2 + \alpha(EU_1^2 + 2U_1) < 2r$ for $H = 1/2$. And the *discount stock price* is *unstable in mean square* for $H > 1/2$.

Remark 1. We may consider mixed Brownian-fractional Brownian market [12] with independent Brownian and fractional Brownian motions, instead of just fractional Brownian markets.

The results will be the same, since

$$\frac{B_t}{t} \to 0$$

almost sure when $t \to +\infty$.

10.4 Stochastic stability of fractional (B,S)-security market in Hu and Oksendal scheme

In this section, we define fractional Brownian market in Hu and Oksendal scheme and state stability almost sure, in mean and mean square for it without and with jumps.

10.4.1 Definition of fractional Brownian market in Hu and Oksendal scheme

Fractional (B,S)-security market in Hu and Oksendal scheme is described by two assets: 1) bond or bank account $B(t)$:

$$dB(t) = rB(t)dt, \quad B(0) = B_0 > 0, \quad 0 \leq t \leq T, \quad r > 0,$$

2) stock $S(t)$, which satisfies the equation

$$dS(t) = \mu S(t)dt + \sigma S(t)dB_t^H, \quad S(0) = S_0 > 0, \tag{10.9}$$

$\mu \in R$ is an appreciate rate, $\sigma > 0$ is a volatility, B_t^H is fBm, $H \in (1/2, 1)$.

Equation (10.9) may be written down in the integral form

$$S(t) = S_0 + \mu \int_0^t S(u)du + \sigma \int_0^t S(u)dB_u^H,$$

where the integral is fractional Ito integral (defined by Wick product, see Section 4.1) with zero mean. Solution of the equation (10.9) has the form

$$S(t) = S_0 \exp\{\sigma B_t^H + \mu t - \frac{1}{2}\sigma^2 t^{2H}\}, \quad t \geq 0. \tag{10.10}$$

10.4.2 Stability almost sure, in mean and mean square of fractional Brownian markets without jumps in Hu and Oksendal scheme

Almost Sure Stability

We remark that $S(t)$ has the following representation (see previous section)

$$S(t) = S_0 \exp\{\sigma B_t^H + \mu t - \frac{1}{2}\sigma^2 t^{2H}\}, \quad H \in (1/2, 1) \quad t \geq 0. \tag{10.11}$$

Using Corollary, Appendix, Section 10.6, we obtain that Lyapunov index for $S(t)$ in (10.11) equals to

$$\lambda_{a.s.} = \lim_{t \to +\infty} \frac{\ln S(t)}{t} = \mu - \lim_{t \to +\infty} \frac{\sigma^2}{2} t^{2H-1} = -\infty,$$

since $H \in (1/2, 1)$.

Similarly, the Lyapunov index for discount stock price equals to

$$\lambda_{a.s.} = \lim_{t \to +\infty} \frac{\ln \frac{S(t)}{B(t)}}{t} = \mu - r - \lim_{t \to +\infty} \frac{\sigma^2}{2} t^{2H-1} = -\infty.$$

Therefore, we have the following result.

Lemma 7. *Stock price* $S(t)$ and *discount stock price* $\frac{S(t)}{B(t)}$ are *stable almost sure* for any $r>0$, $\mu \in R$ and $\sigma>0$, $\quad H \in (1/2,1)$.

Stability in Mean.
From equality (10.3) we obtain that

$$Ee^{\sigma B_t^H} = e^{\frac{\sigma^2 t^{2H}}{2}}.$$

From here and from (10.10) it follows that mean value for stock price $S(t)$ is

$$ES(t) = S_0 e^{\mu t}$$

and Lyapunov index equals to

$$\lambda_{mean} = \lim_{t\to\infty} \frac{\ln S(t)}{t} = \mu.$$

Similarly, for the discount stock price $\frac{S(t)}{B(t)}$ we have

$$E\frac{S(t)}{B(t)} = S_0 e^{(\mu-r)t}$$

and Lyapunov index equals to

$$\lambda_{mean} = \lim_{t\to\infty} \frac{\ln \frac{S(t)}{B(t)}}{t} = \mu - r.$$

In this way, we have the following result.
Lemma 8.
I. Stock price $S(t)$ is *stable in mean* when $\mu<0$.
II. Discount stock price $\frac{S(t)}{B(t)}$ is *stable in mean* when $\mu<r$.

Stability in Mean Square.
From (10.3) we obtain that

$$E(\exp\{\sigma B_t^H\})^2 = \exp\{2\sigma^2 t^{2H}\}.$$

Also, taking into account (10) we have that $E(S(t))^2$ equals to

$$E(S(t))^2 = E[S_0^2 e^{2\sigma B_t^H + 2\mu t - \sigma^2 t^{2H}}] = S_0^2 e^{2\sigma^2 t^{2H}} e^{2\mu t} e^{-\sigma^2 t^{2H}} = S_0^2 e^{2\mu t} e^{\sigma^2 t^{2H}}.$$

The mean value of squared discount stock price $E(fracS(t)B(t))^2$ equals to

$$\begin{aligned} E(\frac{S(t)}{B(t)})^2 &= E[S_0^2 e^{2\sigma B_t^H + 2(\mu-r)t - \sigma^2 t^{2H}}] \\ &= S_0^2 e^{2\sigma^2 t^{2H}} e^{2(\mu-r)t} e^{-\sigma^2 t^{2H}} \\ &= S_0^2 e^{2(\mu-r)t} e^{\sigma^2 t^{2H}}. \end{aligned}$$

The Lyapunov index for $E(S(t))^2$ equals to

$$\lambda_{square} = \lim_{t\to\infty} \frac{\ln E(S(t))^2}{t} = 2\mu + \sigma^2 \lim_{t\to\infty} t^{2H-1} = +\infty,$$

since $H \in (1/2, 1)$.

The similar result about the Lyapunov index is true for $E(\frac{S(t)}{B(t)})^2$.

In this way, we have the following result.

Lemma 9.

Stock price $S(t)$ and discount stock price $\frac{S(t)}{B(t)}$ are unstable in mean square.

10.4.3 Stability almost sure, in mean and mean square of fractional Brownian markets with jumps in Hu and Oksendal scheme

Definition of Fractional Brownian Markets with Jumps in Hu and Oksendal Scheme.

Consider fractional Brownian market with jumps in Hu and Oksendal scheme. Let N_t be Poisson process with intensity $\alpha > 0$ and moments of jumps $\tau_k, \quad k \geq 1$. Let also $(U_k, \quad k \geq 0)$ be sequence of independent identically distributed random variables taking their values from $(-1, +\infty)$. We suppose that on the intervals $[\tau_i, \tau_{i+1}), \quad i \geq 1$ the stock price S_t follows (10.9) and in the moments τ_i S_{τ_i} has jumps, more precisely,

$$S_{\tau_i} - S_{\tau_i -} = S_{\tau_i -} U_i$$

or

$$S_{\tau_i} = (1 + U_i) S_{\tau_i -}. \tag{10.12}$$

The number of jumps will equal N_t on the interval $[0, t]$.

Such process S_t is a solution of stochastic integral equation

$$S_t = S_0 + \int_0^t S_u(\mu du + \sigma dB_u^H) + \sum_{k=1}^{N_t} S_{\tau_k -} U_k$$

with fractional Ito stochastic integral.

In this way, from (10.10) and (10.12) we have that fractional Brownian market with jumps in Hu and Oksendal scheme is defined by the system of two equations:

$$\begin{cases} B_t = B_0 e^{rt}, \quad B_0 > 0, \\ S_t = S_0 e^{\mu t - \frac{1}{2}\sigma^2 t^{2H} + \sigma B_t^H} \prod_{k=1}^{N_t} (1 + U_k), \quad S_0 > 0, \quad H \in (1/2, 1). \end{cases} \tag{10.13}$$

Stability Almost Sure, in Mean and Mean Square of Fractional Brownian Markets with Jumps in Hu and Oksendal Scheme.

From (10.13) it follows that the Lyapunov index for stock price $S(t)$ equals

$$\begin{aligned}\lambda_{a.s.} &:= \lim_{t\to\infty}\tfrac{S_t}{t} = \mu + \lim_{t\to\infty}\tfrac{\sigma B_t^H}{t} - \lim_{t\to\infty}\tfrac{\sigma^2}{2}t^{2H-1} \\ &+ \lim_{t\to\infty}\tfrac{1}{t}\textstyle\sum_{k=1}^{N_t}\ln(1+U_k) \\ &= \mu + \alpha E\ln(1+U_k) - \infty = -\infty.\end{aligned} \tag{10.14}$$

In a similar way, the Lyapunov index for discount stock price $\frac{S(t)}{B(t)}$ equals

$$\begin{aligned}\lambda_{a.s.} &:= \lim_{t\to\infty}\frac{\ln\frac{S_t}{B(t)}}{t} = \mu - r + \lim_{t\to\infty}\tfrac{\sigma B_t^H}{t} - \lim_{t\to\infty}\tfrac{\sigma^2}{2}t^{2H-1} \\ &+ \lim_{t\to\infty}\tfrac{1}{t}\textstyle\sum_{k=1}^{N_t}\ln(1+U_k) \\ &= \mu - r + \alpha E\ln(1+U_k) - \infty = -\infty.\end{aligned}$$

From (10.3), $m=1$, and (10.7) we can find, taking into account (10.13), that

$$ES(t) = S_0 e^{\mu t} e^{\alpha t EU_1}.$$

Hence, the Lyapunov index for mean value of stock price $ES(t)$ here equals

$$\lambda_{mean} = \lim_{t\to\infty}\frac{\ln ES(t)}{t} = \mu + \alpha EU_1. \tag{10.15}$$

Similarly, the Lyapunov index for mean value of discount stock price $E\frac{S(t)}{B(t)}$ is

$$\lambda_{mean} = \lim_{t\to\infty}\frac{\ln E\frac{S(t)}{B(t)}}{t} = \mu - r + \alpha EU_1. \tag{10.16}$$

From (10.3), $m=2$, and (9.8) we can find, taking into account (10.13), that

$$E(S(t))^2 = S_0^2 e^{2\mu t} e^{\sigma^2 t^{2H}} e^{\alpha t(EU_1^2 + 2EU_1)}.$$

Therefore, the Lyapunov index for mean value of squared stock price $E(S(t))^2$ equals

$$\lambda_{square} = \lim_{t\to\infty}\frac{\ln E(S(t))^2}{t} = 2\mu + \alpha(EU_1^2 + 2EU_1) + \lim_{t\to\infty}\sigma^2 t^{2H-1} = +\infty, \tag{10.17}$$

since $H \in (1/2, 1)$.

Similarly, the Lyapunov index for mean value of squared discounted stock price $E(\frac{S(t)}{B(t)})^2$ is

$$\lambda_{square} = \lim_{t\to\infty}\frac{\ln E(\frac{S(t)}{B(t)})^2}{t} = 2(\mu - r) + \alpha(EU_1^2 + 2EU_1) + \lim_{t\to\infty}\sigma^2 t^{2H-1} = +\infty.$$

Summarizing all the above results (see (9.14)-(9.17)) we have the following result.

Lemma 10.

I. *Stock price with jumps* in (10.13) in Hu and Oksendal scheme is *almost sure stable* for any $\mu \in R$, $\sigma > 0$ and $\alpha > 0$. It is *stable in mean* if $\mu + \alpha EU_1 < 0$, *unstable in mean* if $\mu + \alpha EU_1 > 0$, and *unstable in mean square* for any set of parameters included.

II. *Discount stock price with jumps* in Hu and Oksendal scheme is *almost sure stable* for any $\mu \in R$, $\sigma > 0$ and $\alpha > 0$. It is *stable in mean* if $\mu + \alpha EU_1 < r$, *unstable in mean* if $\mu + \alpha EU_1 > r$, and *unstable in mean square* for any set of parameters included.

Remark 2. Stock price $S(t)$ with jumps in Hu and Oksendal scheme is *fractionally stable almost sure* (see Appendix for definition of fractional stability) since

$$\lambda^H_{a.s.} = -\frac{1}{2}\sigma^2 < 0.$$

Here,

$$\begin{aligned}\frac{1}{t^{2H}}\sum_{k=1}^{N_t}\ln(1+U_k) &= \frac{N_t}{t}\frac{t}{N_t}\frac{1}{t^{2H}}\sum_{k=1}^{N_t}\ln(1+U_k)\\ &= \frac{N_t}{t}t^{1-2H}\frac{1}{N_t}\sum_{k=1}^{N_t}\ln(1+U_k) \to 0\end{aligned}$$

almost sure when $t \to \infty$, since $t^{1-2H} \to 0$ for $H \in (1/2, 1)$.

Similarly we can obtain that $\lambda^H_{mean} = 0$ has nothing to say about fractional stability in mean for stock price $S(t)$. Also, $\lambda^H_{square} = \sigma^2 > 0$, and $S(t)$ is fractionally unstable in mean square.

10.5 Stochastic stability of fractional (B,S)-security market in Elliott and van der Hoek scheme

In this section, we define fractional Brownian market in Elliott and van der Hoek scheme and state stability almost sure, in mean and mean square for it without and with jumps.

10.5.1 Definition of fractional Brownian market in Elliott and van der Hoek Scheme

Fractional (B,S)-security market in Elliott and van der Hoek scheme is defined by two assets:

1) bond or bank account $B(t)$

$$dB(t) = rB(t)dt, \quad B(0) = B_0, \quad r > 0;$$

2) stock price $S(t)$ that satisfies the following equation

$$dS(t) = \mu S(t)dt + \sigma S(t)dB_M(t), \quad S(0) = S_0 > 0, \tag{10.18}$$

where $\mu \in R$, $\sigma > 0$, and $B_M(t)$ is a linear combination of fBm on $(S'(R), F)$, which is defined by unique probability measure P that is given by the Bochner–Minlos theorem [3], in contrast to the approach by Hu and Oksendal, where B_t^H is given by measure P_H on $(S'(R),F)$ with fixed Hurst parameter $H \in (1/2,1)$. Here, $M := \sigma_1 M_{H_1} + \ldots + \sigma_m M_{H_m}, \quad 0 < H_k < 1, \quad k = 1,2,\ldots,m, \quad \sigma_k > 0$ is a constant and M_H is a fundamental operator (see [3]):

$$M_H f(x) := \begin{cases} 2\Gamma(H-1/2)\cos(\frac{\pi}{2}(H-1/2))^{-1}\int_R \frac{f(x-t)-f(x)}{|t|^{3/2-H}}dt, & H \in (0,1/2), \\ 2\Gamma(H-1/2)\cos(\frac{\pi}{2}(H-1/2))^{-1}\int_R \frac{f(t)}{|t-x|^{3/2-H}}dt, & H \in (1/2,1), \\ f(x), & H = 1/2, f \in S(R). \end{cases}$$

Process $B_M(t)$ in (10.19) has the form

$$B_M(t) := \sigma_1 B_t^{H_1} + \ldots + \sigma_m B_t^{H_m},$$

where B^H is a fBm with Hurst index $H \in (0,1)$. According to [2], the advantage of this approach consists of the fact that fractional Brownian motions are defined simultaneously for all Hurst indices $H \in (0,1)$.

As it follows from [2], the solution of (10.18) equals the following fractional lognormal process

$$S(t) = S_0 \exp\{\mu t + \sigma B_M(t) - \frac{\sigma^2}{2}\sigma_M^2(t)\}, \tag{10.19}$$

where

$$\sigma_M^2(t) := \sum_{i,j=1}^{m} 2(\sin(\frac{\pi}{2}(H_i+H_j))\Gamma(H_i+H_j+1))^{-1}|t|^{H_i+H_j}.$$

We note, that $B_M(t)$ is a Gaussian process with zero mean and variance $\sigma_M^2(t)$, for all $t \in [0,T]$. Introduce the following notion

$$\hat{H} := \max_{i=1,\ldots,m} H_i.$$

Then

$$\sigma_M^2(t) \le t^{2\hat{H}} \sum_{i,j=1}^{m} 2(\sin(\frac{\pi}{2}(H_i+H_j))\Gamma(H_i+H_j+1))^{-1} := Ct^{2\hat{H}}. \tag{10.20}$$

10.5.2 Stability almost sure, in mean and mean square of fractional Brownian markets without jumps in Elliott and van der Hoek Scheme

Almost Sure Stability

It follows from Lemma and Corollary (see Appendix, Section 10.6) that

$$\frac{B_M(t)}{t} = \sigma_1 \frac{B^{H_1}(t)}{t} + \ldots + \sigma_m \frac{B^{H_m}(t)}{t} \to_{t\to+\infty} 0, \tag{10.21}$$

for any $H_k \in (0,1), \quad \sigma_k > 0, \quad k = 1,2,\ldots,m.$

From (10.20) we may get

$$0 \le \frac{\sigma_M^2(t)}{t} \le \frac{t^{2\hat{H}C}}{t} = t^{2\hat{H}-1C} \to_{t\to+\infty} \begin{cases} 0, & 0<\hat{H}<1/2, \\ +\infty, & \hat{H}>1/2. \end{cases} \tag{10.22}$$

Taking into account (10.21) and (10.22) we obtain that the Lyapunov index for $S(t)$ equals

$$\lambda_{a.s.} = \lim_{t\to\infty} \frac{\ln|S(t)|}{t} = \begin{cases} \mu, & 0<\hat{H}<1/2, \\ -\infty, & \hat{H}>1/2. \end{cases}$$

Similarly, the Lyapunov index for discounted stock price $\frac{S(t)}{B(t)}$ equals

$$\lambda_{a.s.} = \lim_{t\to\infty} \frac{\ln\frac{|S(t)|}{B(t)}}{t} = \begin{cases} \mu - r, & 0<\hat{H}<1/2, \\ -\infty, & \hat{H}>1/2. \end{cases}$$

In this way, we have the following result.

Lemma 11.

I.

1) If $\hat{H} > 1/2$ then stock price is *stable almost sure*;

2) if $0 < \hat{H} < 1/2$ then $S(t)$ is *stable almost sure* for $\mu < 0$ and *unstable* for $\mu > 0$.

II.

1) If $\hat{H} > 1/2$ then the discount stock price $\frac{S(t)}{B(t)}$ is *stable almost sure*;

2) if $0 < \hat{H} < 1/2$ then the discount stock price $\frac{S(t)}{B(t)}$ is *stable almost sure* for $\mu < r$ and *unstable* for $\mu > r$.

Stability in Mean. Since (see [3])

$$Ee^{\sigma B_M(t)} = e^{\frac{\sigma^2}{2}\sigma_M^2(t)}$$

then (see (10.19))

$$ES(t) = S_0 e^{\mu t}. \tag{10.23}$$

Obviously, the Lyapunov index for $ES(t)$ equals μ. Similarly, the Lyapunov index for $E\frac{S(t)}{B(t)}$ equals $\mu - r$.

We have got the following result.

Lemma 12.

I. The stock price $S(t)$ is *stable in mean* when $\mu < 0$, and is *unstable in mean* when $\mu > 0$.

II. The discount stock price $E\frac{S(t)}{B(t)}$ is *stable in mean* when $\mu < r$, and is *unstable in mean* when $\mu > r$.

Stability in Mean Square

From [2] we may find that

$$E[e^{\sigma B_M(t)}]^2 = e^{2\sigma^2 \sigma_M^2(t)},$$

and (see (10.19))

$$E[S(t)]^2 = S_0^2 e^{2\mu t} e^{\sigma^2 \sigma_M^2(t)}. \tag{10.24}$$

Therefore, the Lyapunov index for the mean value of squared stock price $E|S(t)|^2$ equals

$$\lambda_{square} = \lim_{t\to\infty} \frac{\ln E|S(t)|^2}{t} = 2\mu + \sigma^2 \lim_{t\to\infty} \frac{\sigma_M^2}{t} = \begin{cases} 2\mu, & 0 < \hat{H} < 1/2, \\ +\infty, & \hat{H} > 1/2. \end{cases}$$

Similarly, the Lyapunov index for the mean value of discount squared stock price $E(\frac{|S(t)|}{B(t)})^2$ equals

$$\lambda_{square} = \lim_{t\to\infty} \frac{\ln E(\frac{|S(t)|}{B(t)})^2}{t} = 2(\mu - r) + \sigma^2 \lim_{t\to\infty} \frac{\sigma_M^2}{t} = \begin{cases} 2(\mu - r), & 0 < \hat{H} < 1/2, \\ +\infty, & \hat{H} > 1/2. \end{cases}$$

Therefore, we obtain the following result.

Lemma 13.

I.

1) If $0 < \hat{H} < 1/2$, then the stock price $S(t)$ is *stable in mean square* when $\mu < 0$ and *unstable in mean square* when $\mu > 0$;

2) if $\hat{H} > 1/2$ then the stock price $S(t)$ is *unstable in mean square*.

II.

1) If $0 < \hat{H} < 1/2$, then the discount stock price $\frac{S(t)}{B(t)}$ is *stable in mean square* when $\mu < r$ and *unstable in mean square* when $\mu > r$;

2) if $\hat{H} > 1/2$ then the discount stock price $\frac{S(t)}{B(t)}$ is *unstable in mean square*.

10.5.3 Stability almost sure, in mean and mean square of fractional Brownian markets with jumps in Elliott and van der Hoek scheme

Consider fractional Brownian market with jumps in Elliott and van der Hoek scheme. Let N_t be Poisson process with intensity $\alpha > 0$ and moments of jumps $\tau_k, \quad k \geq 1$. Let also $(U_k, \quad k \geq 0)$ be sequence of independent identically distributed random variables taking their values from $(-1, +\infty)$. We suppose that on the intervals $[\tau_i, \tau_{i+1}), \quad i \geq 1$ the stock price S_t follows (10.18) and in the moments τ_i S_{τ_i} has jumps, more precisely,

$$S_{\tau_i} - S_{\tau_i -} = S_{\tau_i -} U_i$$

or

$$S_{\tau_i} = (1 + U_i) S_{\tau_i -}. \tag{10.25}$$

The number of jumps will equal N_t on the interval $[0,t]$.

Such process S_t is a solution of stochastic integral equation (see (10.18))

$$S_t = S_0 + \int_0^t S_u(\mu du + \sigma dB_M(u)) + \sum_{k=1}^{N_t} S_{\tau_k -} U_k.$$

In this way, from (10.19) and (10.23) we have that fractional Brownian market with jumps in Elliott and van der Hoek scheme is defined by the system of two equations:

$$\begin{cases} B_t &= B_0 e^{rt}, \quad B_0 > 0, \\ S_t &= S_0 e^{\mu t - \frac{1}{2}\sigma^2 \sigma_M^2(t) + \sigma B_M(t)} \prod_{k=1}^{N_t} (1 + U_k), \quad S_0 > 0. \end{cases} \tag{10.26}$$

Almost Sure Stability

The Lyapunov index for stock price $S(t)$ in (10.26) equals

$$\lambda_{a.s} = \lim_{t\to\infty} \frac{\ln|S(t)|}{t} = \mu + \alpha E \ln(1+U_1) - \frac{\sigma^2}{2} \lim_{t\to\infty} \frac{\sigma_M^2(t)}{t}$$

$$= \begin{cases} \mu + \alpha E \ln(1+U_k), & 0 < \hat{H} < 1/2, \\ -\infty, & \hat{H} > 1/2. \end{cases}$$

Here we used the Law of Large Numbers for $\frac{1}{t}\sum_{k=1}^{N_t} \ln(1+U_k)$ and limits in (10.21) and (10.22).

Similarly, the Lyapunov index for discount stock price $\frac{S(t)}{B(t)}$ equals

$$\lambda_{a.s} = \lim_{t\to\infty} \frac{\ln\frac{|S(t)|}{B(t)}}{t} = \mu - r + \alpha E \ln(1+U_1) - \frac{\sigma^2}{2} \lim_{t\to\infty} \frac{\sigma_M^2(t)}{t}$$

$$= \begin{cases} \mu - r + \alpha E \ln(1+U_k), & 0 < \hat{H} < 1/2, \\ -\infty, & \hat{H} > 1/2. \end{cases}$$

Thus, we have the following result.

Lemma 14.

I. Stock price $S(t)$ is *stable almost sure* when $0 < \hat{H} < 1/2$; if $\hat{H} > 1/2$, then stock price $S(t)$ is *stable almost sure* when $\mu + \alpha E \ln(1+U_1) < 0$. Otherwise it is *unstable almost sure.*

II. Discount stock price $\frac{S(t)}{B(t)}$ is *stable almost sure* when $0 < \hat{H} < 1/2$; if $\hat{H} > 1/2$, then discount stock price $\frac{S(t)}{B(t)}$ is *stable almost sure* when $\mu + \alpha E \ln(1+U_1) < r$. Otherwise it is *unstable almost sure.*

Stability in Mean

Taking into account (10.7), (10.23), and (10.26) we have the following expression

$$ES(t) = S_0 e^{\mu t} e^{\alpha E U_1 t}.$$

From here we may find the Lyapunov index for $ES(t)$:

$$\lambda_{mean} = \mu + \alpha EU_1.$$

Similarly, the Lyapunov index for $\frac{S(t)}{B(t)}$ equals

$$\lambda_{mean} = \mu - r + \alpha EU_1.$$

Therefore, we have the following result.

Lemma 15.

I. Stock price $S(t)$ is *stable in mean* if $\mu + \alpha EU_1 < 0$. Otherwise it is *unstable in mean.*

II. Discounted stock price $\frac{S(t)}{B(t)}$ is *stable in mean* if $\mu + \alpha EU_1 < r$. Otherwise it is *unstable in mean.*

Stability in Mean Square

Taking into account (10.8), (10.24), and (10.26) we have the following expression

$$E[S(t)]^2 = S_0^2 e^{2\mu t} e^{\sigma^2 \sigma_M^2(t)} e^{\alpha(EU_1^2 + 2EU_1)} t.$$

From here we may find the Lyapunov index for $E[S(t)]^2$:

$$\lambda_{square} = 2\mu + \sigma^2 \lim_{t\to\infty} \frac{\sigma_M^2(t)}{t} + \alpha(EU_1^2 + 2EU_1) =$$

$$\begin{cases} 2\mu + \alpha(EU_1^2 + 2EU_1), & 0 < \hat{H} < 1/2, \\ +\infty, & \hat{H} > 1/2. \end{cases}$$

Similarly, the Lyapunov index for $E[\frac{S(t)}{B(t)}]^2$ equals

$$\lambda_{square} = 2(\mu - r) + \sigma^2 \lim_{t\to\infty} \frac{\sigma_M^2(t)}{t} + \alpha(EU_1^2 + 2EU_1) =$$

$$\begin{cases} 2(\mu - r) + \alpha(EU_1^2 + 2EU_1), & 0 < \hat{H} < 1/2, \\ +\infty, & \hat{H} > 1/2. \end{cases}$$

Therefore, we have the following result.

Lemma 16.

I. If $0 < \hat{H} < 1/2$ then stock price $S(t)$ is *stable in mean square* when $2\mu + \alpha(EU_1^2 + 2EU_1) < 0$. It is *unstable in mean square* if $\hat{H} > 1/2$.

II. If $0 < \hat{H} < 1/2$ then discounted stock price $\frac{S(t)}{B(t)}$ is *stable in mean square* when $2\mu + \alpha(EU_1^2 + 2EU_1) < 2r$. It is *unstable in mean square* if $\hat{H} > 1/2$.

10.6 Appendix

In the Appendix we give some definitions of stochastic stability and some results (Lemma and Corollary) on convergence $t^{-H-\varepsilon'} B_t^H$ when $t \to +\infty$ and $\varepsilon' > 0$.

10.6.1 Definitions of Lyapunov indices and stability

We say that *process* $S(t)$ *is stable almost sure* if $S(t) \to 0$ almost sure, *process* $S(t)$ *is stable in mean* if $ES(t) \to 0$, and *process* $S(t)$ *is stable in mean square* if $E(S(t))^2 \to 0$, when $t \to +\infty$. Otherwise, it is *unstable almost sure, unstable in mean, and unstable in mean square.*

Definition 1. *The Lyapunov index (or exponent)* $\lambda \equiv \lambda(\omega)$ for stochastic process equals

$$\lambda_{a.s.} = \overline{\lim}_{t\to\infty} \frac{\ln|S(t)|}{t} \tag{10.27}$$

almost sure.

In deterministic cases, the value λ is called *index of exponential growth* of function $S(t)$. It is known that λ coincides with lower bound of those values λ for which there exists a constant $N > 0$ such that for all $t \geq 0$:

$$S(t) \leq Ne^{\lambda t}.$$

In such a way, from (10.27) we have the following statements:

if $\lambda < 0$, then

$$S(t) \to 0$$

almost sure, when $t \to \infty$;

if $\lambda > 0$, then

$$S(t) \to +\infty$$

almost sure, when $t \to \infty$.

Definition 2. *The Lyapunov index (or exponent) in mean* equals

$$\lambda_{mean} = \overline{\lim}_{t\to\infty} \frac{\ln|ES(t)|}{t}$$

From here we have that process $S(t)$ is *stable in mean* if $\lambda_{mean} < 0$ and *unstable in mean* if $\lambda_{mean} > 0$.

Definition 3. *The Lyapunov index (or exponent) in mean square* equals

$$\lambda_{square} = \overline{\lim}_{t\to\infty} \frac{\ln E|S(t)|^2}{t}$$

From here we have that process $S(t)$ is *stable in mean square* if $\lambda_{square} < 0$ and *unstable in mean square* if $\lambda_{square} > 0$.

Definition 4. *Fractional Lyapunov index* $\lambda^H_{a.s.}$ is defined as follows

$$\lambda^H_{a.s.} = \overline{\lim}_{t\to\infty} \frac{\ln|S(t)|}{t^{2H}}, \quad H \in (0,1).$$

When $H = 1/2$, then the fractional Lyapunov index coincides with Lyapunov index:

$\lambda_{a.s.} \equiv \lambda_{a.s.}^{H}$.

Definition 5. Process $S(t)$ is *fractionally stable almost sure* if $\lambda_{a.s.}^{H} < 0$, and *fractionally unstable almost sure* if $\lambda_{a.s.}^{H} > 0$.

Example. Since $\frac{B_t^H}{t} \to 0$ almost sure when $t \to \infty$, then $\frac{B_t^H}{t^{2H}} \to 0$ almost sure for $H \in (1/2,1)$ when $t \to \infty$, and B_t^H is fractionally stable almost sure for $H \in (1/2,1)$.

Similar definitions may be obtained for *fractional stability in mean and mean square.*

Remark 3. If one of the Lyapunov indices equals zero, then there is nothing to say about the stability of underlying processes.

10.6.2 Asymptotic property of fractional Brownian motion

Consider an auxiliary construction from [8]. Suppose (T,ρ) is a metric space and $T = \cup_{l=0}^{\infty} C_l$, where the sets C_l are compact. Let $\{x(t), t \in T\}$ be a real separable Gaussian process satisfying the condition:

(i) there exists an increasing function $\sigma = \sigma(h) > 0, \quad \sigma(h) \to 0$, as $h \to 0$ and

$$\sup_{|t-s| \le h} (E(x(t)-x(s))^2)^{1/2} \le \sigma(h).$$

Further, let $\{c(t), t \in T\}$ be a continuous function with $\delta_t = \sup_{t \in C_l} |c(t)| \le 1$, a_l be any point from C_l, $\alpha_l = \sigma(\sup_{t \in C_l} \rho(t, a_t))$, $Z_t = (E(x(a_l))^2)^{1/2}$, $\beta = \sup_{l \ge 0} \frac{\alpha_l}{z_l}$. Let $N_l(u)$ be the smallest number of closed balls of radius u that cover C_l, $H_l(u) = \ln N_l(u)$.

Lemma [8]. Under conditions (i) above and

(ii)

$$\Lambda := \sum_{l=0}^{\infty} \delta_l z_l < \infty;$$

(iii)

$$\int_0^{\alpha_l} (H_l(\sigma^{(-1)}(u)))^{1/2} du < \infty, \quad l \ge 0;$$

(iv)

$$\sum_{l=0}^{\infty} \delta_l \int_0^{\alpha_l^p} (H_l(\sigma^{(-1)}(u)))^{1/2} du < \infty, \quad 0 < p < 1,$$

we have that for any $\gamma > 0, \quad 0 < p < 1$,

$$E \exp\{\gamma \sup |c(t)x(t)|\} \le \Phi(\gamma, p), \tag{10.28}$$

where

$$\Phi(\gamma, p) = 2\exp\{\frac{\gamma^2}{2} A(\gamma, p) + 2\gamma B(\gamma, p)\},$$

$$A(\gamma,p) = \frac{\gamma^2}{1-p}(1+\frac{2p^2}{(1-p)p}),$$

$$B(\gamma,p) = \frac{1}{p(1-p)}\sum_{l=0}^{\infty}\delta_l \int_0^{\alpha_l^p}(H_l(\sigma^{(-1)}(u)))^{1/2}du.$$

Now we apply Lemma to a *fractional Brownian motion* $B_t^H, \quad H \in (0,1)$. Here, $T = R_+, \quad \delta(h) = h^H$, since $(E(B_t^H - B_s^H)^2)^{1/2} = |t-s|^H$. The function $c(t)$ can be chosen as $c(t) = t^{-\alpha}$, $t > 1$, $c(t) = 1$, $0 < t < 1$. Further, we take $C_0 = [0,e]$, $\quad C_l = [e^l, e^{l+1}]$, $l \geq 1$, $\delta_l = e^{\alpha l}$, $a_l = e^l$, $l \geq 1$, $a_0 = e$; $\sup_{t \in C_l}|t - a_l| = e^{l+1} - e^l, \quad l \geq 1, \quad \sup_{t \in C_0}|t - a_0| = e; \quad \alpha_0 = (e^{l+1} - e^l)^H, \quad z_0 = e^H, \quad z_l = e^{Hl}; \quad \beta = \sup \frac{(e^{l+1}-e^l)^H}{e^{Hl}} = (e-1)^H; \quad \Lambda = \sum_{l=0}^{\infty} e^{-\alpha l}e^{Hl} < \infty$ for $\alpha > H$. Since for any C_l, $l \geq 1$ $N_l(u) \leq 1 + \frac{e^{l+1}-e^l}{2\sigma^{(-1)}(u)} = 1 + \frac{e^{l+1}-e^l}{2u^{1/H}}$, then the condition (iii) has a form

$$\int_0^{\alpha_l}(H_l(\sigma^{(-1)}(u)))^{1/2}du = \int_0^{\alpha_l}(\ln\frac{\alpha e^{1/H}}{2u^{1/H}}+1)^{1/2}du$$

$$= \alpha_l \int_0^1 (\ln(\frac{1}{2u^{1/H}}+1))^{1/2}du =: \alpha_l\Delta < \infty. \tag{10.29}$$

It follows from (10.29) that

$$\sum_{l=0}^{\infty}\delta_l \int_0^{\alpha_l^p}(H_l(\sigma^{(-1)}(u)))^{1/2}du \leq \sum_{l=0}^{\infty}\delta_l \int_0^{\alpha_l}(H_l(\sigma^{(-1)}(u)))^{1/2}du = \Delta\sum_{l=1}^{\infty}\delta_l\alpha_l < \infty.$$

Therefore, the conditions (ii)-(iv) hold. This implies that the inequality (3) holds for any $\alpha = H + \varepsilon, \quad \varepsilon > 0$, i.e., for $c(t) = t^{-H-\varepsilon}, \quad t > 1$. In other words, for $c(t) = t^{-H-\varepsilon}, \quad t > 1$, and $c(t) = 1, \quad 0 \leq t \leq 1$.

In this way, $E\exp\{\gamma \sup|c(t)B_t^H|\} < \infty$ for any $\gamma > 0$. Therefore, for any $\varepsilon > 0$ $\theta_\varepsilon := \sup_{t>1} t^{-H-\varepsilon}|B_t^H| < \infty$ almost sure. If we take $\varepsilon' > \varepsilon$ then $t^{-H-\varepsilon'}B_t^H \to 0$ almost sure as $t \to +\infty$. In this way, we have the following result.

Corollary. For $\varepsilon' > \varepsilon$

$$t^{-H-\varepsilon'}B_t^H \to 0$$

almost sure as $t \to +\infty$.

References

[1] Duncan, T. E., Hu, Y., and Pasik-Dunkan, B. Stochastic calculus for fractional Brownian motion. I. Theory. *SIAM J. Control and Optim.* 38, no. 2, 582–612, 2000.

[2] Elliott, R. and Van der Hoek, I. A general fractional white noise theory and applications to finance. *Preprint*, University of Adelaide, 2000.

[3] Feller, W. *An Introduction to Probability Theory and Its Applications*. Wiley & Sons Inc., New York, 2, 1971.

[4] Hu, Y. and Oksendal, B. Fractional white noise analysis and applications to finance, *Preprint*, University of Oslo, 1999.

[5] Kallianpur, G. and Karandikar, R. *Introduction to Option Pricing Theory*. Birkhauser, Boston, 2000, p.269.

[6] Korn, R. and Kraft, H. A stochastic control aproach to portfolio problems with stochastic interest rates. *SIAM J. Control Optim.* 2001, 40, 1250-1269.

[7] Korn, R. and Kraft, H. On the stability of continuous-time portfolio problems with stochastic opportunity set. *Mathematical Finance*, v. 14, no. 3, July 2004, 403-414.

[8] Kozachenko, Y. V. and Vasilik, O. I. On the distribution of suprema of random processes. *Theory of Random Processes*, v. 4(20), no. 1-2, 1998, pp 147-160.

[9] Mandelbrot, B. B. *Fractionals and Scaling in Finance: Discontinuity, Concentration, Risks*. Springer-Verlag, 1997.

[10] Mishura, Y. *Stochastic Calculus for fBM and Related Processes*. Springer, 2008.

[11] Mishura, Y. and Swishchuk, A. Stochastic stability of fractional (B,S)-security markets. *Applied Stat., Financial and Insur. Math.*, 2000, v. 52, n.2.

[12] Sandmann, K. and Sondermann, D. A note on the stability of lognormal interest rate models and the pricing of Eurodollar futures. *Mathematical Finance*, 1997, 7, 119-125.

[13] Sandmann, K. and Sondermann, D. On the stability of lognormal interest rate models. *Working paper B-263*, University Bonn, 1993.

[14] Mishura, Y. and Valkeila, E. On arbitrage in the mixed Brownian-fractional-Brownian market model. *Preprint 261*, May 2000, University of Helsinki.

[15] Shyriaev, A. N. An arbitrage and replication for fractional models: In A. Shiryaev and A. Sulem (editors). Workshop on Mathematical finance, INRIA, Paris, 1998.

[16] Swishchuk, A.V. *Random Evolutions and Their Applications. New Trends*. Kluwer AP, Dordrecht, The Netherlands, 2000, p.315.

[17] Swishchuk, A. and Kalemanova, A. Stochastic stability of interest rates with jumps. *Theory Probability and Mathematical Statistics*, v. 61, 2000.

[18] Zaehle, M. Integration with respect to fractional functions and stochastic integrals. *Probability Theory and Related Fields*, 111, 1997, 333-374.

Chapter 11

Stability of RDS with Jumps in Interest Rate Theory

11.1 Chapter overview

In this chapter, we study the stochastic stability of random dynamical systems arising in the interest rate theory. We introduce different definitions of stochastic stability. Then, the stochastic stability of interest rates for the Black-Scholes, Vasicek, Cox-Ingersoll-Ross models and their generalizations for the case of random jump changes are studied.

11.2 Introduction

The change process of bond values can be described by diverent analytical models. The simplest example of a bond is a bank account with a constant interest rate r. The value of this bond is changing according to the compound percent formula, $dB_t = rB_t dt$; at a moment $t \geq 0$ it equals $B_t = B_0 e^{rt}$, where B_0 is the total amount of the account at the initial time $t = 0$.

The assumption that the interest rate is constant makes the model simpler. However this assumption is realistic only for a short time period.

The bank interest rate is, in fact, varying; it depends on many circumstances, and time is the first of them. The bank interest rate can be given as a deterministic function of time $r = (r_t)_{t \geq 0}$. The formula for the bond value is as follows:

$$B_t = B_0 e^{\int_0^t r_s ds}.$$

It is natural to consider generalizations of analytical bond models which include dependencies of the value on both the time and random circumstances. There are two approaches for doing so: the explicit and implicit approaches.

When applying the explicit approach, one defines the bond value process $B = (B_t)_{t \geq 0}$ directly as the solution of a stochastic differential equation

$$dB_t = B_t[\mu(t, \omega)dt + \sigma(t, \omega)dW_t].$$

In the case of implicit approach, the dependence of B_t on random circumstances appears via an auxiliary stochastic process $r = (r_t(\omega))_{t\geq 0}$ which has the meaning of an instant value of the interest rate and satisfies the following stochastic direrential equation $dr_t = a(t,r_t)dt + b(t,r_t)dW_t$, where W_t is a Wiener process.

Under the condition that $B(t;r_t)$ is smooth enough with respect to t and r and due to the Ito formula, the following representation for the bond value process holds in the form of a stochastic differential: $dB(t,r_t) = B(t,r_t)[\alpha(t,r_t)dt + \beta(t;r_t)dW_t]$. The explicit and implicit approaches for describing the bond value processes are considered, for example, in [2, 4, 7, 10, 12].

11.3 Definition of the stochastic stability

We consider the equation

$$\frac{dx}{dt} = G(x,t,\xi(t,\omega)),\ x(t_0) = x_0, \tag{11.3.1}$$

and assume that $G(0,\xi(t,\omega)) = 0$, where $\xi(t)$ is a stochastic process.

Definition 2.1. The trivial solution $x(t) = 0$ of equation (11.3.1) is called *stable in probability* as $t \geq 0$ if for all $\varepsilon > 0$ and $\delta > 0$ there exists $r > 0$ such that

$$P\{|x(t,\omega,t_0,x_0)| > \varepsilon\} < \delta$$

; for $t \geq 0t_0$ and $|x_0| < r$. Here $x(t,\omega,t_0,x_0)$ is the solution of equation (11.3.1).

Definition 2.2. The trivial solution of equation (11.3.1) is called *asymptotically stable in probability*if it is stable in probability and for all $\varepsilon > 0$ there exists $r = r(\varepsilon)$ such that

$$P\{|x(t,\omega,t_0,x_0)| > \varepsilon\} \to 0, t \to \infty$$

Definition 2.3. The trivial solution of equation (11.3.1) is called *p-stable in probability*if for all $\varepsilon > 0$ there exists $r > 0$ such that

$$E\{|x(t,\omega,t_0,x_0)|^p < \varepsilon, p > 0$$

for $t \geq t_0$ and $|x_0| < r$.

Definition 2.4. The trivial solution of equation (11.3.1) is called *asymptotically p-stable* if it is p-stable and

$$E\{|x(t,\omega,t_0,x_0)|^p < \varepsilon, p > 0$$

for sufficiently small $|x_0|$.

Theorem 11.1 *Let $x(t)$ be the general solution of equation (11.3.1) and*

$$\lambda' = lim_{t\to\infty}\frac{\ln x(t)}{t} \tag{11.3.2}$$

If $\lambda' < 0$, then the trivial solution of equation (11.3.1) is stable in probability. For proof see [9].

11.4 The stability of the Black-Scholes model

We consider the continuous Black-Scholes model of the (B,S) securities market. The bond value process $(S_t)_{t\geq 0}$ is described by the stochastic diifferential equation

$$dS_t = S_t(\mu dt + \sigma dW_t), S_0 \leq 0. \tag{11.4.1}$$

The bond value $(B_t)_{t\geq 0}$ itself is described by the differential equation

$$dB_t = rB_t dt, \quad B_0 \geq 0. \tag{11.4.2}$$

Theorem 11.2

1. *If $\mu < \frac{\sigma^2}{2}$, then the solution $S_t = 0$ of equation (11.4.1) is stable in probability.*
2. *$\mu < -\frac{1}{2}\sigma^2(p-1)$, then the solution $S_t = 0$ of equation (11.4.1) is asymptotically p-stable.*

Proof

1. *We find the Lyapunov index and use Proposition 11.1. The general solution of equation (11.4.1) is written in the form*

$$S_t = S_0 e^{(\mu - \frac{\sigma^2}{2})t + \sigma W_t}. \tag{11.4.3}$$

Then

$$\lambda' = \lim_{t\to+\infty} \frac{\ln|S_t|}{t} = \lim_{t\to+\infty} \frac{\ln(S_0) + (\mu - \sigma^2/2)t + \sigma W_t}{t} = \mu - \sigma^2/2.$$

2. *We find the expression for the p th moment of S_t*

$$\begin{aligned} ES_t^p &= E\left(S_0 e^{p(\mu - \frac{\sigma^2}{2})t + p\sigma W_t}\right) \\ &= S_0 e^{p(\mu \frac{\sigma^2}{2})t} E e^{p\sigma W_t} \\ &= S_0 e^{p(\mu - \frac{\sigma^2}{2})t + \frac{p^2\sigma^2}{2}t} \\ &= S_0^p e^{p(\mu + \frac{\sigma^2}{2}(p-1))t}. \end{aligned}$$

Assertion b) follows from the asymptotic p-stability.

Consider the discount value process of a share

$$X_t = \frac{S_t}{B_t}$$

which is written in the form

$$X_t = \frac{S_0}{B_0} e^{(\mu - r - \frac{\sigma^2}{2})t + \sigma W_t}$$

and is a solution of the equation

$$dX_t = X_t(\mu - r)dt + \sigma dW_t). \tag{11.4.4}$$

As a consequence of Proposition 11.2 we obtain conditions for the stability of the zero solution of this equation.

Corollary.

1. If $\mu < \frac{\sigma^2}{2} + r$, then the solution $X_t = 0$ of equation (11.4.4) is stable in probability.

2. $\mu < r - \frac{1}{2}\sigma^2(p-1)$, then the solution $X_t = 0$ of equation (11.4.4) is asymptotically p-stable.

Remark. In Rendleman and Bartter's model [10] the risk-neutral process for interest rate r_t is

$$dr = \mu r dt + \sigma r dW_t,$$

which means that r follows geometric Brownian motion in (11.4.1). We tried to keep the notation S_t in (11.4.1) consistent with Balck-Scholes analysis for the stock price S_t.

11.5 A model of (B,S)- securities market with jumps

Assume that the share value process is varying according to (11.4.1) on the intervals $[\tau_i, \tau_{i+1}), i = 1, 2, \ldots,$ while, at the random times τ_i, the share values jump, namely

$$S_{\tau_i} - S_{\tau_i^-} = S_{\tau_i} - U_i$$

or

$$S_{\tau_i} = (1 + U_i) S_{\tau_i^-}. \tag{11.5.1}$$

We assume that the total number of jumps on the interval $[0,t]$, denoted by N_t, is a Poisson process with intensity $\lambda > 0$. We also assume that the jumps $(U_i)_{i \geq 0}$ form a sequence of independent identically distributed random variables assuming values in $(-1, \infty)$.

Theorem 11.3 *The share value process described above is of the form:*

$$S_t = S_0 \left(\prod_{i=1}^{N_t} (1 + U_i) \right) e^{\left(\mu - \frac{\sigma^2}{2}\right)t + \sigma W_t}. \tag{11.5.2}$$

Proof *According to (11.4.3) the share value process on the interval* $[0,\tau_1)$ *is of the form:*

$$S_t = S_0 e^{\left(\mu-\frac{\sigma^2}{2}\right)t+\sigma W_t}.$$

Assume that (11.5.2) holds for $t \in [\tau_k, \tau_{k+1})$, *that is,*

$$S_t = S_0\left(\prod_{i=1}^{N_t}(1+U_i)\right)e^{\left(\mu-\frac{\sigma^2}{2}\right)t+\sigma W_t}.$$

In particular,

$$S_{\tau_{k+1}^-} = S_0(1+U_1)\dots(1+U_k)e^{\left(\mu-\frac{\sigma^2}{2}\right)\tau_{k+1}+\sigma W_{\tau_{k+1}}}.$$

Using (11.5.1), we get:

$$\begin{aligned} S_{\tau_{k+1}} &= S_{\tau_{k+1}}e^{\left(\mu-\frac{\sigma^2}{2}\right)(t-\tau_{k+1})+\sigma(W_t-W_{\tau_{k+1}}} \\ &= S_0(1+U_1)\dots(1+U_k)e^{\left(\mu-\frac{\sigma^2}{2}\right)\tau_{k+1}+\sigma W_{\tau_{k+1}}}. \end{aligned}$$

Further, we have on the interval $[\tau_{k+1}, tau_{k+2}]$ *that*

$$\begin{aligned} S_t &= S_{\tau_{k+1}}e^{\left(\mu-\frac{\sigma^2}{2}\right)(t-\tau_{k+1})+\sigma(W_t-W_{\tau_{k+1}}} \\ &= S_0(1+U_1)\dots(1+U_k)e^{\left(\mu-\frac{\sigma^2}{2}\right)t+\sigma W_t} \\ &= S_0\left(\prod_{i=1}^{N_t}(1+U-i)\right)e^{\left(\mu-\frac{\sigma^2}{2}\right)t+\sigma W_t} \end{aligned}$$

Therefore we proved by the mathematical induction principle that the share value process satisfies (11.5.2) for all $t \geq 0$.

The share value process can be represented in the form of a stochastic integral equation, namely:

$$S_t = S_0 + \int_0^t S_u(\mu du + \sigma dW_t) + \sum_{i=1}^{N_t} S_{\tau_i^-}U_i. \tag{11.5.3}$$

Lemma 1. Let $(V_n)_{n\geq 1}$ be a sequence of independent identically distributed non-negative random variables, $EV_n < +\infty$, and $(N_t)_{t\geq 0}$ be a Poisson process, independent of $(V_n)_{n\geq 1}$ and having parameter λ. Then

$$E\left(\prod_{n=1}^{N_t}V_n\right) = e^{\lambda t(E(V1)-1)}$$

for all $t \geq 0$.

Proof *Since $E(V1) = E(V_n)$ for all $n \geq 1$, we have for all $t \geq 0$ that*

$$
\begin{aligned}
E\left(\prod_{n=1}^{N_t} V_n\right) &= E(V_1)P\{N-t=1\} + E(V_1)E(V_2)P\{N_t=2\} + \dots \\
&\quad + E(V_1)\dots E(V_n)P\{N_t=n\} + \dots \\
&= \sum_{n=1}^{\infty} [E(V_1)]^n e^{-\lambda t} \frac{()\lambda t)^n}{n!} = e^{\lambda t(M(V-1)-1)}
\end{aligned}
$$

Theorem 11.4

1. *Let $E \parallel \ln(1+U-1) \parallel < \infty$. The zero solution of equation (11.5.3) is stable in probability if $\mu < \frac{\sigma^2}{2} - \lambda E(\ln(1+U_1)$.*
2. *Let $E \parallel 1+U_1 \parallel^p < +\infty$. Then the zero solution of equation (11.5.3) is asymptotically p-stable if $\mu < -\frac{1}{2}\sigma^2(p-1) - \lambda p^{-1}(E(1+U_1)^p - 1)$.*

Proof *The proof of the proposition follows from (11.5.2) and the following representations for the Lyapunov index λ' and pth moment of the process S_t:*

$$
\begin{aligned}
\lambda' &= \lim_{t\to\infty} \frac{\ln|S_t|}{t} \\
&= \lim_{t\to\infty} \frac{\ln S_0 + \ln\left(\prod_{i=1}^{N_t}(1+U_i)\right) + \sigma W_t + e^{\left(\mu-\frac{\sigma^2}{2}\right)t}}{t} \\
&= \mu - \frac{\sigma^2}{2} + lim_{t\to\infty} \frac{N_t}{t} \cdot \frac{1}{N_t} \sum_{i=1}^{N_t} \ln(1+U_i) = \mu - \frac{\sigma^2}{2} + \lambda E(\ln(1+U_1)),
\end{aligned}
$$

since $\frac{N_t}{t} \to \lambda$ as $t \to \infty$ and $N_t^{-1} \sum_{i=1}^{N_t} \ln(1+U_i) \to E(\ln(1+U_1))$ as $t \to \infty$ by the law of large numbers, and

$$
\begin{aligned}
ES_t^p &= S_0^p E\left(\prod_{i=1}^{N_t}(1+U_i)^p e^{p\left(\mu-\frac{\sigma^2}{2}\right)t + p\sigma W_t}\right) \\
&= S_0^p e^{p\left(p\mu + p(p-1)\frac{\sigma^2}{2} + \lambda(E(1+U_1)^p - 1)t\right)t},
\end{aligned}
$$

by Lemma 11.1.

Now we obtain a corollary for the case of the equation

$$
X_t = \frac{S_0}{B_0} + \int_0^t X_u((\mu - r)d\mu\sigma dW_u) + \sum_{i=1}^{N_t} X_{\tau_i} - U_i, \tag{11.5.4}
$$

whose solution is the discount share value with random jumps:

$$X_t = \frac{S_0}{B_0}\left(\prod_{i=1}^{N_t}(1+U_i)\right) e^{\left(\mu - r - \frac{\sigma^2}{2}\right)t + p\sigma W_t} \tag{11.5.5}$$

Corollary The zero solution of equation (11.5.5) is stable in probability if

$$\mu < \frac{\sigma^2}{2} - E(\ln(1+U_1)) + r$$

and asymptotically p-stable if

$$\mu < -\frac{1}{2}\sigma^2(p-1) - \lambda p^{-1}E(1+U_1)^p - 1) + r.$$

11.6 Vasicek model for the interest rate

The interest rate process in the Vasicek model [12] is described by the following equation

$$dr_t = (c - fr_t)dt + \sigma dW_t, \;\; r_t|_{t=0} = r_0, \tag{11.6.1}$$

where $c \in \mathbf{R}, f > 0$, and $\sigma > 0$.

The following proposition shows that the interest in the Vasicek model is varying about $\frac{c}{f}$ with variance $\frac{\sigma^2}{2f}$.

Theorem 11.5 *The process*

$$r_t = \frac{c}{f}(r_0 - \frac{c}{f})e^{-ft} + \sigma e^{-ft}\int_0^t e^{fs}dW_s$$

is the solution of stochastic differential equation (11.6.1). Its expectation tends to $\frac{c}{f}$ *as* $t \to \infty$ *and variance tends to* $\frac{\sigma^2}{2f}$ *as* $t \to \infty$.

Proof *Using* Ito *formula we change the stochastic process*

$$V_t = e^{ft}r_t$$

in stochastic differential equation (11.6.1) and obtain an equation for V_t*:*

$$dV_t = cedt + \sigma e^{ft}dW_t \tag{11.6.2}$$

$$dV_t = V_0 - \frac{c}{f} + \frac{c}{f}e^{ft} + \sigma\int_0^t e^{fs}dW_s \tag{11.6.3}$$

is the solution of the latter equation.

Coming back to the initial process, we get the solution of equation (11.6.1):

$$r_t = \frac{c}{f}(r_0 - \frac{c}{f})e^{-ft} + \sigma e^{-ft}\int_0^t e^{fs}dW_s \tag{11.6.4}$$

It remains to determine the expectation and variance for the process (11.6.4). The mean value is:

$$Er_t = \frac{c}{f}(r_0 - \frac{c}{f})e^{-ft} \to \frac{c}{f},\ t \to \infty.$$

The second moment is:

$$\begin{aligned}
Er_t^2 &= \frac{c^2}{f^2} + (r_0 - \frac{c}{f})^2 e^{-2ft} + 2\frac{c}{f}(r_0 - \frac{c}{f})e^{-ft} \\
&\quad + \sigma^2 e^{-2ft}\int_0^t e^{2fs}ds \\
&= \frac{c^2}{f^2} + (r_0 - \frac{c}{f})^2 e^{-2ft} + 2\frac{c}{f}(r_0 - \frac{c}{f})e^{-ft} + \frac{\sigma^2}{2f} - \frac{\sigma^2}{2f}e^{-2ft} \\
&\to \frac{c^2}{f^2} + \frac{c^2}{2f}, t \to \infty
\end{aligned}$$

In this way, the limit for the varince is

$$\lim_{t\to+\infty} Var(r_t) = \frac{c^2}{2f}.$$

11.7 The Vasicek model of the interest rate with jumps

Assume in the Vasicek model that the interest process is continuous on time intervals $[\tau_i, \tau_{i+1}), i = 1, 2, \ldots$. At random times τ_i the interest jumps:

$$r_{\tau_i} = (1 + U_i)\tau_{\tau_i^-}.$$

The number of jumps on the interval $[0,t]$, denoted by N_t, is assumed to be a Poisson process with intensity λ. The jumps $(U_i)_{i\geq 0}$ form a sequence of independent identically distributed random variables assuming values in $(-1,\infty)$. Applying results of Sections 11.3 and 11.4, we can represent the interest process in the following form:

$$r_t = \prod_{i=1}^{N_t}(1 + U_t)\left(\frac{c}{f} + \left(r_0 - \frac{c}{f}\right)e^{-ft} + \sigma e^{-ft}\int_0^t e^{fs}dW_s\right) \tag{11.7.1}$$

Let us evaluate the expectation r_t by using Lemma 11.1 and condition $E \parallel U_1 \parallel < +\infty$:

$$\begin{aligned}
Er_t &= e^{\lambda t(E(1+U_1)-1)}\left(\frac{c}{f} + \left(r_0 - \frac{c}{f}\right)e^{-ft}\right) \\
&= e^{(\lambda EU_1 - 1)t}\left(\frac{c}{f}e^{-ft} + r_0 - \frac{c}{f}\right).
\end{aligned}$$

Then

$$\begin{aligned}
\lim_{t\to\infty} Er_t &= \lim_{t\to\infty} e^{\lambda t(E(1+U_1)-1)}\left(\frac{c}{f}+\left(r_0-\frac{c}{f}\right)e^{-ft}\right)\\
&= \lim_{t\to\infty}\frac{cf^{-1}e^{ft}+r_0-cf^{-1}}{e^{(f-\lambda EU_1)t}} = \lim_{t\to\infty}\frac{c}{f-\lambda EU_1}e^{\lambda EU_1 t}\\
&= \begin{cases} 0, & EU_1<0,\\ \frac{c}{f}, & EU_1=0,\\ +\infty, & EU_1>0,\end{cases}
\end{aligned}$$

For the mean square stability, we have

$$\begin{aligned}
Er_t^2 &= e^{\lambda t(E(1+U_1)^2-1)}\left(\frac{c^2}{f^2}+\frac{\sigma^2}{2f}+(r_0-\frac{c}{f})^2e^{-2ft}\right.\\
&\quad \left.-\frac{\sigma^2}{2f}e^{-2ft}+2\frac{c}{f}(r_0-\frac{c}{f})e^{-ft}\right)\\
&= e^{(\lambda(E(1+U_1)^2-1)-2f)t}\\
&\quad\times\left((\frac{c^2}{f^2}+\frac{\sigma^2}{2f}e^{-2ft}+2\frac{c}{f}(r_0-\frac{c}{f})e^{-ft}+(r_0-\frac{c}{f})^2-\frac{\sigma^2}{2f}\right)
\end{aligned}$$

if $E\parallel U_1\parallel^2<+\infty$. Therefore

$$\begin{aligned}
Er_t^2 &= \lim_{t\to\infty} e^{(\lambda t(E(1+U_1)^2-1)-2f)t}\\
&\quad\times\left(\frac{c^2}{f^2}\exp 2ft+2\frac{c}{f}(r_0-\frac{c}{f})e^{ft}+(r_0-\frac{c}{f})^2-\frac{\sigma^2}{2f}\right)\\
&= \lim_{t\to\infty}\frac{\frac{c^2}{f^2}\exp 2ft+2\frac{c}{f}(r_0-\frac{c}{f})e^{ft}+(r_0-\frac{c}{f})^2-\frac{\sigma^2}{2f}}{e^{2f-\lambda(E(1+U_1)^2-1))t}}\\
&= \lim_{t\to\infty}\frac{\left(2\frac{c^2}{f}+\sigma^2\right)\exp 2ft+2c(r_0-\frac{c}{f})e^{ft}}{(2f-\lambda(E(1+U_1)^2-1))e^{(2f-\lambda(E(1+U_1)^2-1))t}}\\
&\quad\times\frac{1}{e^{f-\lambda(E(1+U_1)^2-1))t}}\\
&= \lim_{t\to\infty}\frac{2c^2+f\sigma^2)e^{\lambda(E(1+U_1)^2-1)t}}{(2f-\lambda(E(1+U_1)^2-1))(f-\lambda(E(1+U_1)^2-1))}\\
&= \begin{cases} 0, & E(1+U_1)^2<1,\\ \frac{c^2}{f^2}+\frac{\sigma^2}{2f}, & E(1+U_1)^2=1,\\ +\infty, & E(1+U_1)^2>1,\end{cases}
\end{aligned}$$

In this way, the limit for the variance is:

$$\lim_{t\to+\infty} Var(r_t) = \begin{cases} 0, & E(1+U_1)^2 < 1, \\ \frac{\sigma^2}{2f}, & E(1+U_1)^2 = 1, \\ +\infty, & E(1+U_1)^2 > 1, \end{cases}$$

11.8 Cox-Ingersoll-Ross interest rate model

We consider the following stochastic differential equation that describes the Cox-Ingersoll-Ross interest rate model [5]:

$$dr_t = (c - fr_t)dt + \sigma\sqrt{r_t}dW_t, \tag{11.8.1}$$

where $f > 0, \sigma > 0$, and $c > 0$.

Denote by b_t some one-dimensional Brownian motion on $(\Omega, \mathcal{F}, \mathcal{F}_t^W, P)$, and let $b_0 = 0, r_0 \in \mathbf{R}, V_0 = r_0 - \frac{c}{f}, V_s = V_0 + b(\phi_s^{-1})$ for an increasing function ϕ_s such that

$$\phi_t = \int_0^t a^{-2}(\phi_s, V_s)ds$$

where

$$a(s, V_s) = \sigma e^{fs}\sqrt{e^{-fs}V_s + \frac{c}{f}}$$

and $(\cdot)^{-1}$ is the inverse function of ϕ_s.

Theorem 11.6 *The process*

$$r_t = \frac{c}{f} + e^{-ft}V_0 + e^{-ft}b_{\phi_t^{-1}}$$

is a weak solution of equation (11.8.1).

Proof *Equation (11.8.1) is solved by the change of the variable:*

$$V_t = e^{ft}\left(r_t - \frac{c}{f}\right).$$

By the Ito formula we get the equation:

$$dV_t = \sigma e^{ft}\sqrt{e^{-fs}V_s + \frac{c}{f}}dW_t. \tag{11.8.2}$$

Further, changing the time in the Wiener process [8], we obtain the solution of equation (11.8.2):

$$V_t = V_0 + b_{\phi_t^{-1}} = \bar{V}_t \tag{11.8.3}$$

Let us show that the inverse function ϕ_t^{-1} is equal to

$$\phi_t^{-1} = \int_0^t a^2(\phi_s, V_s)ds.$$

As

$$d\phi_t = a^{-2}(\phi_t, V_t)dt,$$

then

$$a^2(\phi_t, V_t)d\phi_t = dt$$

and

$$t = \int_0^t a^2(\phi_s, V_s)d\phi_s.$$

In this way, substituting ϕ_t^{-1} instead of t in the last equality, we get that the inverse function ϕ_t^{-1} has the form

$$\phi_t^{-1} = \int_0^{\phi_t^{-1}} a^2(\phi_s, V_s)d\phi_s = \int_0^t a^2(\phi_s, V_s)sa.$$

Turning to the initial stochastic process

$$r_t = e^{-ft}V_t + \frac{c}{f}, \quad r_0 = V_0 + \frac{c}{f}$$

we obtain the weak solution of equation (11.8.2):

$$r_t = \frac{c}{f} + e^{-ft}(r_0 - \frac{c}{f}) + e^{-ft}b_{\phi_t^{-1}}. \tag{11.8.4}$$

It is easy to see from (11.8.4) that r_t has the finite expectation and variance. From (11.8.4) we get the proof of the proposition, that is,

$$Er_t = \frac{c}{f} + \left(r_0 - \frac{c}{f}\right)e^{-ft} \to \frac{c}{f}, t \to +\infty$$

Also,

$$\begin{aligned}
Er_t^2 &= \frac{c^2}{f^2} + (r_0 - \frac{c}{f})^2 e^{-2ft} + 2\frac{c^2}{f^2} + (r_0 - \frac{c}{f})e^{-ft} \\
&\quad + e^{-2ft}\int_0^t \sigma^2 e^{-2fs}\left(\frac{c}{f} + e^{-fs}(r_0 - \frac{c}{f})\right)ds \\
&= \frac{c^2}{f^2} + (r_0 - \frac{c}{f})^2 e^{-2ft} + 2\frac{c}{f}(r_0 - \frac{c}{f})e^{-ft} \\
&\quad + e^{-2ft}\left(\frac{c\sigma^2}{2f^2}(e^{2ft} - 1) + \frac{\sigma^2}{f}(r_0 - \frac{c}{f})(e^{ft} - 1)\right) \\
&= \frac{c^2}{f^2} + \frac{c\sigma^2}{2f^2} + e^{-ft}\frac{(r_0 - \frac{c}{f})(\sigma^2 + 2c)}{f} \\
&\quad + e^{-2ft}\left((r_0 - \frac{c}{f})^2 + \frac{\sigma^2}{2f}(\frac{c}{f} - 2r_0)\right);
\end{aligned}$$

Finally,

$$Er_t^2 \to \frac{c^2}{f^2} + \frac{c\sigma^2}{2f^2}.$$

Thus, the limit for the variance is:

$$\lim_{t\to+\infty} Var(r_t) = \frac{c\sigma^2}{2f^2}.$$

11.9 Cox-Ingersoll-Ross model with random jumps

We assume that the random jump process satisfies the conditions of Section 11.5. Consider the Cox-Ingersoll-Ross interest rate process with random jumps:

$$r_t = \left(\prod_{i=1}^{N_t}(1+U_i)\right)\left(\frac{c}{f} + e^{-ft}(r_0 - \frac{c}{f}) + e^{-2ft} + 2\frac{c^2}{f^2} + (r_0 - \frac{c}{f})e^{-ft} b_{\phi_t^{-1}}\right). \tag{11.9.1}$$

Now we evaluate the expectation and variance when $t \to +\infty$. The limit for the mean is equal to:

$$\begin{aligned}
\lim_{t\to\infty} Er_t &= \lim_{t\to\infty} e^{\lambda(EU_1 t}\left(\frac{c}{f} + e^{-ft}\left(r_0 - \frac{c}{f}\right)\right) \\
&= \lim_{t\to\infty} \frac{cf^{-1}e^{ft} + r_0 - cf^{-1}}{e^{(f-\lambda EU_1)t}} = \lim_{t\to\infty} \frac{c}{f - \lambda EU_1} e^{\lambda EU_1 t} \\
&= \begin{cases} 0, & EU_1 < 0, \\ \frac{c}{f}, & EU_1 = 0, \\ +\infty, & EU_1 > 0, \end{cases}
\end{aligned}$$

The limit for the 2nd moment is equal to:

$$\begin{aligned}
Er_t^2 &= \lim_{t\to\infty} e^{\lambda EU_1 t}\left(\frac{c}{f} + e^{-ft}\left(r_0 - \frac{c}{f}\right)\right) \\
&\times \left(\frac{c^2}{f^2} + \frac{c\sigma^2}{2f^2} + \exp{-ft}\frac{(r_0 - \frac{c}{f})(\sigma^2 + 2c)}{f} + exp^{-2ft}\left((r_0 - \frac{c}{f})^2 + \frac{\sigma^2}{2f}(\frac{c}{f} - 2r_0)\right)\right. \\
&= \lim_{t\to\infty} \frac{2c^2 + c\sigma^2)e^{\lambda(E(1+U_1)^2 - 1)t}}{(2f - \lambda(E(1+U_1)^2 - 1))(f - \lambda(E(1+U_1)^2 - 1))} \\
&= \begin{cases} 0, & E(1+U_1)^2 < 1, \\ \frac{c^2}{f^2} + \frac{\sigma^2}{2f}, & E(1+U_1)^2 = 1, \\ +\infty, & E(1+U_1)^2 > 1, \end{cases}
\end{aligned}$$

In this way, the limit for the variance is:

$$lim_{t\to+\infty} Var(r_t) = \begin{cases} 0, & E(1+U_1)^2 < 1, \\ \frac{\sigma^2}{2f}, & E(1+U_1)^2 = 1, \\ +\infty, & E(1+U_1)^2 > 1. \end{cases}$$

11.10 A generalized interest rate model

Letting the volatility to be equal to σr_t^γ where $\gamma \in [\frac{1}{2}, +\infty)$ we get the following generalization of the Cox-Ingersoll-Ross model:

$$dr_t = (c - fr_t)dt + \sigma r_t^\gamma W_t. \tag{11.10.1}$$

As in Section 7 we put

$$a(s, V_s) = \sigma e^{fs} \left(e^{-fs} V_s + \frac{c}{f} \right)^\gamma$$

$$\phi_t = \int_0^t s^{-2}(\phi_s, V_s) ds.$$

Theorem 11.7 *The process*

$$r_t = \frac{c}{f} + \exp^{-ft}(r_0 - \frac{c}{f}) + \exp^{-ft} b_{\phi_t^{-1}}$$

is the weak solution of the stochastic differential equation (11.10.1). Its expectation tends to $\frac{c}{f}$ as $t \to +\infty$ and the second moment tends to

$$\frac{c^2}{f^2} + \frac{\sigma^2 c^{2\gamma}}{2 f^{1+2\gamma}}$$

as $t \to \infty$

Proof *As in the preceding model we change the variable*

$$V_t = e^{ft} \left(r_t - \frac{c}{f} \right)$$

and consider the equation

$$dV_t = \sigma e^{ft} \left(e^{-ft} V_t + \frac{c}{f} \right)^\gamma sW_t. \tag{11.10.2}$$

Using the change of time in the Wiener process we get the solution of equation (11.10.3):

$$V_t = V_0 + b_{\phi^{-1}}. \tag{11.10.3}$$

Turning to the initial variable

$$r_t = e^{-ft}V_t + \frac{c}{f}, \; V_0 = r_0 - \frac{c}{f},$$

we obtain the weak solution of equation (11.10.1):

$$() : r_t = \frac{c}{f} + e^{-ft}\left(r_0 - \frac{c}{f}\right) + e^{-ft}b_{\phi-1}. \tag{11.10.4}$$

Further we determine the behavior of the expectation and the second moment at infinity. The limit for the mean is:

$$Er_t = \frac{c}{f} + \left(r_0 - \frac{c}{f}\right)e^{-ft} \to \frac{c}{f}, \quad t \to +\infty.$$

As long as

$$\begin{aligned} Er_t^2 &= \frac{c^2}{f^2} + (r_0 - \frac{c}{f})^2 e^{-2ft} + 2\frac{c^2}{f^2} + (r_0 - \frac{c}{f})e^{-ft} \\ &+ e^{-2ft}\int_0^t \sigma^2 e^{2fs}\left(\frac{c}{f} + e^{-fs}(r_0 - \frac{c}{f})\right)^{\gamma} ds, \end{aligned}$$

the limit for the second moment is:

$$\begin{aligned} \lim_{t\to\infty} Er_t^2 &= \frac{c^2}{f^2} + \lim_{t\to\infty}\frac{\int_0^t \sigma^2 e^{2fs}(\frac{c}{f} + e^{-fs}(r_0 - \frac{c}{f}))^{2\gamma} ds}{e^{2ft}} \\ &= \frac{c^2}{f^2} + \lim_{t\to\infty}\frac{\sigma^2 e^{2fs}(\frac{c}{f} + e^{-fs}(r_0 - \frac{c}{f}))^{2\gamma}}{e^{2ft}} \\ &= \frac{c^2}{f^2} + \lim_{t\to\infty}\frac{\sigma^2}{2f}\left(\frac{c}{f} + e^{-fs}(r_0 - \frac{c}{f}))^{2\gamma}\right) = \frac{c^2}{f^2} + \frac{\sigma^2 c^{2\gamma}}{2f^{1+2\gamma}}. \end{aligned}$$

Finally, the limit for the variance is:

$$\lim_{t\to+\infty} Var(r_t) = \lim_{t\to\infty}\frac{\sigma^2}{2f}\left(\frac{c}{f} + e^{-fs}(r_0 - \frac{c}{f}))^{2\gamma}\right) = \frac{c^2}{f^2} + \frac{\sigma^2 c^{2\gamma}}{2f^{1+2\gamma}}.$$

11.11 A generalized model with random jumps

The generalized interest rate model with random jumps is described by the process

$$r_t = \left(\prod_{i=1}^{N_t}(1+U_i)\right)\left(\frac{c}{f} + e^{-fs}(r_0 - \frac{c}{f})\right) + e^{-ft}b_{\phi_t-1}.$$

Let us evaluate the expectation and second moment of the generalized interest rate process with jump changes:

$$\begin{aligned}
\lim_{t\to\infty} Er_t &= \lim_{t\to\infty} e^{\lambda(EU_1 t}\left(\frac{c}{f}+e^{-ft}\left(r_0-\frac{c}{f}\right)+e^{-ft}b_{\phi_t^{-1}}\right)\\
&= \lim_{t\to\infty}\frac{cf^{-1}e^{ft}+r_0-cf^{-1}}{e^{(f-\lambda EU_1)t}} = \lim_{t\to\infty}\frac{c}{f-\lambda EU_1}e^{\lambda EU_1 t}\\
&= \begin{cases} 0, & EU_1<0,\\ \frac{c}{f}, & EU_1=0,\\ +\infty, & EU_1>0,\end{cases}
\end{aligned}$$

and

$$\begin{aligned}
\lim_{t\to\infty} Er_t^2 &= \lim_{t\to\infty} e^{\lambda(E(1+U_1)^2-1)t}\\
&\quad \times(\frac{c^2}{f^2}+\exp-2ft(r_0-\frac{c}{f})^2+2\frac{c}{f}(r_0-\frac{c}{f})e^{-ft}\\
&= \lim_{t\to\infty}\frac{2c^2+c\sigma^2)e^{\lambda(E(1+U_1)^2-1)t}}{(2f-\lambda(E(1+U_1)^2-1))(f-\lambda(E(1+U_1)^2-1))}\\
&\quad +e^{-2ft}\int_0^t\sigma^2e^{2fs}\left(e^{-fs}(r_0-\frac{c}{f})+\frac{c}{f}\right)^{2\gamma}ds)\\
&= \begin{cases} 0, & E(1+U_1)^2<1,\\ \frac{c^2}{f^2}+\frac{\sigma^2}{2f}, & E(1+U_1)^2=1,\\ +\infty, & E(1+U_1)^2>1,\end{cases}
\end{aligned}$$

In this way, the limit for the variance is:

$$\lim_{t\to+\infty} Var(r_t)=\begin{cases} 0, & E(1+U_1)^2<1,\\ \frac{\sigma^2}{2f}, & E(1+U_1)^2=1,\\ +\infty, & E(1+U_1)^2>1,\end{cases}$$

References

[1] Aase, K. K., Contingent claims valuation when the security price is a combination of an Ito process and a random point process, *Stoch. Processes Appl.* 28 (1988), 185-200.

[2] Ball, C. A. and Torons, W. N., Bond price dynamics and options. *J. Financial and Quantitative Analysis*, 18 (1983), no. 4, 517-531.

[3] Black, F. and Scholes M., The pricing of options and cooperate liabilities. *J. Political Economy*, 3 (1973), 637-659.

[4] Black, F., Derman, E., and Toy, W., A one-factor model of interest rates and its applications to Treasury bond options. *Financial Anal. J.* (1990), 33-39.

[5] Cox, J. C., Ingersoll, J. E., and Ross, S. A., A theory of the term structure of interest rates, *Econometrica*, 53 (1985), 385-407.

[6] Gikhman, I. I. and Skorokhod, A. V., *Stochastic differential equations, Naukova Dumka, Kiev, 1968; English transl.*, Springer Verlag, Berlin, 1972.

[7] Heath, D., Jarrow, R., and Morton, A., Bond pricing and the term structure of interest rates: a new methodology for contingent claims valuation, *J. Financial and Quantitative Analysis* 25 (1990), no. 3, 419-440.

[8] Ikeda, N. and Watanabe, S., *Stochastic differential equations and diffusion processes,* North Holland Publishing Company, Amsterdam-Oxford-New York, 1981.

[9] Khas'minskii, R. Z., *Stability of systems of differential equations under random perturbations of their parameters,* "Nauka," Moscow, 1969; English transl., Sijthoff and Noordhoff, Alphen aan Rijn, 1980.

[10] Rendleman, R. and Bartter, B., The pricing of options on debt securities, *J. Finan. Quant. Anal.*, 15 (March 1980), 11-24.

[11] Swishchuk, A. and Kalemanova, A., The stochastic stability of interest rates with jumps. *Theory Probab. Math. Stat.*, 2000, v. 61.

[12] Vasicek, O., An equilibrium characterization of the term structure, *J. Financial Economics*, 5 (1977), 177-188.

Chapter 12

Stability of Delayed RDS with Jumps and Regime-Switching in Finance

12.1 Chapter overview

The processes of financial mathematics in conditions of the (B,S)-market is patterned by stochastic differential equations. The complexity of financial relations and reality of markets impose diverse, often complex, conditions that have their influence on dynamics of securities. Taking those influences into account is a serious and actual problem nowadays. It demands applying deep and various mathematical methods and approaches.

The subject of this chapter is stability of trivial solution of stochastic differential delay Ito's equations with Markovian switchings and with Poisson bifurcations. Throughout the work stochastic analogue of second Lyapunov's method is used having been described in [7]. The issues [3] and [1] contain general theory of differential delay equations. Some results on stochastic differential equations are located in [2], [12], and [9]. Survey paper [4] contains general survey of some results on SDDE. The subject of [8] is SDDE with Markovian switchings of their parameters.

12.2 Stochastic differential delay equations with Poisson bifurcations

Let stochastic processes $\{x(t), t \in [-h,T]\} \in R^n$, $\{\phi(\theta), \theta \in [-h,T]\} \in R^n$, scalar process of Brownian motion $\{W(t), t \in [0,T]\}$ and $\{\nu(du,dt), t \in [0,T], u \in [-1,+\infty)\}$ -centralized Poisson measure with parameter $\Pi(du)dt$ [10] be set on the probability space (Ω, Σ, P) and let $\{\phi(\theta)\}$ be independent of the Wiener process and of measure. For all $t \in [-h,T]$ define:

$$x(t) = \begin{cases} \phi(t), & t \in [-h,0]; \\ \phi(0) + \int_o^t f(s,x_s)ds + \int_0^t g(s,x_s)dW(s) + \int_0^t \int_{-1}^{+\infty} ux(s) \cdot \nu(ds,du); \end{cases} \tag{12.2.1}$$

where $f : [0,T] \times D_n[-h,0] \to R^n$, $g : [0,T] \times D_n[-h,0] \to R^n$-are measurable on the set of variables; $x_t = \{x(t+\theta), \theta \in [-h,0]\}$; $D_n[-h,0]$-space of functionals $\{\phi(t), t \in [-h,0]\} \in R^n$ that are right-continuous and have left-sided limits. Let the norm be $\|\phi\| = sup\{|\phi(\theta)|, \theta \in [-h,0]\}$, $phi(0) = x(0) = x$.

Then let $\{x(t), t \geq 0\}$ be called *a solution of SDDE with Poisson jumps*:

$$dx(t) = f(t, x_t)dt + g(t, x_t)dW(t) + \int_{-1}^{\infty} ux(t)\nu(dt, du) \tag{12.2.2}$$

with initial data: $x(t) = \phi(t)$, $t \in [-h, 0]$.

Consider theorem of existness and uniqueness of solution of 12.2.2 [1]:

Theorem 12.1: Let $\{f(t,\phi)\}, \{g(t,\phi)\}$ be continuous and assume Lipshitz's condition:

$$|f(t,\phi) - f(t,\psi)|^2 + |g(t,\phi) - g(t,\psi)|^2 \leq k \cdot \|\phi - \psi\|^2 \tag{12.2.3}$$

for all $\phi, \psi \in D_n$.

If there have been set on the probability space (Ω, Σ, P) Brownian motion $\{W(t), t \in [0,T]\}$, centralized Poisson measure $\{\nu(du, dt)\}$ and, independent of them, process $\{\phi(\theta), \theta \in [-h, 0]\}$ with realizations that don't have any breaks of second type and $E\{\|\phi\|^4\} < \infty$ holds. Then for all $T > 0$ there exists a unique solution of 12.2.2 at $[0, T]$.

12.3 Stability theorems

Assume:

$$(Lv)(s,\phi) = \lim_{t \downarrow 0} \frac{1}{t} \cdot [E\{v(s+t, x_{s+t}(\phi))\} - v(s,\phi)] \tag{12.3.1}$$

Then definition area $D(L)$ of operator L consists of elements of space $C(R_+ \times D_n)$ and hold:

1) For all $(s,\phi) \in R_+ \times D_n$ there exist $\delta > 0$ $C > 0$ such that:

$$\sup_{0<t\leq\delta} \frac{1}{t} |[E\{v(s+t, x_{s+t}(\phi))\} - v(s,\phi)]| \leq C$$

2) In each $(s,\phi) \in R_+ \times D_n$ there exists limit 12.3.1.

Let operator L be called weak infinitesimal operator of process defined by 12.3.1. Let us see which functionals are contained in definition area of L.

Theorem 12.2 [1]: Assume conditions of:

i) uniform boundedness:

$$\sup_{t\geq 0}\{|f(t,0)|^2 + |g(t,0)|^2\} < \infty \tag{12.3.2}$$

ii) Lipschitz, local:

$$\begin{aligned} \forall t \geq 0, r > 0, \{\phi, \psi\} \subset S_r &\equiv \{\phi \in D_n[-h,0] \mid \|\phi\| < r\} : \\ |f(t,\phi) - f(t,\psi)|^2 + |g(t,\phi) - g(t,\psi)|^2 &\leq L_r \cdot \textstyle\int_{-h}^{0} |\phi(\theta) - \psi(\theta)|^2 \cdot d\theta \end{aligned} \tag{12.3.3}$$

Then for functionals:

1) $G(s,\phi(0)) \in D(L)$;

2) $V(s,\phi) = \int_{-h}^{0} l(\theta)\cdot H(s,\phi(\theta),\phi(0))\cdot d\theta \in D(L)$

$$\begin{aligned} 1)\ (LG)(s,\phi) &= \frac{\partial G(s,\phi(0))}{\partial s} + <(\nabla G)(s,\phi(0)), a(s,\phi(0))> \\ &+ \frac{1}{2}sp((\nabla^2 G)(s,\phi(0))\cdot g(s,\phi)\cdot g^T(s,\phi)) \\ &+ \int_{-1}^{+\infty} [G(s,\phi+\phi(0)\cdot y) - G(s,\phi(0)) \\ &- <(\nabla G)(s,\phi(0)), \phi(0)\cdot y>]\Pi(dy) \end{aligned} \tag{12.3.4}$$

$$\begin{aligned} 2)\ (LV(s,\phi)) &= l(0)\cdot H(s,\phi(0),\phi(0)) - l(-h)\cdot H(s,\phi(-h),\phi(0)) \\ &- \int_{-h}^{0} \frac{dl(\theta)}{d\theta}\cdot H(s,\phi(\theta),\phi(0))d\theta \\ &+ \int_{-h}^{0} l(\theta)\cdot (L_2 H)(s,\phi(\theta),\phi(0))d\theta, \end{aligned} \tag{12.3.5}$$

where: L_2 acts at $H(s,\phi(\theta),\phi(0))$ by the first and the third arguments by the rule 12.3.4 with fixed second argument, $<\cdot,\cdot>$ is inner product; $G(s,\phi(0))$ has continuous second derivatives w.r.t. $\phi(0)$; l has a continuous derivative on some open set containing $[-r,0]$; $H(s,\alpha,beta)$, H'_s, H'_β, $H''_{\beta\beta}$ are continuous in s, α and β; ∇ stands for derivative w.r.t. $\phi(0)$.

Consider definition of stability of solution of 12.2.1.
Assume:

$$f(t,0) = g(t,0) = 0 \tag{12.3.6}$$

Then let trivial solution of 12.2.1 be called *stable* if for all $\varepsilon > 0$ there exists $\delta > 0$ such that for all $s \geq 0$ from $\phi \in S_\delta = \{\phi \in D_n \| \|\phi\| < \delta\}$ we impose $x_t(s,\phi) \in S_\varepsilon = \{\phi \in D_n \| \|\phi\| < \varepsilon\}$ for all $t \geq s \geq 0$.

Trivial solution is called *asymptotically stable* if it is stable and for all $s \geq 0$ there exists $\delta > 0$ such that: $\lim_{t\to\infty} E\{|x_t(s,\phi)|^2\} = 0$ for all $\phi \in S_\delta$.

We will investigate the stability of trivial solution of 12.2.1 by stochastic analogue of second method of Lyapunov.

Let us consider a set:

$$W := \{V \in C(R_+ \times D_n) |\ c_1\cdot|\phi(0)|^2 \leq V(s,\phi) \leq c_2\cdot \int_{-h}^{0} |\phi(\theta)|^2 d\theta\}$$

for some $c_1 > 0, c_2 > 0$ and for all $s \in R_+, \phi \in D_n$.

Theorem 12.3 [2]: Assume conditions 12.3.3 and 12.3.6 and existence of functional $V \in W \cap D(L)$ such that $LV \leq -f$, $f \in W$. Then trivial solution of 12.2.1 is asymptotically stable.

12.3.1 *Stability of delayed equations with linear Poisson jumps and Markovian switchings*

Let $\{\Omega, \mathcal{F}, \{\mathcal{F}_t\}_{t\geq 0}, P\}$ be a complete probability space with filtration satisfying the usual conditions (i.e., it is right continuous and $\mathcal{F}_0$ contains all P-null sets). Define stochastic processes $\{x(t), t \in [-h,T]\} \in R^n$, $\{\phi(\theta), \theta \in [-h,0]\} \in R^n$, scalar Brownian motion $\{W(t), t \in [0,T]\}$, centralized Poisson measure $\{\nu(dy,dt), t \in [0,T], y \in [-1,+\infty)\}$ with parameter $\Pi(dy)dt$ [10] and for all $t \in [-h,T]$:

$$\begin{aligned} x(t) &= \phi(t), \quad t \in [-h,0]; \phi(0) + \int_o^t [a(r(s))x(s) + \mu(r(s))x(s-\tau)]ds \\ &+ \int_0^t \sigma(r(s))x(s-\rho)dW(s) + \int_0^t \int_{-1}^{+\infty} yx(s) \cdot \nu(dy,ds). \end{aligned} \tag{12.3.7}$$

Here $a(\cdot), \mu(\cdot), \sigma(\cdot)$ are matrix maps with dimension $n \times n$ acting from set $S = \{1,2,...,N\}$; $\tau > 0$, and $\rho > 0$; $\{r(t), t \in [0,+\infty)\}$ is Markovian chain taking values at the set S with generator $\Gamma = (\gamma_{ij})_{N\times N}$:

$$P(r(t+\delta) = j | r(t) = i) = \begin{cases} \gamma_{ij}\delta + o(\delta), i \neq j \\ 1 + \gamma_{ii}\delta + o(\delta), i = j \end{cases}$$

where $\delta > 0, \gamma_{ij} \geq 0$ and $\gamma_{ii} = -\sum_{i\neq j} \gamma_{ij}$. Assume $r(\cdot)$ is independent of $W(\cdot)$.

We call $x(t)$ a solution of stochastic differential delay equation with linear Poisson jumps and with Markovian switchings:

$$\begin{aligned} dx(t) &= [a(r(t))x(t) + \mu(r(t))x(t-\tau)]dt + \sigma(r(t))x(t-\rho)dW(t) \\ &+ \int_{-1}^{\infty} yx(t)\nu(dy,dt) \end{aligned} \tag{12.3.8}$$

Theorem 12.4:
Equation 12.3.8 has unique solution $x(t)$ of 12.3.7 for $t \in [-h,+\infty)$.

Proof There exists a sequence $\{\tau_k\}_{k\geq 0}$ of stopping times that $0 = \tau_0 < \tau_1 < ... < \tau_k \to \infty$ and $r(t)$ is constant at each interval $[\tau_k, \tau_{k+1})$:

$$\forall k \geq 0 : r(t) = r(\tau_k) \quad \tau_k \leq t < \tau_{k+1}.$$

Consider equation 12.3.8 at $t \in [0, \tau_1 \wedge T]$, that is:

$$\begin{aligned} dx(t) &= [a(r(0))x(t) + \mu(r(0))x(t-\tau)]dt + \\ &+ \sigma(r(0))x(t-\rho)dW(t) + \int_{-1}^{\infty} yx(t)\nu(dy,dt) \end{aligned} \tag{12.3.9}$$

with initial data $x(t) = \phi(t), t \in [-h,0]$, where ϕ is a continuous process. Using theorem 12.1 we know that equation 12.3.9 has unique right-continuous solution at $[-h, \tau_1 \wedge T]$ with the property: $E[\sup_{-h\leq s\leq \tau_1\wedge T} |x(s)|^2] < +\infty$. At the moment $\tau_1 \wedge T$ we have:

$$x(\tau_1 \wedge T) - x(\tau_1 \wedge T-) = Y_1 x(\tau_1 \wedge T-),$$

where Y_1 is a random variable with distribution $\Pi(dy)$.

Further we consider 12.3.8 at $t \in [\tau_1 \wedge T, \tau_2 \wedge T]$:

$$\begin{aligned} dx(t) &= [a(r(\tau_1 \wedge T))x(t) + \mu(r(\tau_1 \wedge T))x(t-\tau)]dt + \\ &+ \sigma(r(\tau_1 \wedge T))x(t-\rho)dW(t) + \textstyle\int_{-1}^{\infty} yx(t)\nu(dy,dt) \end{aligned} \quad (12.3.10)$$

with initial data assigned by function $x_{\tau_1 \wedge T}$ defined: $x_{\tau_1 \wedge T}(\theta) = x(\theta + \tau_1 \wedge T)$ being a solution of 12.3.9. At the moment $\tau_2 \wedge T$ we have:

$$x(\tau_2 \wedge T) - x(\tau_2 \wedge T-) = Y_2 x(\tau_2 \wedge T-),$$

where Y_2 has the distribution $\Pi(dy)$ and independent on Y_1. We know that 12.3.10 has a unique right-continuous solution at $[\tau_1 \wedge T - h, \tau_2 \wedge T]$. Continuing this procedure we obtain a unique solution $x(t)$ at $[-h, T]$. For T is arbitrarily defined $x(t)$ exists and is unique at $[-h, +\infty]$. Q.E.D.

Assign $C^{2,1}(R^n \times R_+ \times S; R_+)$ the family of all non-negative functions $V(x,t,i)$ at $R^n \times R_+ \times S$ which have continuous second derivatives by x and have continuous derivative by t. For $V \in C^{2,1}(R^n \times R_+ \times S; R_+)$ introduce an operator $LV: R^n \times R_+ \times S \to R$ by rule:

$$\begin{aligned} LV(x,t,i) &= V_t(x,t,i) + V_x(x,t,i) \cdot [a(i)x(t) + \mu(i)x(t-\tau)] \\ &+ \tfrac{1}{2} tr[x^T(t-\rho)\sigma^T(i)V_{xx}(x,t,i)\sigma(i)x(t-\rho)] \\ &+ \textstyle\sum_{j=1}^{N} \gamma_{ij} V(x,t,j) + \int_{-1}^{\infty} [V(x+yx,t,i) - V(x,t,i) \\ &- V_x(x,t,i)yx]\Pi(dy) \end{aligned} \quad (12.3.11)$$

where:

$$V_t(x,t,i) = \frac{\partial V}{\partial t}(x,t,i); \quad V_x(x,t,i) = (\frac{\partial V}{\partial x_1}(x,t,i), \ldots, \frac{\partial V}{\partial x_n}(x,t,i));$$

$$V_{xx}(x,t,i) = (\frac{\partial^2 V}{\partial x_i \partial x_j}(x,t,i))_{n \times n}.$$

Let us introduce here generalized Ito's formula [8].

For $V \in C^{2,1}(R^n \times R_+ \times S; R_+)$ and for all stopping moments $0 \le \rho_1 < \rho_2 < \infty$:

$$EV(x(\rho_2), \rho_2, r(\rho_2)) = EV(x(\rho_1), \rho_1, r(\rho_1)) + E \int_{\rho_1}^{\rho_2} LV(x(s), s, r(s))ds, \quad (12.3.12)$$

whether expectations of all intergrals exist.

Consider the stability theorem.

Theorem 12.5: Assume p, c_1, c_2 are positive integers and $\lambda_1 > 0$ and there exists function $V(x,t,i) \in C^{2,1}(R^n \times R_+ \times S; R_+)$ such that

$$c_1\|x\|^p \le V(x,t,i) \le c_2\|x\|^p \quad (12.3.13)$$

and

$$LV(x,t,i) \le -\lambda_1\|x\|^p \quad (12.3.14)$$

for all $(x,t,i) \in R^n \times R_+ \times S$. Then

$$\lim_{t\to\infty} sup\frac{1}{t}ln(E|x(t)|^p) \leq -\gamma, \tag{12.3.15}$$

where $x(t)$ is solution of 12.3.8 with initial data $\phi(t)$, $\gamma > 0$ is defined by: $\gamma = \lambda_1/c_2$.

In other words, the trivial solution of 12.3.8 is p-exponentially stable and p-exponent of Lyapunov no greater then $-\gamma$.

Proof: Fix $\phi(t) \in C[-h,0]$ is a continuous process. Define $U(x,t,i) = e^{\gamma t}V(x,t,i)$. By Ito's formula:

$$\begin{aligned} EU(x(t),t,r(t)) &= EU(x(0),0,r(0)) + E\int_0^t LU(x(s),s,r(s))ds \\ &= EU(x(0),0,r(0)) + E\int_0^t e^{\gamma t}[\gamma V(x(s),s,r(s)) \\ &+ LV(x(s),s,r(s))]ds \leq c_2 \cdot E\|\phi\|^p \\ &+ (\gamma \cdot c_2 - \lambda_1) \cdot E\int_0^t e^{\gamma s}|x(s)|^p ds. \end{aligned}$$

For $\gamma c_2 = \lambda_1$ we have: $EU(x(t),t,r(t)) \leq c_2 E\|\phi\|^p$.

But $EU(x(t),t,r(t)) \geq c_1 \cdot e^{\gamma t} \cdot E\|x(t)\|^p$. So

$$E\|x(t)\|^p \leq \frac{E\|\phi\|^p}{c_1} c_2 e^{-\gamma t}$$

and demanded inequality follows here. Q.E.D.

12.4 Application in finance

In financial mathematics, equation 12.2.1 describes the process of stock cost S_t with linear Poisson jumps. In this chapter we will consider an equation:

$$dS(t) = [aS(t) + \mu S(t-\tau)]dt + \sigma S(t-\rho)dW(t) + \int_{-1}^{\infty} yS(t)\nu(dt,dy) \tag{12.4.1}$$

Parameters μ σ have a sense of coefficients of growth and volatility.

Equation 12.4.1 without jump component was named after Ito. Solutions of this equation are continuous, but solutions of 12.4.1 at finite interval have finite quantity of breaks of the first type.

Besides stocks, primary securities also include bonds. Values of bond prices are determined and independent of outside factors. It has such a look:

$$dB(t) = [bB(t) + \nu B(t-\beta)]dt \tag{12.4.2}$$

Within conditions of (B,S)-market consider new terms:

Discounted cost of stock : $S_t^* = S_t/B_t$, where S_t and B_t are described by 12.4.1 and 12.4.2.

Capital of holder of securities: $X_t = \alpha_t S_t + \beta_t B_t$, where: (α_t, β_t) is the portfolio of the holder, i.e., quantity of stocks and bonds which holder holds at the time t.

Further we will use self-financing strategies, i.e., portfolio for which: $S_t d\alpha_t + B_t d\beta_t = 0$. Hence, capital X_t satisfies an equation:

$$dX_t = \alpha_t dS_t + \beta_t dB_t.$$

Whether trivial solution of 12.4.1 is stable in the sense of Lyapunov, then according to the definition of stochastic stability it can be treated such as: $S = (S_t)$ is taking values close to 0 with probability 1. The latter is undesirable in the sense of finance. That is, obtaining such restrictions at the parameters we obtain conditions of taking advantage for the holder of securities.

Theorem 12.6: For the functionals G and V we have:

1) $G(s,\phi(0)) \in D(L)$;

2) $V(s,\phi(0)) = \int_{-h}^{0} l(\theta)\cdot H(s,\phi(\theta),\phi(0))d\theta \in D(L)$

where: L is infinitesimal operator.

And:

$$\begin{aligned} 1)\ (LG)(s,\phi) &= \tfrac{\partial G(s,\phi(0))}{\partial s} + < \tfrac{\partial G}{\partial \phi}(s,\phi(0)), a\phi(0) + \mu\phi(-\tau) > \\ &+ \tfrac{1}{2}\tfrac{\partial^2 G}{\partial \phi^2}\sigma^2\phi^2(-\rho) \\ &+ \int_{-1}^{\infty}[G(s,\phi(0)+\phi(0)y) - G(s,\phi(0)) \\ &- \tfrac{\partial G}{\partial \phi}(s,\phi(0))y\phi(0)]\Pi(dy); \end{aligned}$$

$$\begin{aligned} 2)\ (LV(s,\phi)) &= l(0)\cdot H(s,\phi(0),\phi(0)) - l(-h)\cdot H(s,\phi(-h),\phi(0)) \\ &- \int_{-h}^{0} \tfrac{dl(\theta)}{d\theta}\cdot H(s,\phi(\theta),\phi(0))d\theta \\ &+ \int_{-h}^{0} l(\theta)\cdot(L_2 H)(s,\phi(\theta),\phi(0))d\theta \end{aligned}$$

where: L_2 acts at $H(s,\phi(\theta),\phi(0))$ by the first and the third arguments by the rule 1) with fixed second argument.

This thoerem is the consequence of theorem 12.2.

12.5 Examples

1. Consider equations:

$$\begin{cases} dS(t) = [aS(t) + \mu S(t-\tau)]dt + \sigma S(t-\rho)dW(t) \\ dB(t) = [bB(t) + \nu B(t-\beta)]dt. \end{cases} \quad (12.5.1)$$

Let us find restrictions at a,μ,σ,b,ν that trivial solution of 12.5.1 is stochastically stable.

Let us take the functional of Lyapunov:

$$V(x) = \frac{x^2(0)}{2} + A\int_{-\tau}^{0} x^2(\theta)d\theta + B\int_{-\beta}^{0} x^2(\theta)d\theta; \quad where: A > 0, B > 0 \quad (12.5.2)$$

Consider the first equation. The infinitesimal operator has the look:

$$\begin{aligned} LV(x) &= x(0)[ax(0)+\mu x(-\tau)]+\tfrac{\sigma^2}{2}x^2(-\rho)+A[x^2(0)-x^2(-\tau)] \\ &+ B[x^2(0)-x^2(-\rho)]=(A+B+a)x^2(0) \\ &+ (\tfrac{\sigma^2}{2}-B)x^2(-\rho)-Ax^2(-\tau)+\mu x(0)x(-\tau). \end{aligned}$$

Write $LV(x)$ as square form with relations to $[x(0),x(-\tau),x(-\rho)]$.

$$LV(x)=(x(0),x(-\tau),x(-\rho))\cdot\begin{pmatrix} A+B+a & \mu/2 & 0 \\ \mu/2 & -A & 0 \\ 0 & 0 & \sigma^2/2-B \end{pmatrix}\cdot\begin{pmatrix} x(0) \\ x(-\tau) \\ x(-\rho) \end{pmatrix}$$

For square form to be negative defined it is sufficient that all main minors of dimension k have the sign $(-1)^k$. So it is sufficient to satisfy the inequalities:

$$\begin{aligned} A+B+a<0, &\qquad (A+B+a)(-A)-\tfrac{\mu}{2}\cdot\tfrac{\mu}{2}>0, \\ -A<0, &\qquad (-A)(\tfrac{\sigma^2}{2}-B)>0, \\ \tfrac{\sigma^2}{2}-B<0, &\qquad [(A+B+a)(-A)-\tfrac{\mu}{2}\cdot\tfrac{\mu}{2}](\tfrac{\sigma^2}{2}-B)<0. \end{aligned}$$

This system of inequalities is equivalent to:

$$\begin{aligned} A+B+a<0, &\qquad \tfrac{\sigma^2}{2}-B<0, \\ (A+B+a)(-A)> &\qquad \tfrac{\mu^2}{4}. \end{aligned} \tag{12.5.3}$$

The set of $\{A,B\}$ satisfying 12.5.3 is nonempty if:

$$a+\frac{\sigma^2}{2}+|\mu|<0 \tag{12.5.4}$$

Then there exist positive A,B such that the functional V is positively defined and LV is negative defined. It is equivalent to conditions of theorem 12.5 to be performed.

That is, (12.5.4) is condition of stability of trivial solution of (12.5.6).

For the second equation of (12.5.1) we have:

$$b+|\nu|<0. \tag{12.5.5}$$

2. Consider the process of discounted cost of stock: $S_t^*=S_t/B_t$.

For the investigation of stability of trivial portfolio we will use a vector: $X_t=\begin{pmatrix} B_t \\ S_t^* \end{pmatrix}$.

It is easy to see that X_t is satisfying an equation:

$$dX_t=\begin{pmatrix} bB_t+\nu B_{t-\beta} \\ (a-b)S_t^*+\mu S_{t-\tau}^*\frac{B_{t-\tau}}{B_t}-\nu S_t^*\frac{B_{t-\beta}}{B_t}dt \end{pmatrix}+\begin{pmatrix} 0 \\ \sigma\frac{B_{t-\rho}}{B_t}S_{t-\rho}^*dW_t. \end{pmatrix} \tag{12.5.6}$$

Let us write a functional of Lyapunov:

$$V(x) = x_1^2(0)x_2^2(0)/2 + A\int_{-\tau}^{0} x_1^2(\theta)x_2^2(\theta)d\theta + B\int_{-\rho}^{0} x_1^2(\theta)x_2^2(\theta)d\theta;\ A,B > 0$$

$$\begin{aligned}(LV)(x) &= x_1(0)x_2^2(0)[bx_1(0) + \nu x_1(-\beta)] + x_1^2(0)x_2(0)[(a-b)x_2(0)\\
&+ \mu x_2(-\tau)x_1(-\tau)/x_1(0) - \nu x_2(0)x_1(-\beta)/x_1(0)]\\
&+ \tfrac{1}{2}x_1^2(0)\sigma^2 x_2^2(-\rho)x_1^2(-\rho)/x_1^2(0)\\
&+ A[x_1^2(0)x_2^2(0) - x_1^2(-\tau)x_2^2(-\tau)] + B[x_1^2(0)x_2^2(0) - x_1^2(-\rho)x_2^2(-\rho)]\\
&= bx_1^2(0)x_2^2(0) + (a-b)x_1^2(0)x_2^2(0) + \mu x_1(0)x_2(0)x_1(-\tau)x_2(-\tau)+\\
&+ \tfrac{1}{2}\sigma^2 x_1^2(-\rho)x_2^2(-\rho) + Ax_1^2(0)x_2^2(0) - Ax_1^2(-\tau)x_2^2(-\tau)\\
&+ Bx_1^2(0)x_2^2(0) - Bx_1^2(-\rho)x_2^2(-\rho)\\
&= (A+B+a)(x_1(0)x_2(0))^2 + \mu(x_1(0)x_2(0))(x_1(-\tau)x_2(-\tau))\\
&- A(x_1(-\tau)x_2(-\tau))^2 + (\tfrac{1}{2}\sigma^2 - B)(x_1(-\rho)x_2(-\rho))^2.\end{aligned}$$

Let us consider a substitution: $\eta(\theta) := x_1(\theta)x_2(\theta)$.

Then:

$$LV(x) = (a+A+B)\eta^2(0) + \mu\eta(0)\eta(-\tau) - A\eta^2(-\tau) + (\sigma^2/2 - B)\eta^2(-\rho)$$

Write LV as square form with accordance to $[\eta(0), \eta(-\tau), \eta(-\rho)]$:

$$\Lambda = \begin{pmatrix} a+A+B & \mu/2 & 0 \\ \mu/2 & -A & 0 \\ 0 & 0 & \sigma^2/2 - B \end{pmatrix}.$$

Notice that we were also considering the form of such a type in Example 1. Thus the condition of negative definition has the look:

$$a + |\mu| + \sigma^2/2 < 0 \tag{12.5.7}$$

That is, (12.5.7) is a condition of stability of the trivial solution of (12.5.6). Since S_t^* is component of vector of solution so it is also stable within condition (12.5.7). Notice that this condition is imposed only on the parameters of change of the stock's cost.

3. Let us consider the capital process:

$$X_t = \beta_t B_t + \gamma_t S_t$$

Let strategy (β_t, γ_t) be called self-financing if $B_t d\beta_t + S_t d\gamma_t = 0$. Thus $dX_t = \beta_t dB_t + \gamma_t dS_t$.

For investigation of stability consider a system:

$$\begin{cases} X_t = \beta_t B_t + \gamma_t S_t, \\ Y_t = -\beta_t B_t + \gamma_t S_t. \end{cases} \tag{12.5.8}$$

From 12.5.8 we have:

$$\begin{cases} dX_t = \beta_t dB_t + \gamma_t dS_t \\ dY_t = -\beta_t dB_t + \gamma_t dS_t \end{cases}$$

or:

$$\begin{cases} dX_t &= [b\beta_t B_t + \nu\beta_t B_{t-\beta} + a\gamma_t S_t + \mu\gamma_t S_{t-\tau}]dt + \sigma\gamma_t S_{t-\rho} dW_t \\ dY_t &= [-b\beta_t B_t - \nu\beta_t B_{t-\beta} + a\gamma_t S_t + \mu\gamma_t S_{t-\tau}]dt + \sigma\gamma_t S_{t-\rho} dW_t. \end{cases}$$

For simplicity impose $\beta = \rho = \tau = 1$.

Then:

$$\begin{aligned} dX_t &= \tfrac{1}{2}[(a+b)X_t + (\nu\tfrac{\beta_t}{\beta_{t-1}} + \mu\tfrac{\gamma_t}{\gamma_{t-1}})X_{t-1} \\ &+ (a-b)Y_t + (-\nu\tfrac{\beta_t}{\beta_{t-1}} + \mu\tfrac{\gamma_t}{\gamma_{t-1}})Y_{t-1}]dt \\ &+ \tfrac{1}{2}\sigma\tfrac{\gamma_t}{\gamma_{t-1}}(X_{t-1} + Y_{t-1})dW_t \\ dY_t &= \tfrac{1}{2}[(a-b)X_t + (-\nu\tfrac{\beta_t}{\beta_{t-1}} + \mu\tfrac{\gamma_t}{\gamma_{t-1}})X_{t-1} \\ &+ (a+b)Y_t + (\nu\tfrac{\beta_t}{\beta_{t-1}} + \mu\tfrac{\gamma_t}{\gamma_{t-1}})Y_{t-1}]dt \\ &+ \tfrac{1}{2}\sigma\tfrac{\gamma_t}{\gamma_{t-1}}(X_{t-1} + Y_{t-1})dW_t. \end{aligned}$$

Lyapunov's functional:

$$V(x,y) = x^2(0) + y^2(0) + A\int_{-1}^{0} x^2(\theta) + B\int_{-1}^{0} y^2(\theta)$$

Now we have:

$$\begin{aligned} LV(x,y) &= x(0)[(b+a)x(0) + (\nu\tfrac{\beta_t}{\beta_{t-1}} + \mu\tfrac{\gamma_t}{\gamma_{t-1}})x(-1) \\ &+ (-b+a)y(0) + (-\nu\tfrac{\beta_t}{\beta_{t-1}} + \mu\tfrac{\gamma_t}{\gamma_{t-1}})y(-1)] \\ &+ y(0)[(-b+a)x(0) + (-\nu\tfrac{\beta_t}{\beta_{t-1}} + \mu\tfrac{\gamma_t}{\gamma_{t-1}})x(-1) \\ &+ (b+a)y(0) + (\nu\tfrac{\beta_t}{\beta_{t-1}} + \mu\tfrac{\gamma_t}{\gamma_{t-1}})y(-1)] \\ &+ \tfrac{\sigma^2}{2}\tfrac{\gamma_t^2}{\gamma_{t-1}^2}(x(-1) + y(-1))^2 + A[x^2(0) - x^2(-1)] + B[y^2(0) - y^2(-1)]. \end{aligned}$$

Considering it as a square form of $[x(0), y(0), x(-1), y(-1)]$ we impose the matrix with dimension 4×4. Analyzing the values of main minors we find conditions on a, b, μ, ν, σ. That is, we reduce the stability problem to a linear algebra task.

4. In item 1 we imposed conditions on the coeficients of growth and volatility. But parameters τ, ρ, β could be arbitrary. So let us find restrictions on them.

For equation:

$$dx(t) = -bx(t-r)dt, \quad where\ b > 0 \tag{12.5.9}$$

consider an integral:

$$-b\int_{t-2r}^{t-r} x(\theta - r)d\theta = \int_{t-r}^{t} x\prime(t)d\theta = x(t) - x(t-r)$$

Then 12.5.9 is equivalent to:

$$x\prime(t) = -bx(t) - b^2\int_{t-2r}^{t-r} x(\theta)d\theta \tag{12.5.10}$$

The initial data is a function at $[-2r, 0]$. Let us consider 12.5.10 according to a stability problem. Impose:

$$V(x) = x^2(0)/2 + A\int_{-r}^{0} x^2(\theta)d\theta + B\int_{-2r}^{-r}(\theta + 2r)x^2(\theta)d\theta$$

The infinitesimal operator:

$$\begin{aligned} LV(x) &= x(0)[-bx(0) - b^2\textstyle\int_{-2r}^{-r} x(\theta)d\theta] + A[x^2(0) - x^2(-r)] \\ &+ B[(-r+2r)x^2(-r) - (-2r+2r)x^2(-2r) - \textstyle\int_{-2r}^{-r} x^2(\theta)d\theta] \\ &= (A-b)x^2(0) - b^2x(0)rx(\theta) \\ &+ (Br - A)x^2(-r) - Brx^2(\theta), \;\; \theta \in (-2r, -r). \end{aligned}$$

Considering it as a square form impose:

$$0 < br < 1 \tag{12.5.11}$$

5. Consider an equation with jump component:

$$dS_t = (aS_t + \mu S_{t-\tau})dt + \sigma S_{t-\rho}dW_t + \int_{-1}^{+\infty} yS_t\nu(dt, dy) \tag{12.5.12}$$

Similar to preceding:

$$V(x) = x^2(0)/2 + A\int_{-\tau}^{0} x^2(\theta) + B\int_{-\rho}^{0} x^2(\theta)$$

$$\begin{aligned} LV(x) &= x(0)(ax(0) + \mu x(-\tau)) + \tfrac{1}{2}\sigma^2x^2(-\rho) + \tfrac{1}{2}\textstyle\int_{-1}^{+\infty}[(x(0) + yx(0))^2 \\ &- x^2(0) - 2x(0)yx(0)]\Pi(dy) + A[x^2(0) - x^2(-\tau)] + B[x^2(0) - x^2(-\rho)] \end{aligned}$$

$$\Lambda = \begin{pmatrix} a + A + B + \frac{1}{2}\int_{-1}^{+\infty} y^2\Pi(dy) & \mu/2 & 0 \\ \mu/2 & -A & 0 \\ 0 & 0 & \sigma^2/2 - B \end{pmatrix}.$$

Condition of stability has the look:

$$a + \frac{1}{2}\int_{-1}^{+\infty} y^2\Pi(dy) + |\mu| + \sigma^2/2 < 0 \tag{12.5.13}$$

Remark. Stability of stochastic Ito differential delay equations (sidde) is studied in [7]. Stability of sidde with jumps is studied in [11]. Stability of sdde with Markovian switchings is investigated in [8]. Novelty of this chapter is to prove the theorems of existence, uniqueness, and stability for sdde with jumps and Markovian switchings. The results connected with applications of stability theorems with delay to mathematical finance are also new.

References

[1] Bellman, R., and Cook, K. *Differential Delay Equations*. Moscow, Nauka, 1978 (In Russian).

[2] Gihman, I.I., and Skorohod, A.V. *Stochastic Differential Equations*. Moscow, Nauka, 1968 (In Russian).

[3] Hale, J. *Theory of Functional-Differential Equations*. Moscow, 1986 (In Russian).

[4] Ivanov, A.F., and Swishchuk, A.V. Stochastic differential delay equations and stochastic stability: a survey of some results. *SITMS Research Report 2/99*, University of Ballarat, Australia, Jan. 1999.

[5] Khasminsky, R. *Stochastic Stability of Differential Equations*. Moscow, Nauka, 1969 (In Russian).

[6] Kolmanovsky, V., and Nosov, V. *Stability of Functional-Differential Equations*. Moscow, Nauka, 1972 (In Russian).

[7] Kushner, H. On the stability of processes defined by stochastic difference-differential equations. *J. Diff. Eq.*, 4(1968), 424-443.

[8] Mao, X., Matasov, A., Piunovski, A. Stochastic differential delay equations with Markovian switchings. *Bernoulli*, 6(1), 2000, pp.73-90.

[9] Skorohod, A.V. *Asymptotical Methods in the Theory of Stochastic Differential Equations*. *Naukova Dumka*, Kyiv, 1989.

[10] Swishchuk, A. V., and Kazmerchuk, Yu. I. Stability of stochastic Ito equations with delay, Poisson jumps and Markov switchings with applications to finance, *Theory Probab. and Mathem. Statis.*, 2001, v. 64 (translated by AMS, N64, 2002).

[11] Tsarkov, M.L., Sverdan, M.L., and Yasynsky, V.K. Stability in stochastic modelling of the complex dynamical systems. *Preprint,* Institute of Mathematics, Kiev, 1996.

[12] Vatanabe, S., and Ikeda, T. *Stochastic Differential Equations and Diffusion Processes*. Kodansha, Tokyo, 1982.

Chapter 13

Optimal Control of Delayed RDS with Applications in Economics

13.1 Chapter overview

This chapter is devoted to the study of optimal control of random delayed dynamical systems and their applications. By using the Dynkin formula and solution of the Dirichlet-Poisson problem (see Chapter 4), the Hamilton-Jacobi-Bellman (HJB) equation and the inverse HJB equation are derived. Application is given to a stochastic model in economics.

13.2 Introduction

In our presentation at the Conference on Stochastic Modelling of Complex Systems SMOCS05 [7] the following controlled stochastic differential delay equation (SDDE) was introduced:

$$x(t) = x(0) + \int_0^t a(x(s-1),u(s))ds + \int_0^t b(x(s-1),u(s))dw(s),$$

where $x(t) = \phi(t), \quad t \in [-1,0]$, is a given continuous process, $u(t)$ is a control process, and $w(t)$ is a standard Wiener process.

We presented the Dynkin formula and solution of the Dirichlet–Poisson problem for the SDDE. These results can be obtained from the relevant results about the Dynkin formulas and boundary value problems for multiplicative operator functionals of Markov processes [12]. By using the Dynkin formula and the solution of the Dirichlet-Poisson problem, the Hamilton-Jacobi-Bellman (HJB) and the inverse HJB equations have been stated. We have also found the stochastic optimal control and optimal performance for the SDDE. The results there have been presented without proof.

In the present chapter, we give a complete proof of two theorems from the talk: Theorem 1 (HJB equation) and Theorem 2 (inverse of the HJB equation) about the stochastic optimal control. For the definitions related to the stochastic optimal control and stochastic optimal performance see [10]. Application is given to a stochastic model in economics, a Ramsey model [3, 10] that takes into account the delay and randomness in the production cycle.

The Ramsey model is described by the equation

$$dK(t) = [AK(t-T) - u(K(t))C(t)]\,dt + \sigma(K(t-T))dw(t)$$

where K is the capital, C is the production rate, u is a control process, A is a positive constant, σ is a standard deviation of the "noise." The "initial capital"

$$K(t) = \phi(t), \quad t \in [-T, 0],$$

is a continuous bounded positive function. For this stochastic economic model the optimal control is found to be $u_{\min} = K(0) \cdot C(0)$, and the optimal performance is

$$\begin{aligned} J(K, u_{\min}) &= \tfrac{K^2(0)}{2} + \tfrac{K^2(0) \cdot C^2(0)}{2} + \textstyle\int_{-T}^{0} \phi^2(\theta)\,d\theta = \\ &= \tfrac{K^2(0)}{2}(1 + C^2(0)) + \textstyle\int_{-T}^{0} \phi^2(\theta)\,d\theta. \end{aligned}$$

By time rescaling, the delay T can be normalized to $T = 1$, which will be our assumption in the theoretical considerations that follow. The obtained results are valid however for general delay $T > 0$.

13.3 Controlled stochastic differential delay equations

13.3.1 Assumptions and existence of solutions

Below we recall some basic notions and facts from [2, 5, 7, 9] necessary for subsequent exposition in this chapter. Let $x(t), t \in [-1, \infty)$ be a stochastic process, $\mathcal{F}_{\alpha\beta}(x)$ be a minimal σ-algebra with respect to which $x(t)$ is measurable for every $t \in [\alpha, \beta]$. Let $w(t), t \in [-1, \infty)$ be a Wiener process with $w(0) = 0$, and let $\mathcal{F}_{\alpha\beta}(dw)$ be a minimal Borel σ-algebra such that $w(t) - w(s)$ is measurable for all t, s with $\alpha \le t \le s \le \beta$. Let $u(t) \in \mathcal{U}, t \in [-1, \infty)$ be a stochastic process whose values can be chosen from the given Borel set $\mathcal{U}$ and such that $u(t)$ is $\mathcal{F}_{\alpha\beta}(u)$-adapted for all $t \in [\alpha, \beta]$.

Let $\mathcal{C}$ denote the metric space of all continuous functions defined on the interval $[-1, 0]$ with the standard norm $|h| = \sup_{-1 \le t \le 0} |h(t)|$. One also has the notation $h_t(s) := h(t+s), s \in [-1, 0]$. If $h(t)$ is continuous for $t \ge -1$ then $h_t \in \mathcal{C}$. For definitions, notations, and basics of the deterministic differential delay equations, see e.g., [4].

Let $a(\cdot,\cdot), b(\cdot,\cdot)$ be continuous functionals defined on $\mathcal{C} \times \mathcal{U}$. A stochastic process $x(t)$ is called a solution of the stochastic differential delay equation

$$dx(t) = a(x_t, u(t))dt + b(x_t, u(t))dw(t), \quad t \in [0, \infty) \tag{13.3.1}$$

if

$$\mathcal{F}_{-1t}(x) \vee \mathcal{F}_{0t}(dw) \vee \mathcal{F}_{0t}(u)$$

is independent of $\mathcal{F}_{t\infty}(dw)$ for every $t \in [0,\infty)$. Here $\mathcal{F}_{-1t}(x) \vee \mathcal{F}_{0t}(dw) \vee \mathcal{F}_{0t}(u)$ stands for the minimal σ-algebra containing $\mathcal{F}_{-1t}(x)$, $\mathcal{F}_{0t}(dw)$, and $\mathcal{F}_{0t}(u)$, and

$$x(t) - x(s) = \int_s^t a(x_r, u(r))dr + \int_s^t b(x_r, u(r))dw(r),$$

where the last integral is the Ito integral.

Equation (13.3.1) is meant in the integral form

$$x(t) = x(0) + \int_0^t a(x_s, u(s))ds + \int_0^t b(x_s, u(s))dw(s) \qquad (13.3.2)$$

with the initial condition $x(t) = \phi(t), t \in [-1,0]$, where $\phi \in \mathcal{C}$ is a given continuous function. Therefore, we assume that the processes $\phi(t), t \in [-1,0], w(t)$ and $u(t), t \geq 0$, are defined on the probability space $(\Omega, \mathcal{F}, \mathcal{F}_t, P)$ and $\mathcal{F}_t := \mathcal{F}_{-1t}(x) \vee \mathcal{F}_{0t}(dw) \vee \mathcal{F}_{0t}(u)$.

Let the following conditions be satisfied for equation (13.3.2).

A.1 $a(\phi, u)$ and $b(\phi, u)$ are continuous real-valued functionals defined on $\mathcal{C} \times \mathcal{U}$;

A.2 $\phi \in \mathcal{C}$ is continuous with probability 1 in the interval $[-1,0]$, independent of $w(s), s \geq 0$, and $E|\phi(t)|^4 < \infty$;

A.3 $\forall \phi, \psi \in \mathcal{C}$:

$$|a(\phi,u) - a(\psi,u)| + |b(\phi,u) - b(\psi,u)| \leq K \int_{-1}^{0} |\phi(\theta) - \psi(\theta)| d\theta, \qquad (13.3.3)$$

with $|a(0,u)| + |b(0,u)| \leq M$ for some $M, K > 0$ and all $u \in \mathcal{U}$.

Under assumptions A.1-A.3 the solution $x(t)$ of the initial value problem (13.3.2) exists and is a unique stochastic continuous process [2, 5, 9]. The function x_t is a Markov process. The solution can be viewed at time $t \geq 0$ as an element x_t of the space $\mathcal{C}$, or as a point $x(t)$ in $\mathbf{R}$. We shall use both interpretations in this paper, as appropriate.

13.3.2 Weak infinitesimal operator of Markov process $(x_t, x(t))$

In the case of stochastic differential delay equations the solution $x(t)$ is not Markovian. However, we can Markovianize it by considering the pair $(x_t, x(t)) := (x(t+s), x(t))$, $s \in [-1,0]$, i.e., the path of the process from $t-1$ till t and the value of the process at t. The pair is a strong Markov process to which we can apply the weak infinitesimal generator (see e.g., [1]).

A real valued functional $J(x_t, x(t))$ on $\mathcal{C} \times \mathbf{R}$ is said to be in the *domain* of A^u, the weak infinitesimal operator (w.i.o.), if the limit

$$\lim_{t \to 0+}((E^u_{x,x(0)} J(x_t, x(t)) - J(x, x(0)))/t) = q(x, x(0), u),$$
$$x = x_0 = \phi \in \mathcal{C},\ u \in \mathcal{U}$$

exists pointwise in $\mathcal{C} \times \mathcal{U}$, and

$$\lim_{t\to 0+} \sup_{x,u} |E^u_{x,x(0)} q(x_t, x(t), u) - q(x, x(0), u)| = 0.$$

Here $x_t := x_t(\theta) = x(t+\theta), \theta \in [-1,0]$, is in $\mathcal{C}$ and E^u_x is the expectation under the conditional probability with respect to x and u. We set $A^u J(x, x(0)) := q(x, x(0), u)$.

For an open and bounded set $H \times G \subset \mathcal{C} \times \mathbf{R}$ denote by $\tilde{A}^u_{H\times G}$ the w.i.o. of $(\tilde{x}_t, \tilde{x}(t)) := (x_t, x(t))$ stopped at $\tau_{H\times G} := \inf\{t : (x_t, x(t)) \notin H \times G\}$ [1].

Let $F : \mathbf{R} \to \mathbf{R}$ be continuous and bounded on bounded sets and set $J(x_t, x(t)) := F(x(t))$. Then if $F \in D(\tilde{A}^u_G)$ and $\tilde{A}^u_G F = q$ is bounded on bounded sets, the restriction of F to G is in $D(\tilde{A}^u_G)$, and

$$\begin{aligned} \tilde{A}^u_G J(x) &= L^u F(x(0)) = q(x(0), u) := \\ &= F'(x(0)) a(x(0), u) + F''(x(0)) \tfrac{1}{2} b^2(x(0), u) \end{aligned} \tag{13.3.4}$$

where $u = u(0)$ (see [8]).

It is not simple to completely characterize the domain of the weak infinitesimal operator of either processes ϕ or $x(t)$. For example, in the case of $J(x) = x(-1)$ the operator is not necessarily in $D(\tilde{A}^u_G)$, since $x(t)$ can be not differentiable.

It is possible to study functionals $J(x(0))$ whose dependence on $\phi \in \mathcal{C}$ is in the form of an integral. For example, let the above conditions be satisfied for the functional

$$J_\phi(x(0)) := \int_{-1}^{0} F(\phi(s), x(0))\, ds,$$

where $F : \mathcal{C} \times \mathbf{R} \to \mathbf{R}$ is continuous. Let in addition $F(\phi, x)$, $F'_x(\phi, x)$, $F''_{xx}(\phi, x)$ be continuous in ϕ, x. Then $J_\phi(x) \in D(\tilde{A}^u_G)$ and

$$\begin{aligned} \tilde{A}^u_G J_\phi(x(0)) &= q(x(0)) = F(\phi(0), x(0)) - F(\phi(-1), x(0)) + \\ &+ \int_{-1}^{0} L^u F(\phi(s), x(0))\, ds, \end{aligned} \tag{13.3.5}$$

where the operator L^u is defined by (13.3.4) and acts on F as a function of $x(0)$ only (see [8]).

13.3.3 Dynkin formula for SDDEs

Let $x(t)$ be a solution of the initial value problem (13.3.2). For the strong Markov process $(x_t, x(t))$ consider the functional

$$J(x_t, x(t)) := \int_{-1}^{0} F(x(t+\theta), x(t))\, d\theta.$$

From (13.3.5) we obtain the following Ito formula for the functional J :

$$\begin{aligned} & J(x_t, x(t)) = J(x, x(0)) + \int_0^t F(x(s), x(s))\, ds - \\ - & \int_0^t F(x(s-1), x(s))\, ds + \int_0^t \int_{-1}^{0} L^u F(x(t+\theta), x(s))\, d\theta ds \\ + & \int_0^t \int_{-1}^{0} \sigma(x(s-1)) F'_x(x(t+\theta), x(s))\, d\theta dw(s). \end{aligned}$$

Let τ be a stopping time for the strong Markov process $(x_t, x(t))$ such that $E_{x,x(0)}|\tau| < \infty$. Then we have the following Dynkin formula [12]

$$\begin{aligned} & E_{x,x(0)}J(x_\tau, x(\tau)) = J(x, x(0)) + E_{x,x(0)} \int_0^\tau F(x(s), x(s))\,ds \\ - & \int_0^\tau F(x(s-1), x(s))ds + E_{x,x(0)} \int_0^\tau \int_{-1}^0 L^u F(x_s(\theta), x(s)) d\theta ds \\ = & J(x, x(0)) + E_{x,x(0)} \int_0^\tau \tilde{A}^u_G J(x_s, x(s))\,ds, \end{aligned} \tag{13.3.6}$$

where $\tilde{A}^u_G$ is defined by (13.3.5).

13.3.4 Solution of Dirichlet-Poisson problem for SDDEs

Let $G \subset \mathbf{R}$ and $H \subset \mathcal{C}$ be bounded open sets, and $\partial(H \times G)$ be the regular boundary of the set $H \times G$. Let $\psi(x, x(0))$ be a given function continuous on the closure of the set $H \times G$ and bounded on $\partial(H \times G)$. Let function $F(x, x(0), u) \in C(\mathcal{C} \times \mathbf{R} \times \mathcal{U})$ be such that

$$E_{x,x(0)} \left[\int_{-1}^0 \int_0^{\tau_{H\times G}} |F(\phi(\theta), x(s), u(s))|\,ds d\theta \right] < \infty \quad \forall\,(x, x(0)) \in H \times G,$$

where $\tau_{H\times G} = \inf\{t : (x_t, x(t)) \notin H \times G\}$ is the exit time from the set $H \times G$.

Define

$$\begin{aligned} J(x, x(0), u) \;:=\; & E_{x,x(0)} \left[\int_{-1}^0 \int_0^{\tau_{H\times G}} F(x_s(\theta), x(s), u(s))\,ds d\theta \right] \\ + \; & E_{x,x(0)} \left[\psi(x_{\tau_G}, x(\tau_G)) \right], \quad (x, x(0)) \in H \times G. \end{aligned}$$

Then [12]

$$\tilde{A}^u J(x, x(0), u) = -\int_{-1}^0 F(\phi(\theta), x, u)\,d\theta, \quad \text{in} \quad H \times G \quad \forall u \in \mathcal{U}$$

and

$$\lim_{t \uparrow \tau_G} J(x_t, x(t), u) = \psi(x_{\tau_G}, x(\tau_G)) \quad \forall (x, x(0)) \in H \times G.$$

13.3.5 Statement of the problem

We assume that the cost function is given in the form

$$\begin{aligned} J(x, x(0), u) \;:=\; & E_{x,x(0)} \left[\int_{-1}^0 \int_0^{\tau_{H\times G}} F(x_s(\theta), x(s), u(s))\,ds d\theta + \right. \\ + \; & \left. \psi(x_{\tau_{H\times G}}, x(\tau_{H\times G})) \right], \end{aligned} \tag{13.3.7}$$

where ψ, F and $\tau_{H\times G}$ are as in Section 13.3.4. In particular, $\tau_{H\times G}$ can be a fixed time t_0. We assume that $E_{x,x(0)}|\tau_{H\times G}| < \infty, \forall (x, x(0)) \in H \times G$. Similar cost functions are considered in [10] for systems without the delay.

The problem is as follows. For each $(x, x(0)) \in H \times G$ find a number $J^*(x, x(0))$ and a control $u^* = u^*(x, x(0)), \omega)$ such that

$$J^*(x, x(0)) := \inf_u \{J(x, x(0), u)\} = J(x, x(0), u^*),$$

where the infimum is taken over all $\mathcal{F}_t$-adapted processes $u(t) \in \mathcal{U}$. Such a control u^*, if it exists, is called an *optimal control* and $J^*(x, x(0))$ is called the *optimal performance*.

13.4 Hamilton–Jacobi–Bellman equation for SDDEs

We consider only Markov controls $u(t) := u(x_t, x(t))$. For every $v \in \mathcal{U}$ define the following operator

$$\begin{aligned}(A^v J)(x, x(0)) &= F(x(0), x(0), v(0)) - F(x(-1), x(0), v(0)) \\ &+ \int_{-1}^{0} L^v F(\phi(\theta), x(0), v(0))\, d\theta, \quad v(0) := v(x, x(0)),\end{aligned} \tag{13.4.1}$$

where operator L^v is given by (13.3.4), and let

$$J(x, x(0)) := \int_{-1}^{0} F(\phi(\theta), x(0), v(0))\, d\theta.$$

With $x(t)$ being the solution of equation (13.3.2), for each control u the pair $(x_t, x(t))$ is an Ito diffusion with the infinitesimal generator $(AJ)(x, x(0)) = (A^u J)(x, x(0))$.

Theorem 1. (HJB-equation)
Let

$$J^*(x, x(0)) = \inf\{J(x, x(0), u) | u := u(x, x(0)) \text{ - Markov control}\}. \tag{13.4.2}$$

Suppose that $J \in C^2(H \times G)$ and the optimal control u^* exists. Then

$$\inf_{v \in \mathcal{U}} \left[\int_{-1}^{0} F(\phi(\theta), x, v)\, d\theta + (A^v J^*)(x, x(0)) \right] = 0, \quad \forall (x, x(0)) \in H \times G, \tag{13.4.3}$$

and

$$J^*(x, x(0)) = \psi(x, x(0)), \quad \forall (x, x(0)) \in \partial(H \times G),$$

where F and ψ are as in (13.3.7), and operator A^v is given by (13.4.1).

The infimum in (13.4.2) is achieved when $v = u^*(x, x(0))$, where u^* is optimal. In other words,

$$\int_{-1}^{0} F(\phi(\theta), x, u^*)\, d\theta + (A^{u^*} J^*)(x, x(0)) = 0, \quad \forall (x, x(0)) \in H \times G, \tag{13.4.4}$$

which is equation (13.4.3).

Proof Now we proceed to prove (13.4.3). Fix $(x, x(0)) \in H \times G$ and choose a Markov control process u. Let $\alpha \leq \tau_{H \times G}$ be a stopping time. By using the strong

Markov property of $(x_t, x(t))$ we obtain for $J(x, x(0), u)$:

$$
\begin{aligned}
& E_{x,x(0)}\left[J(x_\alpha, x(\alpha), u)\right] \\
& = E_{x,x(0)}\left[E_{x_\alpha,x(\alpha)}\left[\int_{-1}^{0}\int_{0}^{\tau_{H\times G}} F(x_s(\theta), x(s), u(s))\,ds d\theta + \right.\right. \\
+ \quad & \left.\left.\psi(x_{\tau_{H\times G}}, x(\tau_{H\times G}))\right]\right] \\
= \quad & E_{x,x(0)}\left[E_{x,x(0)}\left[S_\alpha\left(\int_{-1}^{0}\int_{0}^{\tau_{H\times G}} F(x_s(\theta), x(s), u(s))\,ds d\theta + \right.\right.\right. \\
& \left.\left.\left. + \psi(x_{\tau_{H\times G}}, x(\tau_{H\times G}))\right) / F_\alpha\right]\right] \\
& = E_{x,x(0)}\left[E_{x,x(0)}\left[\int_{-1}^{0}\int_{0}^{\tau_{H\times G}} F(x_s(\theta), x(s), u(s))\,ds d\theta + \right.\right. \\
+ \quad & \left.\left.\psi(x_{\tau_{H\times G}}, x(\tau_{H\times G}))\right] / F_\alpha\right] \\
= \quad & E_{x,x(0)}\left[\int_{-1}^{0}\int_{0}^{\tau_{H\times G}} F(x_s(\theta), x(s), u(s))\,ds d\theta + \right. \\
+ \quad & \left.\psi(x_{\tau_{H\times G}}, x(\tau_{H\times G})) - \int_{-1}^{0}\int_{0}^{\alpha} F(x_s(\theta), x(s), u(s))\,ds d\theta\right] \\
= \quad & J(x, x(0), u) - E_{x,x(0)}\left[\int_{-1}^{0}\int_{0}^{\alpha} F(x_s(\theta), x(s), u(s))\,ds d\theta\right],
\end{aligned}
$$

where S_α is a shift operator (see, eg., [1]). Therefore

$$
\begin{aligned}
J(x, x(0), u) \quad = \quad & E_{x,x(0)}\left[\int_{-1}^{0}\int_{0}^{\alpha} F(x_s(\theta), x(s), u(s))\,ds d\theta\right] + \\
+ \quad & E_{x,x(0)}\left[J(x_\alpha, x(\alpha), u(\alpha))\right].
\end{aligned} \tag{13.4.5}
$$

Now let $V \subset H \times G$ be of the form $V := \{(y, y(0)) \in H \times G : |(y, y(0)) - (x, x(0))| < \varepsilon\}$. Let $\alpha = \tau_V$ be the first exit time of the pair $(x_t, x(t))$ from V.

Suppose the optimal control u^* exists. For every $v \in \mathcal{U}$ choose:

$$
u = \begin{cases} v(x, x(0)), & \text{if} \quad (x, x(0)) \in V \\ u^*(x, x(0)), & \text{if} \quad (x, x(0)) \in H \times G \setminus V. \end{cases} \tag{13.4.6}
$$

Then $J^*(x_\alpha, x(\alpha)) = J(u^*, x_\alpha, x(\alpha))$, and by combining (13.4.5) and (13.4.6), we obtain

$$
\begin{aligned}
J^*(x, x(0)) \leq J(x, x(0), v) = E_{x,x(0)}\left[\int_{-1}^{0}\int_{0}^{\alpha} F(x_s(\theta), x(r), v(r))\,d\theta dr\right] \\
+ E_{x,x(0)}\left[J(x_\alpha, x(\alpha), v)\right].
\end{aligned}
$$

By Dynkin formula (13.3.6) we have

$$
E_{x,x(0)}\left[J(x_\alpha, x(\alpha), v)\right] = J(x, x(0)) + E_{x,x(0)}\left[\int_{0}^{\alpha} A^v J(x_r, x(r), v)\,dr\right],
$$

where A^v is defined by (13.4.1). By substituting the latter into the previous inequality we obtain

$$
\begin{aligned}
J^*(x, x(0)) \qquad & \leq E_{x,x(0)}\left[\int_{-1}^{0}\int_{0}^{\alpha} F(x_s(\theta), x(s), v(s))\,ds\right] + J(x, x(0)) \\
+ \quad & E_{x,x(0)}\left[\int_{0}^{\alpha} A^v J(x_r, x(r), v)\,dr\right],
\end{aligned}
$$

or

$$E_{x,x(0)}\left[\int_{-1}^{0}\int_{0}^{\alpha}F(x_r(\theta),x(r),v(r))\,drd\theta+\int_{0}^{\alpha}A^{v}J(x_r,x(r),v)\,dr\right]\geq 0.$$

Therefore,

$$\begin{aligned} E_{x,x(0)} & \quad \left[\int_{-1}^{0}\int_{0}^{\alpha}F(x_r(\theta),x(r),v(r))\,d\theta dr+\right.\\ & + \left.\int_{0}^{\alpha}(A^{v}J)(x_r,x(r),v)\,dr\right]/E_{x,x(0)}[\alpha]\geq 0. \end{aligned}$$

By letting $\varepsilon\to 0$ we derive

$$\int_{-1}^{0}F(x,x(0),v)\,d\theta+(A^{v}J)(x,x(0),v)\geq 0,$$

which combined with (13.4.4) gives (13.4.3).

Theorem 2. (Converse of the HJB-equation)
Let g be a bounded function in $C^2(H\times G)\cap C(\partial(H\times G)))$. Suppose that for all $u\in\mathcal{U}$ the inequality

$$\int_{-1}^{0}F(x,x(0),u)\,d\theta+(A^{u}g)(x,x(0))\geq 0,\quad (x,x(0))\in H\times G$$

and the boundary condition

$$g(x,x(0))=\psi(x,x(0)),\quad (x,x(0))\in\partial(H\times G) \tag{13.4.7}$$

are satisfied. Then $g(x,x(0))\leq J(x,x(0),u)$ for all Markov controls $u\in\mathcal{U}$ and for all $(x,x(0))\in H\times G$.

Moreover, if for every $(x,x(0))\in H\times G$ there exists u^0 such that

$$\int_{-1}^{0}F(x,x(0),u^0)\,d\theta+(A^{u^0}g)(x,x(0))=0, \tag{13.4.8}$$

then u^0 is a Markov control, $g(x)=J(x,x(0),u^0)=J^*(x,x(0))$, and therefore u^0 is an optimal control.

Proof Assume that g satisfies hypotheses (13.4.7) and (13.4.8). Let u be a Markov control. Then $A^uJ\geq-\int_{-1}^{0}F(x,x(0),u)\,d\theta$ for all u in $\mathcal{U}$, and we have by Dynkin formula (13.3.6)

$$\begin{aligned} E_{x,x(0)}\left[g(x_{\tau_r},x(\tau_r))\right] &= g(x,x(0))+E_{x,x(0)}\left[\int_{0}^{\tau_r}(A^{u}g)(x_s,x(s))\,ds\right]\geq\\ \geq g(x,x(0)) &- E_{x,x(0)}\left[\int_{-1}^{0}\int_{0}^{\tau_r}F(x_s(\theta),x(s),u(s))\,d\theta ds\right], \end{aligned}$$

where

$$\tau_r:=\min\{r,\tau_{H\times G},\inf\{t>0:|x_t|\geq r\}\},\, r>0.$$

By taking the limit as $\tau_r \to +\infty$ this gives

$$
\begin{aligned}
& g(x,x(0)) \leq \\
& E_{x,x(0)}\left[\int_{-1}^{0}\int_{0}^{\tau_r} F(x_s(\theta),x(s),u(s))\,d\theta ds + \psi(x_{\tau_r},x(\tau_r))\right] \leq \\
& \leq \lim_{\tau_r\to\infty}\left[E_{x,x(0)}\left[\int_{-1}^{0}\int_{0}^{\tau_r} F(x_s(\theta),x(s),u(s))\,d\theta ds + \psi(x_{\tau_r},x(\tau_r))\right]\right] \\
= \; & E_{x,x(0)}\left[\int_{-1}^{0}\int_{0}^{\tau_{H\times G}} F(x_s(\theta),x(s),u(s))\,d\theta ds + \psi(x_{\tau_{H\times G}},x(\tau_{H\times G}))\right] \\
= \; & J(x,x(0),u),
\end{aligned}
$$

which proves the first assertion of the theorem. If u^0 is such that (13.4.8) holds, then the above calculation gives the equality. This completes the proof.

Remark. The Hamilton–Jacobi–Bellman (HJB) equation and the inverse HJB equation are classical results in the optimization theory. Theorems 1 and 2 above provide their extensions to the case of stochastic differential delay equations considered in this chapter. Both statements assume the existence of the optimal control u_0 as a hypothesis. The existence of the optimal control in a general setting is an important and difficult problem by itself. Under certain conditions on functions a,b,F,ϕ and the boundary of the set $H \times G$, and with the compactness of the set of control values, one can show, by using related general results from nonlinear PDEs, that a smooth function J satisfying equation (13.4.5) and boundary condition $J^*(x,x(0)) = \phi(x,x(0))$ exists. Then by applying a measurable selection theorem one can find a measurable function u^* satisfying equation (13.4.6) for almost all points in $H \times G$. For more details of this possible approach to tackle the existence problem see, for example, A. Bensoussan and J.L. Lions “Applications of Variational Inequalities to Stochastic Control,” North-Holland, 1982 and N.V. Krylov “Controlled Diffusion Processes,” Springer-Verlag, 1980. We plan to address this general problem of existence of optimal control in our future research. In the next section we show the existence of the optimal control and find it explicitly for the Ramsey SDDE model with a given cost function.

13.5 Economics model and its optimization

13.5.1 Description of the model

In 1928 F.R. Ramsey introduced an economics model describing the rate of change of capital K and labor L in a market by a system of ordinary differential equations [10]. With P and C being the production and consumption rates, respectively, the model is given by the system

$$
\frac{dK(t)}{dt} = P(t) - C(t), \qquad \frac{dL(t)}{dt} = a(t)L(t), \tag{13.5.1}
$$

where $a(t)$ is the rate of growth of labor (population).

The production, capital and labor are related by the Cobb–Douglas formula,

$P(t) = AK^{\alpha}(t)L^{\beta}(t)$, where A, α, β are some positive constants [3]. In certain circumstances the dependence of P on K and L is linear, i.e., $\alpha = \beta = 1$, which will be our assumption throughout this section. We shall also assume that the labor is constant, $L(t) = L_0$, which is true for certain markets or relatively short time intervals of several years. Therefore, the production rate and the capital are related by $P(t) = BK(t)$, where $B = AL_0$. Another important assumption we make is that the production rate is subject to small random disturbances, i.e., $P(t) = BK(t) +$ "noise." System (13.5.1) then results in the equation

$$\frac{dK(t)}{dt} = BK(t) + \text{"noise"} - C(t),$$

which can be rewritten in the differential form as

$$dK(t) = [BK(t) - C(t)]\,dt + \sigma(K(t))\,dw(t),$$

where $w(t)$ is a standard Wiener process, $\sigma(K)$ is a given (small) real function, characteristic of the noise.

The original model of Ramsey is based on the assumption of instant transformation of the investments. This can be accepted as satisfactory in only very rough models. In reality the transformation of the invested capital cannot be accomplished instantly. A certain essential period of time is normally required for this transformation, such as the length of the production cycle in many economical situations. Therefore, a more accurate assumption is that the rate of change of capital K at present time t depends on the investment that was made at time $t - T$, where T is the cycle duration required for the creation of working capital. This leads to the following delay differential equation

$$dK(t) = [BK(t-T) - C(t)]\,dt + \sigma(K(t-T))\,dw(t).$$

Our next assumption is that the consumption rate C can be controlled by the available amount of the capital, i.e., it is of the form $C(t)\,u(K(t))$, where $u(\cdot)$ is a control. By normalizing the delay to $T = 1$ (by time rescaling) one arrives at the equation

$$dK(t) = [BK(t-1) - u(K(t))C(t)]\,dt + \sigma(K(t-1))\,dw(t). \qquad (13.5.2)$$

The initial investment of the capital K is naturally represented in equation (13.5.2) by a given initial function ϕ

$$K(t) = \phi(t), \quad t \in [-1, 0]. \qquad (13.5.3)$$

Therefore, we propose to study a modified Ramsey model with delay and random perturbations given by the system (13.5.2)–(13.5.3).

13.5.2 Optimization calculation

Usually one wants to minimize the investment capital under the assumption of labor being constant. Let us choose the following cost function

$$J(K,u) = \frac{K^2(0)}{2} + \int_{-1}^{0} \phi^2(\theta)\,d\theta + \frac{u^2(0)}{2}.$$

The operator $A^u J$ has the following form

$$\begin{aligned} A^u J &= \tfrac{K^2(0)}{2} + \phi^2(0) + \tfrac{u^2(0)}{2} - \left[\tfrac{K^2(0)}{2} + \phi^2(-1) + \tfrac{u^2(0)}{2}\right] \\ &+ \left[K(0)\cdot(B\cdot K(0) - u(0)\cdot C(0)) + \tfrac{1}{2}\sigma^2(K(0))\right], \end{aligned}$$

since

$$\begin{aligned} &F(K(0),K(0),u(0)) = \tfrac{K^2(0)}{2} + \phi^2(0) + \tfrac{u^2(0)}{2}, \\ &F(K(0),K(-1),u(0)) = \tfrac{K^2(0)}{2} + \phi^2(-1) + \tfrac{u^2(0)}{2}, \\ &L^u J(K,u) = K(0)\left(B\cdot K(0) - u(0)\cdot C(0) + \tfrac{1}{2}\sigma^2(K(0))\right). \end{aligned}$$

From (13.4.1) we obtain the following HJB-equation

$$\begin{aligned} \inf_u \quad & \left[\tfrac{K^2(0)}{2} + \textstyle\int_{-1}^0 \phi^2(\theta)\,d\theta + \tfrac{u^2(0)}{2} + \phi^2(0) - \phi^2(-1) + B\cdot K^2(0)\right. \\ + \quad & \left. u(0)\cdot K(0)\cdot C(0) + \tfrac{1}{2}\sigma^2(K(0))\right] = 0, \end{aligned}$$

or equivalently,

$$\begin{aligned} \inf_u \quad & \left[u^2(0) - 2K(0)C(0)u(0) + (2\phi^2(0) - 2\phi^2(-1) + 2\textstyle\int_{-1}^0 \phi^2(\theta)\,d\theta\right. \\ & \left. + K^2(0)(1+2B) + \sigma^2(K(0)))\right] = 0. \end{aligned}$$

Let

$$\begin{aligned} 4K^2(0)C^2(0) \quad & \geq 4(2\phi^2(0) - 2\phi^2(-1) + 2\textstyle\int_{-1}^0 \phi^2(\theta)\,d\theta \\ + \quad & K^2(0)(1+2B) + \sigma^2(K(0))), \end{aligned}$$

or

$$K^2(0)\cdot(C^2(0) - 3 - 2B) \geq 2\int_{-1}^0 \phi^2(\theta)\,d\theta - 2\phi^2(-1) + \sigma^2(K(0)),$$

since $K(0) = \phi(0)$. Hence, the infimum is achieved when

$$u(0) = -\left(-\frac{2K(0)\cdot C(0)}{2}\right) = K(0)\cdot C(0).$$

Therefore $u_{\min} = K(0)\cdot C(0)$ and

$$\begin{aligned} J(K,u_{\min}) &= \tfrac{K^2(0)}{2} + \tfrac{K^2(0)\cdot C^2(0)}{2} + \textstyle\int_{-1}^0 \phi^2(\theta)\,d\theta \\ &= \tfrac{K^2(0)}{2}(1 + C^2(0)) + \textstyle\int_{-1}^0 \phi^2(\theta)\,d\theta. \end{aligned}$$

Note that in the case of general delay $T > 0$ in model (13.5.2)–(13.5.3) the last expression for J remains valid with the integration range $[-T,0]$.

References

[1] Dynkin, E.B., *Markov Processes,* Vols 1-2, Die Grundlehreu der Math. Wissenschaften 121-122, Springer-Verlag, 1965.

[2] Fleming, W. and Nisio, M., *On the existence of optimal stochastic control.* J. Math. Mech. 15, No.5 (1966), 777-794.

[3] Gandolfo, G., *Economic Dynamics.* Springer-Verlag, 1996, 610 pp.

[4] Hale, J.K., *Theory of Functional Differential Equations,* Springer-Verlag: Applied Mathematical Sciences 3 (1977), 365 pp.

[5] Ito, K. and Nitio, M., *On stationary solutions of a stochastic differential equation.* J. Math. Kyoto Univ. 4-1 (1964), 1–70.

[6] Ivanov, A.F. and Swishchuk, A.V., *Stochastic delay differential equations and stochastic stability: a survey of some results.* SITMS Research Report 2/99, University of Ballarat, January 1999, 22 pp.

[7] Ivanov, A.F. and Swishchuk, A.V., *Optimal control of stochastic differential delay equations with application in economics.* Abstracts of SMOCS05, July 10-16, 2005, Daydream Island, Australia.
(http://www.conferences.unimelb.edu.au/smocs05/abstracts.htm)

[8] Ivanov, A. F. and Swishchuk, A. V., Optimal control of stochastic differential delay equations with application in economics, *International Journal of Qualitative Theory of Differential Equations and Applications*, Vol. 2, No. 2 (2008), pp. 201–213.

[9] Kushner, H., *On the stability of processes defined by stochastic difference-differential equations.* J. Differential Equations 4 (1968), 424–443.

[10] Oksendal, B., *Stochastic Differential Equations. An Introduction with Applications.* Springer-Verlag, 1992, 224 pp.

[11] Ramsey, F.P., A mathematical theory of savings. *Economic J.* 38 (1928), 543–549.

[12] Swishchuk, A. V., *Random Evolutions and their Applications,* Kluwer Academic Publishers, Dordrecht, v. 408, (1997), 215 p.

Chapter 14

Optimal Control of Vector Delayed RDS with Applications in Finance and Economics

14.1 Chapter overview

This chapter is devoted to the study of RDS arising in optimal control theory for vector stochastic differential delay equations (SDDEs) and its applications in mathematical finance and economics. By using the Dynkin formula and solution of the Dirichlet-Poisson problem, the Hamilton-Jacobi-Bellman (HJB) equation and the converse HJB equation are derived. Furthermore, applications are given to an optimal portfolio selection problem and a stochastic Ramsey model in economics.

14.2 Introduction

This chapter is devoted to the RDS arising in optimal control theory for vector stochastic differential delay equations (SDDEs) with applications in finance and economics. We adopt the name stochastic differential delay equations as in [9] instead of stochastic functional differential equations as in [15]. We believe that SDDEs are useful dynamical models to understand the behavior of natural processes that take into consideration the influence of past events on the current and future states of the system [1, 9, 11]. This view is especially appropriate in the study of financial variables, since predictions about their evolution take strongly into account the knowledge of their past [7, 15].

The SDDEs are very important objects that have many applications. One of the problems in the theory of SDDEs is the study of optimal control that also has many applications including finance. The main idea in finance is to find the optimal portfolio of an investor to maximize his wealth or cost function. In this way, the SDDEs with controlled parameters are the main object of investigation of this chapter.

The chapter is organized in the following way. In Section 14.3 we present the basic spaces, the norms, properties, and notation which we are going to work with in the following sections and formulation of the problem that is the goal of this work. In

Section 14.4 we state the results on existence and uniqueness of the solution of the SDDEs. We proved that the pair of processes, one with delayed parameter and another one as the solution of the SDDEs, is a strong Markov process. With this result in hand we can define and calculate the weak infinitesimal generator of the Markov process and apply the theory of controlled Markov processes to the solution of our optimization problem. We found the sufficient conditions for the optimality of the solution and derived the Hamilton-Jacobi-Bellman equation (HJB) equation and the converse of the HJB equation. The HJB equation has been studied by many authors. See, for example, [2, 9, 13]. Also, Reference [8] surveys many results in this area. In Section 14.5, the results obtained in Section 14.4 are applied to two problems: an optimal portfolio selection problem and an economics problem's model describing the rate of change of capital and labor in a market. In both cases the optimal control has been found in explicit form.

14.3 Preliminaries and formulation of the problem

Let $a>0$, U be a closed set of $\mathbf{R}^m$ and $\left(\Omega,\mathcal{F},(\mathcal{F}_t)_{0\leq t\leq a},\mathbf{P}\right)$ be a complete filtered probability space. Assume also that each $\mathcal{F}_t$ contains all the sets of measure zero in $\mathcal{F}$. Let $r>0$, $J:=[-r,0,]$ and $T:=[0,a]$.

We denote by $V:=L^2([-r,0],\mathbf{R}^n)$, $H:=L^2([0,a],\mathbf{R}^n)$, with respective norms and inner products $\|\cdot\|_V$, $\langle\cdot\rangle_V$, and $\|\cdot\|_H$, $\langle\cdot\rangle_H$. Assume $\mu: V\times\mathbf{R}^n\times U\to\mathbf{R}^n$, and $\sigma: V\times\mathbf{R}^n\times U\to\mathbf{R}^{n\times d}$ are measurable. Now, we consider the following stochastic differential delay equations (SDDEs)

$$S(t)=\begin{cases}\vec{x}+\int_0^t\mu(S_s,S(s),u(s))ds+\int_0^t\sigma(S_s,S(s),\mathbf{u}(s))dW(s), t\in T\\ \phi(t),\quad -r\leq t<0,\end{cases}\tag{14.3.1}$$

where ϕ is an initial path in V, $\vec{x}$ an initial vector in $\mathbf{R}^n$ and $W(t)$ is an $\mathcal{F}_t-$ adapted $d-$dimensional Brownian motion, and $\mathbf{u}(s)$ is defined below.

See also, e.g., [2] for stochastic control of SDDEs. The solution $\{S(t)\}_{-r\leq t\leq a}$ of (14.3.1) is an $n-$dimensional stochastic process. Its *segment process* $\{S_t: t\in T\}$ is defined by

$$S_t(\omega)(s):=S(t+s,\omega)\text{ for } s\in J.\tag{14.3.2}$$

The function $\mathbf{u}(t)=\mathbf{u}(S_t,S(t))$ will be called *Markov control law*. A Markov control law $\mathbf{u}: V\times\mathbf{R}^n\to U$ is *admissible* if it is a Borel measurable function and satisfies $|\mathbf{u}(\phi,\vec{x})-\mathbf{u}(\eta,\vec{y})|^2\leq K\left\{|\vec{x}-\vec{y}|^2+\|\phi-\eta\|_V^2\right\}$ with constant K and for some constant K_1 and for all $\phi\in V$, $\vec{x},\vec{y}\in\mathbf{R}^n$ holds $|\mathbf{u}(\phi,\vec{x})|^2\leq K_1\left\{1+|\vec{x}|^2+\|\phi\|_V^2\right\}$. We denote by $\mathcal{U}$ the set of all admissible Markov control laws.

Let $G\subseteq V\times\mathbf{R}^n$ be an open connected subset and $\Gamma:=\partial(G)$ be the boundary of the set G. Let $\psi(\cdot,\cdot)$ be a function continuous on the closure of the set G and bounded on Γ. Let $L(\cdot,\cdot,\cdot)$ be a function continuous on $C(G\times U)$ such that

$$E^{(\phi,\vec{x})}\left[\int_0^{\tau_G}|L(S_t,S(t),\mathbf{u}(t))|dt\right]<\infty\ \forall(\phi,\vec{x})\in G. \tag{14.3.3}$$

where τ_G is the first exit time from the set G, and $E^{\phi,\vec{x}}$ the expectation under the conditional probability with respect to $\left(\phi,\vec{x}\right)$ and $\mathbf{u}$, Markov control.
Now we are given a *cost function*(or *performance criterion*)

$$\begin{aligned}J(\phi,\vec{x},\mathbf{u}):=\ & E^{(\phi,\vec{x})}\left[\int_0^{\tau_G}L(S_t,S(t),\mathbf{u}(t))dt+\right.\\ & \left.+\ \psi(S_{\tau_G},S(\tau_G))\right].\end{aligned} \tag{14.3.4}$$

For simplicity, we assume that interest rate $r=0$ and there is no discount factor. Of course, all calculations below can be performed with a discount factor as well.

The problem is, for each $\left(\phi,\vec{x}\right)\in G$ to find the number $\Phi(\phi,\vec{x})$ and a control $\mathbf{u}^\star=\mathbf{u}^\star(t,\omega)$ such that

$$\Phi(\phi,\vec{x}):=\inf_{\mathbf{u}(t)}J(\phi,\vec{x},\mathbf{u})=J(\phi,\vec{x},\mathbf{u}^\star) \tag{14.3.5}$$

where the infimum is taken over all $\mathcal{F}_t-$ adapted processes $\mathbf{u}(t)$ with values in U. If such a control $\mathbf{u}^\star$ exists it is called an *optimal control* and Φ is called the *optimal performance*.

See [11,13]. We denote by $B_b(V\times\mathbf{R}^n)$ the Banach space of all real bounded Borel functions, endowed with the sup norm.

14.4 Controlled stochastic differential delay equations

Given the Markov control $\mathbf{u}$ and a function $g(\phi,\vec{x},u)$, we use the notation

$$g^{\mathbf{u}}\left(\phi,\vec{x}\right)=g(\phi,\vec{x},\mathbf{u}).$$

Then (14.3.1) can be written as

$$S(t)=\begin{cases}\vec{x}+\int_0^t\mu^{\mathbf{u}}(S_s,S(s))ds+\int_0^t\sigma^{\mathbf{u}}(S_s,S(s))dW(s), & t\in T\\ \phi(t),\ -r\le t<0,\end{cases} \tag{14.4.1}$$

Theorem 14.1 *Let $\phi:\Omega\to V$ such that $E\left[\|\phi\|_V^2\right]<+\infty$ and $\vec{x}:\Omega\to\mathbf{R}^n$ such that $E\left[\|\vec{x}\|^2\right]<+\infty$ and $\mathcal{F}_0$ mensurable. Assume that there exists a constant C such that*

$$\begin{aligned}\left\|\mu(\phi,\vec{x},u)\right.&\left.-\ \mu(\eta,\vec{x_1},u_1)\right\|^2+\left\|\sigma(\phi,\vec{x},u)-\sigma(\eta,\vec{x_1},u_1)\right\|^2\le\\ &\le C\left[\left\|\vec{x}-\vec{x_1}\right\|^2+|u-u_1|^2+\|\phi-\eta\|_V^2\right]\end{aligned} \tag{14.4.2}$$

and

$$|\mu(\phi,\vec{x},u)|^2+|\sigma(\phi,\vec{x},u)|^2\leq C(1+|\vec{x}|^2+\|\phi\|_V^2+|u|^2). \tag{14.4.3}$$

for all $\phi,\ \eta\in V,\ \vec{x},\ \vec{x_1}\in\mathbf{R}^n,\ u,u_1\in U.$

Then we have a unique measurable solution $S(t)$ *to (14.4.1) with continuous trajectories* $\{(S_t,S(t)),\ t\in T\}$ *adapted to* $(\mathcal{F}_t)_{t\in T}$.

Proof *We prove by using the standard method of successive approximations (see [12], page 227).*

We can still assure the existence and uniqueness of the solution to (14.4.1) under weaker conditions.

Theorem 14.2 *Let* $\phi:\Omega\to V$ *such that* $E\left[\|\phi\|_V^2\right]<+\infty$ *and* $\vec{x}:\Omega\to\mathbf{R}^n$ *such that* $E\left[\|\vec{x}\|^2\right]<+\infty$ *and* $\mathcal{F}_0$*−mensurable. Assume that there exists a constant C such that*

$$|\mu(\phi,\vec{x},u)|^2+|\sigma(\phi,\vec{x},u)|^2\leq C(1+|\vec{x}|^2+\|\phi\|_V^2+|u|^2). \tag{14.4.4}$$

for all $\phi\in V,\ \vec{x}\in\mathbf{R}^n,\ u\in U.$

And for each N there exists C_N *for which*

$$\begin{aligned}\left\|\mu(\phi,\vec{x},u)\right. & \left.-\mu(\eta,\vec{x_1},u_1)\right\|^2+\left\|\sigma(\phi,\vec{x},u)-\sigma(\eta,\vec{x_1},u_1)\right\|^2\leq\\ & \leq C_N\left[\left\|\vec{x}-\vec{x_1}\right\|^2+|u-u_1|^2+\|\phi-\eta\|_V^2\right]\end{aligned} \tag{14.4.5}$$

for all $\phi,\ \eta\in V,\ \vec{x},\ \vec{x_1}\in\mathbf{R}^n$, *with* $\|\vec{x}\|\leq C_N,\ \|\vec{x_1}\|\leq C_N\ u,u_1\in U.$

Then we have a unique measurable solution $S(t)$ *to (14.4.1) with continuous trajectories* $\{(S_t,S(t)),\ t\in T\}$ *adapted to* $(\mathcal{F}_t)_{t\in T}$.

Proof *See [6], Theorem 3, page 45.*

Remark. Let $0\leq t_1\leq t\leq T,\ \phi\in V,\ \vec{x}:\Omega\to\mathbf{R}^n$ such that $E\left[\|\vec{x}\|^2\right]<+\infty$ with $\phi,\ \vec{x}\ \mathcal{F}_{t_1}$−mensurable. We can solve the following equation at time t_1

$$\begin{cases} S(t) & =\vec{x}+\int_{t_1}^t\mu^{\mathbf{u}}(S_s,S(s))ds+\int_{t_1}^t\sigma^{\mathbf{u}}(S_s,S(s))dW(s),t\in[t_1,T]\\ S(t) & =\phi(t-t_1),t\in[t_1-r,t_1).\end{cases} \tag{14.4.6}$$

We denote by $S(\cdot,t_1,\phi,\vec{x})$ the solution of (14.4.6).

Moreover, the solution has properties similar to the solutions of stochastic differential equations.

Theorem 14.3 *Under the assumptions of Theorem 14.1 (or 14.2), there exists $C(a,r) > 0$ such that, for arbitrary $\phi, \eta : \Omega \to V$ such that $E\left[\|\phi\|_V^2\right]$, $E\left[\|\eta\|_V^2\right] < +\infty$, and $x, y : \Omega \to \mathbf{R}^n$ such that $E\left[\| \vec{x} \|^2\right]$, $E\left[\| \vec{y} \|^2\right] < +\infty$ and $\mathcal{F}_0$ mensurable. Let $0 \le t_1 \le t \le a$, then*

$$E\left(\|(S(\cdot,t_1,\phi,\vec{x}))_t\|_V^2 + |S(t,t_1,\phi,\vec{x})|^2\right) \le C(a,r)E(\|\phi\|_V^2) + E\left(|\vec{x}|^2\right) \quad (14.4.7)$$

$$\begin{aligned} \sup_{s\in[t_1,a]} E\left(|S(s,\right. & \left. t_1,\phi,\vec{x}) - (S(s,t_1,\eta,\vec{y}))|\right) + \\ & + E\left(\|(S(\cdot,t_1,\phi,\vec{x}))_t - (S(\cdot,t_1,\eta,\vec{y}))_t\|_V^2\right) \le \\ & C(a,r)E\left(\|\phi-\eta\|_V^2\right) + E\left(|\vec{x} - \vec{y}|^2\right). \end{aligned} \quad (14.4.8)$$

$$\begin{aligned} E\left(|S(t,t_1,\phi,\vec{x}) - \right. & \left. S(t,t_1,\phi,\vec{x})|\right) + \\ & + E\left(\|(S(\cdot,t_1,\phi,\vec{x}))_t - (S(\cdot,t_1,\eta,\vec{y}))_t\|_V^2\right) \le \\ & C(a,r,\phi,\vec{x})|t-t_1|^2, \end{aligned} \quad (14.4.9)$$

where we denoted by $S(\cdot,t_1,\phi,\vec{x})$ the solution of (14.4.6).

Proof *The proof is using ideas similar to the case of where there is no delay (see [4], Theorem 9.1) and to the proof of Theorem 3.1 page 41 of [12]*

Let $A \in \mathbf{B}(\mathbf{R}) \otimes \mathcal{B}(V)$, we define the transition probability

$$\begin{aligned} p\left(t_1,(\phi,\vec{x}),t,A\right) := & \ \mathbf{P}\left(\left((S(\cdot,t_1,\phi,\vec{x}))_t, S(t,t_1,\phi,\vec{x})\right) \in A\right) = \\ & = E\left[1_A(S(\cdot,t_1,\phi,\vec{x})_t, S(t,t_1,\phi,\vec{x}))\right]. \end{aligned}$$

We will show now, following [12], that the process $(S_t, S(t))$, $t \in T$, is a Markov process with transition probability $p\left(t_1,(\phi,\vec{x}),t,A\right)$.

Lemma. Assume that $S(t)$ is the solution to (14.4.1). Then $(S_t, S(t))$ will be a Markov process with transition probability $p\left(t_1,(\phi,\vec{x}),t,A\right)$, $0 \le t_1 \le t \le a$. and $A \in \mathcal{B}\mathbf{R}^n) \otimes \mathcal{B}(V)$.

Proof *Denote by G^t the $\sigma-$algebra generated by $W(s) - W(t)$ for $t \le s$. We observe that G^t and $\mathcal{F}_t$ are independent. We observe that $S(t) = S(t,t_1,S_{t_1},S(t_1))$ for $t > t_1$, because both are solutions of the equation:*

$$\begin{cases} Z(t) & = Z(t_1) + \int_{t_1}^t \mu^{\mathbf{u}}(Z_s,Z(s))ds + \int_{t_1}^t \sigma^{\mathbf{u}}(Z_s,Z(s))dW(s), \ t_1 \le t \le T \\ Z(t) & = S(t-t_1), t \in [t_1 - r, t_1). \end{cases} \quad (14.4.10)$$

Let $B \in \mathcal{F}_{t_1}$. Since

$$\begin{aligned}\int_B 1_A((\ S_t, S(t)))d\mathbf{P}(\omega) &= \int_\Omega 1_A((S_t, S(t)))1_B d\mathbf{P}(\omega) = \\ &= \int_\Omega 1_A((S(\cdot,t_1,S_{t_1},S(t_1)))_t, S(t,t_1,S_{t_1},S(t_1)))1_B d\mathbf{P}(\omega) = \\ &= \int_\Omega 1_A((S(\cdot,t_1,S_{t_1},S(t_1)))_t, S(t,t_1,S_{t_1},S(t_1))))d\mathbf{P}(\omega)\int_\Omega 1_B d\mathbf{P}(\omega) = \\ &= \int_B \mathbf{P}\left(t_1,(\phi,\overrightarrow{x}),t,A\right)d\mathbf{P}(\omega)|_{x=S(t_1),\phi=S_{t_1}},\end{aligned}$$

thus we have that $\mathbf{P}\left((S_t,S(t))\in A|\mathbf{F}_{t_1}\right) = p(t_1,(S_{t_1},St_1),t,A)$. To see that $\mathbf{P}\left((S_t,S(t))\in A|(S_{t_1},S(t_1))\right) = p(t_1,(S_{t_1},S(t_1)),t,A)$, firstly we prove that $P(t_1,.,t,A)$ is measurable for fixed t,t_1,A, since $(S_{t_1},S(t_1))$ is measurable with respect to $\sigma-$algebra generated by $(S_{t_1},S(t_1))$ we finish the proof.

With similar arguments we can prove the following theorem. See, for example, [3], Theorem 9.8.

Theorem 14.4 *Let $S(t) := S(t,t_1,\phi,\overrightarrow{x})$ be the solution to (14.4.6). For arbitrary $f \in B_b(V\times\mathbf{R}^n)$ and $0\le t_1\le t\le a$,*

$$E\left[f(S_t,S(t))|\mathcal{F}_{t_1}\right] = E\left[f(S_t,S(t))|(S_{t_1},S(t_1))\right] \tag{14.4.11}$$

Now, following [4] we will prove that the solutions to (14.4.1) are a strong Markov process.

Theorem 14.5 *(The strong Markov property) Let $S(t)$ as in Theorem 14.4, f in $B_b(V\times\mathbf{R}^n)$, τ a stopping time with respect to $\mathcal{F}_t$, $\tau<\infty$ a.s. Then*

$$E^{(\phi,\overrightarrow{x})}\left[f(S_{\tau+h},S(\tau+h))|\mathcal{F}_\tau\right] = E^{(S_\tau,S(\tau))}f(S_h,S(h)) \tag{14.4.12}$$

for all $h\ge 0$.

Proof *We prove (14.4.12) as in [3], Theorem 9.14 page 255 using the properties of Theorem 14.3.*

For every $f\in B_b(V\times\mathbf{R}^n)$ and $(\phi,\overrightarrow{x})\in V\times\mathbf{R}^n$ let

$$P_t f(\phi,\overrightarrow{x}) := E^{(\phi,\overrightarrow{x})}\left(f(S_t,S(t))\right).$$

Definition. The weak infinitesimal operator of P_t (or of $(S_t,S(t))$), $\mathbf{A^u} = \mathbf{A}^\mathbf{u}_S$, is defined by

$$\mathbf{A^u} f(\phi,\overrightarrow{x}) := \lim_{h\to 0} h^{-1}\left[P_h f(\phi,\overrightarrow{x}) - f(\phi,\overrightarrow{x})\right]. \tag{14.4.13}$$

The set of functions f such that the limit (14.4.13) exists in $(\phi,\overrightarrow{x})$ is denoted by $\mathcal{D}_{\mathbf{A^u}}(\phi,\overrightarrow{x})$ and $\mathcal{D}_{\mathbf{A^u}}$ denotes the set of functions such that the limit (14.4.13) exists for all $(\phi,\overrightarrow{x})$.

Let e_j for $j = 1,\dots,d$ be the canonical basis of $\mathbf{R}^d$ for $(\phi,\vec{x}) \in V \times \mathbf{R}^n$ let

$$\widehat{\phi^{\vec{x}}}(t) := \begin{cases} \vec{x}, & t \in T \\ \phi(t), & t \in [-r,0). \end{cases} \tag{14.4.14}$$

Then, for each $s \in J, t \in T$,

$$\widehat{\phi_t^{\vec{x}}}(s) = \widehat{\phi^{\vec{x}}}(s+t) = \begin{cases} \vec{x}, & t+s \geq 0 \\ \phi(t), & t+s < 0. \end{cases} \tag{14.4.15}$$

Denote by $\mathbf{S}_t$ for $t \in T$ the weakly continuous contraction semigroup of the shift operators defined on $C_b(V \times \mathbf{R}^n)$ (see [12], Chapter 4) by

$$\mathbf{S}_t(f)(\phi,\vec{x}) := f(\widehat{\phi_t^{\vec{x}}},\vec{x}) \text{ for } f \in C_b(V \times \mathbf{R}^n)$$

Denote by $\mathbf{S}$ the weak infinitesimal operator of $\mathbf{S}_t$ with domain $D(\mathbf{S})$ and $D(\mathbf{S}) \subset C_b^0 = \{f \in C_b(V \times \mathbf{R}^n) : S_t \text{ is strongly continuous}\}$. We note, that $\mathbf{S}_t(f)(\phi,\vec{x})$ has an explicit expression:

Remark. (See [16] Section 9)

$$\mathbf{S}f(\phi,\vec{x}) = \int_{-r}^{0} \frac{\partial f}{\partial \phi}(\phi,\vec{x})(s)d\phi(s)$$

for any $(\phi,\vec{x}) \in V \times \mathbf{R}^n$.

Now we have a formula for the weak infinitesimal operator $\mathbf{A}^{\mathbf{u}}$ similar to the formula in the no delay case: this is a sum of differential operators, and this formula depends of the coefficients $\mu^{\mathbf{u}}$ and $\sigma^{\mathbf{u}}$.

Theorem 14.6 *Let $S(t)$ be the solution to (14.4.1). Suppose $f \in C_b^2(V \times \mathbf{R}^n)$, belongs to the domain of $\mathbf{A}^{\mathbf{u}}$, $\sigma^i \in C_b^2(V \times \mathbf{R}^n \times U;\mathbf{R}^n)$ (where σ^i are the vector columns of σ) and $\mu \in C_b^1(V \times \mathbf{R}^n \times U;\mathbf{R}^n)$. Assume that $\phi \in V$, $\vec{x} \in \mathbf{R}^n$. Let $e_j : j = 1,\dots,d$ be a normalized basis of $\mathbf{R}^d$. Then*

$$\begin{aligned} \mathbf{A}^{\mathbf{u}}f(\phi,\vec{x}) = \; & \mathbf{S}f(\phi,\vec{x}) + \tfrac{\partial f}{\partial \vec{x}}(\phi,\vec{x})\mu^{\mathbf{u}}(\phi,\vec{x}) + \\ & + \tfrac{1}{2}\textstyle\sum_j^n \tfrac{\partial^2 f}{\partial \vec{x}^2}(\phi,\vec{x}) \left[(\sigma^{\mathbf{u}}(\phi,\vec{x}))e_j \otimes (\sigma^{\mathbf{u}}(\phi,\vec{x}))e_j\right]. \end{aligned} \tag{14.4.16}$$

Proof *Is consequence of Lemma 9.3 of [16].*

Remark. Let $\mathbf{L}$ denote the differential operator given by the right-hand side of (14.4.16). The Theorem 14.6 above says that $\mathbf{A}^{\mathbf{u}}$ and $\mathbf{L}$ coincide on $f \in C_b^2(V \times \mathbf{R}^n)$.

Lemma (Dynkin formula). Let $S(t)$ be the solution of (14.4.1). Let $f \in C_b^2(V \times \mathbf{R}^n)$, τ is a stopping time such that $E^{(\phi,\vec{x})}[\tau] < \infty$, with $(\phi,\vec{x}) \in V \times \mathbf{R}^n$ then

$$E^{(\phi,\vec{x})}[f(S_\tau, S(\tau))] = f(\phi,\vec{x}) + E^{(\phi,\vec{x})}\left[\int_0^\tau \mathbf{A^u} f(S_s, S(s))ds\right]. \quad (14.4.17)$$

Proof *From Dynkin [4], corollary of Theorem 5.1.*

Definition. Let $S(t)$ be the solution of (14.4.1). The characteristic operator $\mathcal{A}^{\mathbf{u}} = \mathcal{A}_S^{\mathbf{u}}$ of $(S_t, S(t))$ is defined by

$$\mathcal{A}^{\mathbf{u}} f(\phi,\vec{x}) := \lim_{U \downarrow (\phi,\vec{x})} \frac{E^{(\phi,\vec{x})}[f(S_{\tau_U}, S(\tau_U))] - f(\phi,\vec{x})}{E^{\phi,\vec{x}}[\tau_U]} \quad (14.4.18)$$

where the $U^{'}$s are open sets U_k decreasing to the point $(\phi,\vec{x})$, in the sense that $U_{k+1} \subset U_k$ and $\bigcap_k U_k = (\phi,\vec{x})$, and $\tau_U = \inf\{t > 0;\ (S_t, S(t)) \notin U\}$. We denote by $\mathcal{D}_{\mathcal{A}^{\mathbf{u}}}$ the set of functions f such that the limit (14.4.18) exists for all $(\phi,\vec{x}) \in V \times \mathbf{R}^n$ (and all $\{U_k\}$.) If $E^{(\phi,\vec{x})}[\tau_U] = \infty$ for all open $U \ni (\phi,\vec{x})$, we define $\mathcal{A}^{\mathbf{u}} f(\phi,\vec{x}) = 0$.

Theorem 14.7 *Let $f \in C^2(V \times \mathbf{R}^n)$. Then $f \in \mathcal{D}_{\mathcal{A}^{\mathbf{u}}}$ and*

$$\mathcal{A}^{\mathbf{u}} f = \mathbf{L} f. \quad (14.4.19)$$

Proof *See [13], Theorem 7.5.4.*

Theorem 14.8 *Let $\psi \in C(\partial(G))$ be bounded and let $g \in C(G)$ satisfy*

$$E^{(\phi,\vec{x})}\left[\int_0^{\tau_G} |g(S_t, S(t))|dt\right] < \infty\ \forall\ \in G. \quad (14.4.20)$$

Define

$$\begin{aligned} w(\phi,\vec{x}) &= E^{(\phi,\vec{x})}[\psi(S_{\tau_G}, S(\tau_G))] + \\ &+ E^{(\phi,\vec{x})}\left[\int_0^{\tau_{V_1 \times O}} g(S_t, S(t))dt\right], \ (\phi,\vec{x}) \in G. \end{aligned} \quad (14.4.21)$$

Then

a)

$$\mathcal{A}^{\mathbf{u}} w = -g \, in G \quad (14.4.22)$$

and

$$\lim_{t \uparrow \tau_G} w(S_t, S(t)) = \psi(S_{\tau_G}, S(\tau_G)) \ a.s., \quad (14.4.23)$$

b) Moreover, if there exists a function $w_1 \in C^2(G)$ and a constant C such that

$$|w_1(\phi,\vec{x})| < C\left(1 + E^{(\phi,\vec{x})}\left[\int_0^{\tau_G} |g(S_t, S(t))|dt\right]\right),\ \left(\phi,\vec{x}\right) \in G, \quad (14.4.24)$$

and w_1 satisfies (14.4.22) and (14.4.23), then $w_1 = w$.

Proof *The proof follows arguments similar to [13] Theorem 9.3.3.*

Let $M : V \times \mathbf{R}^n \times U \to \mathbf{R}$, such that $E^{(\phi,\vec{x})} \int_0^{\tau_G} |M^{\mathbf{u}}(S_t, S(t))|dt < \infty$, we consider the equation

$$(\mathbf{A}^{\mathbf{u}} f + M^{\mathbf{u}})(\phi, \vec{x}) = 0, \ \ (\phi, \vec{x}) \in G \tag{14.4.25}$$

with boundary data

$$f(\phi, \vec{x}) = \psi(\phi, \vec{x}) \text{ with } (\phi, \vec{x}) \in \partial^* G. \tag{14.4.26}$$

Here $\partial^*(G)$ denotes a closed subset of $\partial(G)$ such that $\mathbf{P}^{(\phi,\vec{x})}((S_{\tau_G}, S\tau_G) \notin \partial^*(G),\ \tau_G < \infty) = 0$ for each $(\phi, \vec{x}) \in G$.

Lemma. Let $S(t)$ be the solution to (14.4.1), f in $C^2(G)$, with F continuous and bounded. Suppose that $\mathbf{P}^{(\phi,\vec{x})}(\tau_G < \infty) = 1$ for each $(\phi, \vec{x}) \in G$.

(a) If $(\mathbf{A}^{\mathbf{u}} f + M^{\mathbf{u}})(\phi, \vec{x}) \geq 0$ for all $(\phi, \vec{x}) \in G$, then

$$f(\phi, \vec{x}) \leq E^{(\phi,\vec{x})} \left\{ \int_0^{\tau_G} M^{\mathbf{u}}(S_t, S(t))dt + f(S_{\tau_G}, S(\tau_G)) \right\}, \ \ (\phi, \vec{x}) \in G \tag{14.4.27}$$

(b) If f is a solution of (14.4.25) and (14.4.26) for all $(\phi, \vec{x}) \in G$,

where $E^{(\phi,\vec{x})} \int_0^{\tau_G} |M^{\mathbf{u}}(S_t, S(t))| < \infty$, then

$$f(\phi, \vec{x}) = E^{(\phi,\vec{x})} \left\{ \int_0^{\tau_G} M^{\mathbf{u}}(S_t, S(t))dt + \Psi(S_{\tau_G}, S(\tau_G)) \right\}, \ \ (\phi, \vec{x}) \in G \tag{14.4.28}$$

Proof *(a) From the Dynkin formula*

$$\begin{aligned} f(\phi, \vec{x}) = \ & E^{(\phi,\vec{x})} f(S_{\tau_G}, S(\tau_G) + \\ & -E^{(\phi,\vec{x})} \{ \textstyle\int_0^{\tau_G} \mathbf{A}^{\mathbf{u}} f(S_t, S(t))dt \} \leq \\ & \leq E^{(\phi,\vec{x})} \{ \textstyle\int_0^{\tau_G} M^{\mathbf{u}}(S_t, S(t))dt + f(S_{\tau_G}, S(\tau_G)) \} \end{aligned}$$

(b) Since $M^{\mathbf{u}} = -\mathbf{A}^{\mathbf{u}} f$ satisfies the condition integrability, we get (b) as in (a).
For $v = \mathbf{u}(S_t, S(t))$, let

$$\mathbf{A}^v f(S_t, S(t)) := \mathbf{A}^{\mathbf{u}} f(S_t, S(t))$$

The dynamic programming equation is:

$$0 = \inf_{v \in U} \left[(\mathbf{A}^v f + L^v)(\phi, \vec{x}) \right], \ (\phi, \vec{x}) \ in \ G, \tag{14.4.29}$$

with the boundary data

$$f(\phi, \vec{x}) = \psi(\phi, \vec{x}) \ \ (\phi, \vec{x}) \in \partial^*(G), \tag{14.4.30}$$

and L as in (14.3.3).

We assume that

$$L(\phi, \vec{x}, v) \geq c > 0 \tag{14.4.31}$$

for some constant c.

Theorem 14.9 *(Sufficient conditions for optimality) Let f be a solution of (14.4.29)-(14.4.30) such f is in $C^2(G)$, and f is bounded and continuous in $\overline{G}$. Then:*
(a) $f(\phi, \vec{x}) \leq J(\phi, \vec{x}, \mathbf{u})$ for any $\mathbf{u} \in \mathcal{U}$ and $(\phi, \vec{x}) \in G$.
(b) If $\mathbf{u}^\star \in \mathcal{U}$, $J(\phi, \vec{x}, \mathbf{u}^\star) < \infty$ and

$$\mathbf{A}^{\mathbf{u}^\star} f(\phi, \vec{x}) + L^{\mathbf{u}^\star}(\phi, \vec{x}) = \inf_{v \in U} \left[(\mathbf{A}^v f + L^v)(\phi, \vec{x}) \right] \tag{14.4.32}$$

for all $(\phi, \vec{x}) \in G$, then $f(\phi, \vec{x}) = J(\phi, \vec{x}, \mathbf{u}^\star)$. Thus $\mathbf{u}^\star$ is an optimal control, for all choices of initial data $(\phi, \vec{x}) \in G$.

Proof *(a). It is sufficient to consider those $\mathbf{u}$ for which $J(\phi, \vec{x}, \mathbf{u}) < \infty$. The Chebishev inequality, (14.4.31) and the boundedness of ψ on $\partial^*(G)$, implies that $\mathbf{P}^{(\phi, \vec{x})}(\tau_G < \infty) = 1$. For each $v \in U$, $(\phi, \vec{x}) \in G$, $0 \leq (\mathbf{A}^{\mathbf{u}} f + L^{\mathbf{u}})(\phi, \vec{x})$. We conclude the proof by using the Lemma (14.4) replacing M by $L^{\mathbf{u}}$.*
(b) The condition (14.3.3) implies that

$$E^{(\phi, \vec{x})} \int_0^{\tau_G} |M[(S_t, S(t))]| dt < \infty.$$

For $\mathbf{u} = \mathbf{u}^\star$, we get $\mathbf{A}^{\mathbf{u}} f + L^{\mathbf{u}}(\phi, \vec{x}) = 0$. Then, using Lemma (14.4)(b), we have $f(\phi, \vec{x}) = J(\phi, \vec{x}, \mathbf{u}^\star)$.

Definition. A point $(\phi, \vec{x}) \in \partial(G)$ is called **regular** for G (with respect to $(S_t, S(t))$) if

$$\mathbf{P}^{(\phi, \vec{x})}(\tau_G = 0) = 1.$$

Otherwise the point $(\phi, \vec{x})$ is called **irregular**.

Theorem 14.10 *(The Hamilton–Jacobi–Bellman (HJB) equation) Suppose that $\mathbf{P}^{(\phi, \vec{x})}(\tau_G < \infty) = 1$ for each $(\phi, \vec{x}) \in G$. Define*

$$\Phi(\phi, \vec{x}) = \inf_{\mathbf{u}} \left\{ J^{\mathbf{u}}(\phi, \vec{x});\ \mathbf{u} \textit{ Markov control} \right\}$$

Suppose that $\Phi \in C^2(G) \bigcap C(\overline{G})$ is bounded and that an optimal Markov control $\mathbf{u}^\star$ exists and that $\partial(G)$ is regular for $(S_t^{\mathbf{u}^\star}, S^{\mathbf{u}^\star}(t))$. Then

$$\inf_{v \in U} \left\{ L^v(\phi, \vec{x}) + \mathbf{A}^v \Phi(\phi, \vec{x}) \right\} = 0 \ \forall\ (\phi, \vec{x}) \in G \tag{14.4.33}$$

and

$$\Phi(\phi,\vec{x}) = \psi(\phi,\vec{x})\ \forall\ (\phi,\vec{x}) \in \partial(G). \tag{14.4.34}$$

The infimum in (14.4.33) is obtained if $v = \mathbf{u}^\star(\phi,\vec{x})$ *where* $\mathbf{u}^\star(\phi,\vec{x})$ *is optimal. Equivalently*

$$L^{\mathbf{u}^\star(\phi,\vec{x})}(\phi,\vec{x}) + (\mathbf{A}^{\mathbf{u}^\star(\phi,\vec{x})}\Phi)(\phi,\vec{x}) = 0\ \forall\ (\phi,\vec{x}) \in G. \tag{14.4.35}$$

Proof *Since* $\mathbf{u}^\star(\phi,\vec{x})$ *is optimal, we obtain*

$$\begin{aligned}\Phi(\phi,\vec{x}) = \quad & J^{\mathbf{u}^\star}(\phi,\vec{x}) = E^{(\phi,\vec{x})}\left[\int_0^{\tau_G} L(S_t,S(t),u(t))dt + \right.\\ & \left. + \psi(S_{\tau_G},S(\tau_G))\right].\end{aligned} \tag{14.4.36}$$

If $(\phi,\vec{x}) \in \partial(G)$ *then* $\tau_G = 0$ *a.s. and we get (14.4.34). From (14.4.36) and Theorem 14.8 we obtain (14.4.35).*
The proof is completed if we prove (14.4.33). Following [9], fix $(\phi,\vec{x}) \in G$ *and choose a Markov control* $\mathbf{u}$*. Let* $\alpha \leq \tau_G$ *be a bounded stopping time. Since*

$$J^{\mathbf{u}}(\phi,\vec{x}) = E^{(\phi,\vec{x})}\left[\int_0^{\tau_G} L^u(S_t,S(t))dt + \psi(S_{\tau_G},S(\tau_G))\right]$$

using the Theorem 14.5 and the properties of the shift operator θ*. (see [12] sections 7.2 and 9.3) we have*

$$\begin{aligned}E^{(\phi,\vec{x})}[J^{\mathbf{u}}(S_\alpha,S(\alpha))] &= E^{(\phi,\vec{x})}\left[E^{(S_\alpha,S(\alpha))}\left[\int_0^{\tau_G} L^{\mathbf{u}}(S_t,S(t))dt + \psi(S_{\tau_G},S(\tau_G))\right]\right]\\ &= E^{(\phi,\vec{x})}\ \left[E^{(\phi,\vec{x})}\left[\theta_\alpha\left(\int_0^{\tau_G} L^u(S_t,S(t))dt + \psi(S_{\tau_G},S(\tau_G))\right)|\mathcal{F}_\alpha\right]\right]\\ &= E^{(\phi,\vec{x})}\ \left[E^{(\phi,\vec{x})}\left[\theta_\alpha\left(\int_\alpha^{\tau_G} L^u(S_t,S(t))dt + \psi(S_{\tau_G},S(\tau_G))\right)|\mathcal{F}_\alpha\right]\right]\\ &= E^{(\phi,\vec{x})}\ \left[\int_0^{\tau_G} L^u(S_t,S(t))dt + \psi(S_{\tau_G},S(\tau_G)) - \int_0^\alpha L^u(S_t,S(t))dt\right]\\ &= J^{\mathbf{u}}(\phi,\vec{x})\ - E^{(\phi,\vec{x})}\left[\int_0^\alpha L^u(S_t,S(t))dt\right].\end{aligned}$$

Then

$$J^{\mathbf{u}}(\phi,\vec{x}) = E^{(\phi,\vec{x})}\left[\int_0^\alpha L^u(S_t,S(t))dt\right] + E^{(\phi,\vec{x})}[J^{\mathbf{u}}(S_\alpha,S(\alpha))] \tag{14.4.37}$$

Now, we consider $W \subset G$ *and* $\alpha := \inf\{t \geq 0; (S_t,S(t)) \notin W\}$*. Suppose an optimal control* $\mathbf{u}^\star(\phi,\vec{x})$ *exists, let* $v \in U$ *arbitrary we define*

$$\mathbf{u}(\eta,\vec{y}) = \begin{cases} v & \text{if } (\eta,\vec{y}) \in W \\ \mathbf{u}^\star(\eta,\vec{y}) & \text{if } (\eta,\vec{y}) \in G\setminus W \end{cases}$$

Then

$$\Phi(S_\alpha,S(\alpha)) = J^{\mathbf{u}}(S_\alpha,S(\alpha)) = J^{\mathbf{u}^\star}(S_\alpha,S(\alpha)). \tag{14.4.38}$$

From this, (14.4.37) and using the Dynkin formula we obtain

$$\begin{aligned}\Phi(\phi,\vec{x}) &\leq J^{\mathbf{u}}(\phi,\vec{x}) = E^{(\phi,\vec{x})}\left[\int_0^\alpha L^v(S_t,S(t))dt\right] + E^{(\phi,\vec{x})}\left[\Phi(S_\alpha,S(\alpha))\right] \\ &= E^{(\phi,\vec{x})}\left[\int_0^\alpha L^v(S_t,S(t))dt\right] + \Phi(\phi,\vec{x}) + \\ &+ E^{(\phi,\vec{x})}\left[\int_0^\alpha \mathbf{A}^v\Phi(S_t,S(t))dt\right],\end{aligned} \tag{14.4.39}$$

therefore

$$E^{(\phi,\vec{x})}\left[\int_0^\alpha (L^v(S_t,S(t)) + \mathbf{A}^v\Phi(S_t,S(t)))\,dt\right] \geq 0.$$

Thus

$$\frac{E^{(\phi,\vec{x})}\left[\int_0^\alpha (L^v(S_t,S(t)) + \mathbf{A}^v\Phi(S_t,S(t)))\,dt\right]}{E^{(\phi,\vec{x})}[\alpha]} \geq 0.$$

Taking in account that $L^v(\cdot)$ and $\mathbf{A}^v(\cdot)$ are continuous, we obtain
$L^v(\phi,\vec{x}) + \mathbf{A}^v(\phi,\vec{x}) \geq 0$, *as* $W \downarrow (\phi,\vec{x})$.
From this and (14.4.35) we obtain (14.4.33).

Remark. The HJB equation has been studied by many authors. See, for example, [2, 9, 13]. Also, the paper [8] surveys many results in this area.

14.5 Examples: optimal selection portfolio and Ramsey model

14.5.1 An optimal portfolio selection problem

Let $S(t)$ denote the wealth of a person at time t. The person has two investments. Let $P(t)$ be an investment *risk free*:

$$dP(t) = kP(t)dt, \quad P(0) = P_0.$$

And the another investment is a *risky* one:

$$dS_1(t) = \mu S_1(t)dt + \sigma S_1(t)dW(t), \quad S(0) = S_0.$$

At each instant t an investor can choose what fraction $u(t)$ of this wealth he will invest in the risky asset, then investing $1-u(t)$ in the risk free asset. Suppose that the past has influence over the wealth $Z(t)$ of the investor; thus we have the following SDDE (we suppose that we have a self-financing portfolio [13]):

$$\begin{aligned}dZ(t) = \;& u(t)\mu Z(t)\tfrac{1}{1+\|Z_t\|}dt + u(t)\sigma S(t)\tfrac{1}{1+\|Z_t\|}dW(t) + \\ &+(1-u(t))\left(Z(t)\tfrac{1}{1+\|Z_t\|}\right)kdt \\ &= (\mu u(t) + k(1-u(t)))\left(Z(t)\tfrac{1}{1+\|Z_t\|}\right)dt + \sigma u(t)\left(Z(t)\tfrac{1}{1+\|Z_t\|}\right)dW(t),\end{aligned}$$

and $(Z_0, Z(0)) = (\phi, x)$ with $\|\phi\| > 0$ and $x > 0$. By the Theorem 14.2 there is a solution $S(t)$ with initial condition (ϕ, x).

Assume that $0 \leq u(t) \leq 1$, and $\psi : [0,\infty) \to [0,\infty)$, $\psi(0) = 0$, the problem is to find $\Xi(\phi,x)$ and a control $\mathbf{u}^\star = \mathbf{u}^\star(Z_t, Z(t))$, $0 \leq \mathbf{u}^\star \leq 1$, such that

$$\Xi(\phi,x) = \sup\{J^{\mathbf{u}}(\phi,x) : \ \mathbf{u} \text{ Markov control, } 0 \leq \mathbf{u} \leq 1\} = J^{\mathbf{u}^\star}(\phi,x),$$

where $J^{\mathbf{u}}(\phi,x) = E^{(\phi,x)}\left[\psi(Z^{\mathbf{u}}_{\tau_G}, Z^{\mathbf{u}}(\tau_G))\right]$ and τ_G is the first exit time from

$$G = \left\{(\phi,x) \in V \times \mathbf{R} : \ x, \|\phi\| > 0 \text{ and } \frac{(\mu-k)^2}{2\sigma^2(1-p)} + \frac{k}{1+\|\phi\|} = 0\right\}.$$

We observe that

$$\Xi = -\inf\{-J^{\mathbf{u}}(\phi,x)\} = -\inf\left\{E^{(\phi,x)}\left[-\psi(Z^{\mathbf{u}}_{\tau_G}, Z^{\mathbf{u}}(\tau_G))\right]\right\}$$

so $-\Xi$ coincides with the solution Φ of the problem (14.3.5), but with ψ replaced by $-\psi$ and $L = 0$. Thus, we see that the equation (14.4.33) for Φ gets for Ξ the form

$$\sup_v\{(\mathbf{A}^v f)(\phi,x)\} = 0, \text{ for } (\phi,x) \in G;$$

and

$$f(\phi,x) = -\psi(\phi,x) \text{ for } (\phi,x) \in \partial G$$

We suppose that f is a smooth function. From (14.4.16) the differential operator $\mathbf{A}^v$ has the form

$$(\mathbf{A}^v f)(\phi,x) = \frac{\partial f}{\partial x}(\phi,x)(\mu v + k(1-v))\left(x\frac{1}{1+\|\phi\|}\right) + \frac{1}{2}\frac{\partial^2 f}{\partial x^2}(\phi,x)\sigma^2 v^2\left(\frac{x}{1+\|\phi\|}\right)^2 + \\ +\mathbf{S}f(\phi,x).$$

Therefore, for each (ϕ,x) we try to find the value $v = (\phi,x)$ which maximizes the function

$$m(v) = ((\mu-k)v + k))\left(x\frac{1}{1+\|\phi\|}\right)\frac{\partial f}{\partial x}(\phi,x) + \sigma^2 v^2 \frac{1}{2}\left(x\frac{1}{1+\|\phi\|}\right)^2 \frac{\partial^2 f}{\partial x^2}(\phi,x) + \\ +\mathbf{S}f(\phi,x). \tag{14.5.1}$$

If $\frac{\partial f}{\partial x} > 0$ and $\frac{\partial^2 f}{\partial x^2} < 0$, the solution is

$$v = \mathbf{u}(\phi,x) = -\frac{(\mu-k)\left(x\frac{1}{1+\|\phi\|}\right)\frac{\partial f}{\partial x}}{\sigma^2\left(x\frac{1}{1+\|\phi\|}\right)^2\frac{\partial^2 f}{\partial x^2}} = -\frac{(\mu-k)\frac{\partial f}{\partial x}}{\sigma^2\left(x\frac{1}{1+\|\phi\|}\right)\frac{\partial^2 f}{\partial x^2}} \tag{14.5.2}$$

We replace this in (14.5.1) we obtain the following boundary value problem

$$-\frac{(\mu-k)^2}{2\sigma^2\frac{\partial^2 f}{\partial x^2}}\left(\frac{\partial f}{\partial x}(\phi,x)\right)^2+kx\frac{\partial f}{\partial x}(\phi,x)+$$

$$+\ \mathbf{S}f(\phi,x)=0 \tag{14.5.3}$$

$$f\ (\ \phi,x)=-\psi(\phi,x) \text{ for } (\phi,x)\in\partial G \tag{14.5.4}$$

Consider $\psi(\phi,x)=x^p$ where $0<p<1$.

We try to find a solution of (14.5.3) and (14.5.4) of the form

$$f(\phi,x)=x^p.$$

We note, that if $p>1$, then f is convex and when $0<p<1$, then f is a concave function. Substituting in (14.5.3) and using the definition of $\mathbf{S}$ we obtain $\frac{(\mu-k)^2}{2\sigma^2(1-p)}+k\frac{1}{1+\|\phi\|}=0.$

Using (14.5.2) we obtain the optimal control

$$\mathbf{u}^\star(\phi,x)=\frac{\mu-k}{\sigma^2(1-p)}(1+\|\phi\|).$$

If $0<\frac{\mu-k}{\sigma^2(1-p)}(1+\|\phi\|)<1$, this $\mathbf{u}^\star$ is the solution to the problem.

14.5.2 Stochastic Ramsey model in economics

Another application is given to a stochastic model in economics, a Ramsey model, that takes into account the influence of past and randomness in the production cycle.

In this example we use the notation of subsection 4.1 of [9]. In 1928 F.R. Ramsey introduced an economics model describing the rate of change of capital K and labor L in a market by a system of ordinary differential equations [14]. With P and C being the production and consumption rates, respectively, the model is given by

$$\frac{dK(t)}{dt}=P(t)-C(t),\quad \frac{dL(t)}{dt}=a(t)L(t), \tag{14.5.5}$$

where $a(t)$ is the rate of growth of labor (population).

The production, capital and labor are related by the Cobb–Douglas formula, $P(t)=AK^\alpha(t)L^\beta(t)$, where A, α, β are some positive constants. In certain circumstances the dependence of P on K and L is linear, i.e., $\alpha=\beta=1$, which will be our assumption throughout this section. We shall also assume that the labor is constant, $L(t)=L_0$, which is true for certain markets for relatively short time intervals of several years. Therefore, the production rate and the capital are related by $P(t)=BK(t)$, where $B=AL_0$. Another important assumption we make is that the production rate is subject to small random disturbances, i.e., $P(t) = BK(t) + \text{"}noise.\text{"}$ System (14.5.5) then results in

$$\frac{dK(t)}{dt} = P(t) + \text{"}noise\text{"} - C(t)$$

we can rewrite this equation in the following form

$$dK(t) = [BK(t) - C(t)]\,dt + \frac{K(t)}{1 + \|K_t\|} dW(t)$$

where $W(t)$ is a real Wiener process.

The original model of Ramsey is based on the assumption of instant transformation of the investments, we suppose that the past has influence over the investments in the following way

$$dK(t) = \left[B\frac{K(t)}{1 + \|K_t\|} - C(t)\right] dt + \frac{K(t)}{1 + \|K_t\|} dW(t)$$

Our next assumption is that C is constant and that the consumption rate C can be controlled by the available amount of the capital, i.e., it is of the form $u(K(t))C$, where u is a control. Thus we get the following stochastic differential delay equation

$$dK(t) = \left[B\frac{K(t)}{1 + \|K_t\|} - u(K(t))C\right] dt + \frac{K(t)}{1 + \|K_t\|} dW(t) \qquad (14.5.6)$$

and $(K_0, K(0)) = (\phi, x)$ with $\|\phi\| > 0$ and $x > 0$.

We note, that control u only depends here on $K(t)$. Of course, we could assume that it depends on K_t and $K(t)$ as well that changes operator $\mathbf{A}^v$ below with obvious changes in the calculations that follow. By the Theorem 14.2 there is a solution $K(t)$ with initial condition (ϕ, x).

The problem is to find $\Phi(\phi, x)$ and a control $\mathbf{u}^\star = \mathbf{u}^\star(K(t))$ such that

$$\Phi(\phi, x) = \inf\{J^{\mathbf{u}}(\phi, x) : \mathbf{u}(t) = \mathbf{u}(K(t)) \text{ Markov control}\} = J^{\mathbf{u}^\star}(\phi, x)$$

where $J^{\mathbf{u}}(\phi, x) = E^{(\phi,x)}\left[\int_0^{\tau_G} \frac{K(s)^2}{2} + \frac{\mathbf{u}(s)^2}{2} ds\right]$ and τ_G is the first exit time from

$$G = \left\{(\phi, x) \in V \times \mathbf{R} : \, x, \|\phi\| > 0, \text{ and } -\frac{4B}{1 + \|K_t\|} - \frac{2}{(1 + \|K_t\|)^2} = k = constant\right\}.$$

The HJB equation (14.4.33) for Φ gets the form

$$\inf_v \left\{(\mathbf{A}^v f)(\phi, x) + \frac{x^2}{2} + \frac{v^2}{2}\right\} = 0, \text{ for } (\phi, x) \in G; \qquad (14.5.7)$$

and

$$f(\phi, x) = 0 \text{ for } (\phi, x) \in \partial G.$$

From (14.4.16) the differential operator $\mathbf{A}^v$ has the form

$$(\mathbf{A}^v f)(\phi,x) = \frac{\partial f}{\partial x}(\phi,x)(B\frac{x}{1+\|\phi\|} - vC) + \frac{1}{2}\frac{\partial^2 f}{\partial x^2}(\phi,x)\frac{x^2}{(1+\|\phi\|)^2} + \\ +\mathbf{S}f(\phi,x).$$

We try for each (ϕ,x) to find the value $v = \mathbf{u}(\phi,x)$ which minimizes the function

$$h(v) = \frac{\partial f}{\partial x}(\phi,x)(B\frac{x}{1+\|\phi\|} - vC) + \frac{1}{2}\frac{\partial^2 f}{\partial x^2}(\phi,x)\frac{x^2}{(1+\|\phi\|)^2} + \\ +\mathbf{S}f(\phi,x) + \frac{x^2}{2} + \frac{v^2}{2}. \tag{14.5.8}$$

If $\frac{\partial f}{\partial x} > 0$ and $C > 0$ the solution is

$$v = C\frac{\partial f}{\partial x}. \tag{14.5.9}$$

If we substitute this in (14.5.7) we obtain the following boundary value problem

$$\frac{\partial f}{\partial x}(B\frac{x}{1+\|\phi\|} - \frac{C}{2}\frac{\partial f}{\partial x}C) + \frac{1}{2}\frac{\partial^2 f}{\partial x^2}\frac{x^2}{(1+\|\phi\|)^2} + \\ +\mathbf{S}f(\phi,x) + \frac{x^2}{2} + \frac{(\frac{C}{2}\frac{\partial f}{\partial x})^2}{2} = 0 \tag{14.5.10}$$

$$f(\phi,x) = 0 \text{ for } (\phi,x) \in \partial G \tag{14.5.11}$$

We try a solution of the form $f(\phi,x) = \frac{x^2}{c_1}$, where c_1 is a constant. We substitute this in (14.5.10) and using the definition of $\mathbf{S}$ we obtain

$$\frac{2x^2}{c_1}B\frac{1}{1+\|\phi\|} + \frac{1}{c_1}\frac{x^2}{(1+\|\phi\|)^2} + \frac{x^2}{2} = 0$$

From here, $c_1 = -\frac{4B+2}{1+\|\phi\|}$. Substituting in (14.5.9) we obtain the optimal control

$$\mathbf{u}^\star(\phi,x) = \frac{2xC}{c_1} = -\frac{2xC(1+\|\phi\|)}{4B+2}.$$

References

[1] Arriojas, M., Hu, T., Pap, Y. and Mohammed, S.-E. A delayed Black-Scholes formula. *Journal of Stochastic Analysis and Applications*, Vol. 25, No. 2 (2007), 471-492.

[2] Chang, M.-H. *Stochastic Control of Hereditary Systems and Applications.* Springer-Verlag, v. 59, 2008.

[3] Da Prato, G. and Zabczyk, J. *Stochastics Equations in Infinite Dimensions*, Cambridge University Press, 1992.

[4] Dynkin, E. B. *Markov Process,* Vol. I, Die Grundlehreu der Math. Wissenschaften, Springer-Verlag, 1965.

[5] Fleming, W. H. and Rishel, R. W., *Deterministic and Stochastic Control,* Springer-Verlag, 1975.

[6] Gihman, I. I., and Skorohod, A. I. *Stochastic Differential Equations*, Springer-Verlag, 1972.

[7] Hu, Y. and Øksendal, B. Fractional white noise calculus and applications to finance. *Infinite Dimensional Analysis, Quantum Probability and Related Topics*, Vol. 6, No. 1 (2003), 1-32.

[8] Ivanov, A., Kazmerchuk, Y. and Swishchuk, A. Theory, stochastic stability and applications of stochastic delay differential equations: a survey of recent results. *Differ. Equat. and Dynam. Syst. J.*, 11, 1-2, (2003) 55–115 .

[9] Ivanov, A. F. and Swishchuk, A. V. Optimal control of stochastic differential delay equations with applications in economics. *International Journal of Qualitative Theory of Differential Equations and Applications*, Vol. 2, No. 2 (2008), 201–213.

[10] Karatzas, I. and Shreve, S. E. *Brownian Motion and Stochastic Calculus*, Second Edition Springer, NY, 1991.

[11] Kushner, H. J. On the stability of process defined by stochastic difference-differential equations, *Journal of Differential Equations*, 4, (1968) 424–443 .

[12] Øksendal, B., *Stochastic Differential Equations. An Introduction with Applications* Springer-Verlag, Sixth Ed. 2003.

[13] Ramsey, F.P. A mathematical theory of savings. *The Economic Journal*, Vol. 38, No. 152 (1928), 543–549.

[14] Schoenmakers, J., and Kloeden, P. Robust option replication for a Black and Scholes model extended with nondeterministic trends. *Journal of Applied Mathematics and Stochastic Analysis*, 12:2 (1999), 113–120.

[15] Yan, F., and Mohammed, S.-E. A stochastic calculus for systems with memory. *Stochastic Analysis and Applications*, 23:3 (2005) 613–657.

Chapter 15

RDS in Option Pricing Theory with Delayed/Path-Dependent Information

15.1 Chapter overview

The analogue of Black-Scholes formula for vanilla call option price in conditions of (B,S)-securities market with delayed information is derived. A special case of a continuous version of GARCH is considered. The results are compared with the results of Black and Scholes.

15.2 Introduction

In the early 1970s, Black and Scholes (1973) [4] made a major breakthrough by deriving pricing formulas for vanilla options written on the stock. Their model and its extensions assume that the probability distribution of the underlying cash flow at any given future time is lognormal. This assumption is not always satisfied by real-life options as the probability distribution of an equity has a fatter left tail and thinner right tail than the lognormal distribution (see Hull (2000)), and the assumption of constant volatility σ in a financial model (such as the original Black–Scholes model) is incompatible with derivatives prices observed in the market.

The above issues have been addressed and studied in several ways, such as
i) Volatility is assumed to be a deterministic function of the time: $\sigma \equiv \sigma(t)$ (see [40]);
ii) Volatility is assumed to be a function of the time and the current level of the stock price $S(t)$: $\sigma \equiv \sigma(t, S(t))$ (see [23]); the dynamics of the stock price satisfies the following stochastic differential equation:

$$dS(t) = \mu S(t)dt + \sigma(t, S(t))S(t)dW_1(t),$$

where $W_1(t)$ is a standard Wiener process;
iii) The time variation of the volatility involves an additional source of randomness represented by $W_2(t)$ and is given by

$$d\sigma(t) = a(t, \sigma(t))dt + b(t, \sigma(t))dW_2(t),$$

where $W_2(t)$ and $W_1(t)$ (the initial Wiener process that governs the price process) may be correlated (see [6]), [24]);

iv) The volatility depends on a random parameter x such as $\sigma(t) \equiv \sigma(x(t))$, where $x(t)$ is some random process (see [16], [19], [36], [37], [38]).

In the approach (i), the volatility coefficient is independent of the current level of the underlying stochastic process $S(t)$. This is a deterministic volatility model, and the special case where σ is a constant reduces to the well-known Black–Scholes model that suggests changes in stock prices are lognormally distributed. But the empirical test by Bollerslev (1986) [5] seems to indicate otherwise. One explanation for this problem of a lognormal model is the possibility that the variance of $\log(S(t)/S(t-1))$ changes randomly. This motivated the work of [9], where the prices are analyzed for European options using the modified Black–Scholes model of foreign currency options and a random variance model. In their works the results of [24], [33] and [39] were used in order to incorporate randomly changing variance rates.

In the approach (ii), several ways have been developed to derive the corresponding Black-Scholes formula: one can obtain the formula by using stochastic calculus and, in particular, Ito's formula (see [31], for example). In the book by Cox and Rubinstein (1985) [11], an alternative approach was developed: the Black–Scholes formula is interpreted as the continuous-time limit of a binomial random model. A generalized volatility coefficient of the form $\sigma(t,S(t))$ is said to be *level-dependent*. Because volatility and asset price are perfectly correlated, we have only one source of randomness given by $W_1(t)$. A time and level-dependent volatility coefficient makes the arithmetic more challenging and usually precludes the existence of a closed-form solution. However, the *arbitrage argument* based on portfolio replication and a completeness of the market remains unchanged.

The situation becomes different if the volatility is influenced by a second "non-tradable" source of randomness. This is addressed in approaches (iii) and (iv) and one usually obtains a *stochastic volatility model*, which is general enough to include the deterministic model as a special case. The concept of stochastic volatility was introduced by [24], and subsequent development includes the following works [39], [25], [33], [35] and [20]. We also refer to [17] for an excellent survey on level-dependent and stochastic volatility models. We should mention that the approach (iv) is taken by, for example, in [19].

There is yet another approach connected with stochastic volatility, namely, the uncertain volatility scenario (see [6]). This approach is based on the uncertain volatility model developed in [2], where a concrete volatility surface is selected among a candidate set of volatility surfaces. This approach addresses the sensitivity question by computing an upper bound for the value of the portfolio under arbitrary candidate volatility, and this is achieved by choosing the local volatility $\sigma(t,S(t))$ among two extremal values $\sigma_{\min}$ and $\sigma_{\max}$ such that the value of the portfolio is maximized locally.

The assumption made implicitly by Black and Scholes (1973) [4] is that the historical performance of the (B,S)-securities markets can be ignored. In particular, the so-called Efficient Market Hypothesis implies that all information available is already reflected in the present price of the stock and the past stock performance gives no information that can aid in predicting future performance. However, some statistical studies of stock prices (see [34], and [1]) indicate the dependence on past returns. For example, [27] obtained a diffusion approximation result for processes satisfying some equations with past-dependent coefficients, and they applied this result to a model of option pricing, in which the underlying asset price volatility depends on the past evolution to obtain a generalized (asymptotic) Black–Scholes formula. Reference [21] suggested a new class of nonconstant volatility models, which can be extended to include the aforementioned level-dependent model and share many characteristics with the stochastic volatility model. The volatility is nonconstant and can be regarded as an endogenous factor in the sense that it is defined in terms of the *past behavior* of the stock price. This is done in such a way that the price and volatility form a multi-dimensional Markov process.

Reference [7] studied the pricing of a European contingent claim for the (B,S)-securities markets with a hereditary price structure in the sense that the rate of change of the unit price of the bond account and rate of change of the stock account S depend not only on the current unit price but also on their historical prices. The price dynamics for the bank account and that of the stock account are described by a linear functional differential equation and a linear stochastic functional differential equation, respectively. They show that the rational price for a European contingent claim is independent of the mean growth rate of the stock. Later [8] generalized the celebrated Black-Scholes formula to include the (B,S)-securities market with hereditary price structure.

Other alternatives to the Black–Scholes model include models where a company's equity is assumed to be an option of its assets (see [18]); models of (B,S,X)-securities market with Markov or semi-Markov stochastic volatility $\sigma(x(t))$ (see [19]), and [36]); models of fractional (B,S)-securities markets with Hurst index $H > 1/2$, $H \in (0,1)$ or combined with the assumption of Markov or semi-Markov volatility (see [22], [15], [16]).

Clearly related to our work is the work by [30] devoted to the derivation of a delayed Black-Scholes formula for the (B,S)-securities market using PDE approach. In their paper, the stock price satisfies the following equation:

$$dS(t) = \mu S(t-a)S(t)dt + \sigma(S(t-b))S(t)dW(t),$$

where a and b are positive constants and σ is a continuous function, and the price of the option at time t has the form $F(t, S(t))$.

The subject of our work is the study of stochastic delay differential equations (SDDE), which arise in the pricing of options for security markets with delayed response. We show that a continuous-time equivalent of GARCH arises as a stochastic volatility model with delayed dependence on the stock value. We derive an analogue of Ito's lemma for a general type of SDDEs and we obtain an integro-differential equation for a function of option price with boundary conditions specified according to the type of option to be priced. This equation is solved using a simplifying assumption and the graph of the closed-form solution is shown on Figure 15.1. An implied volatility plot is generated to demonstrate the difference between the Black-Scholes model and our model (see Figure 15.2). Hobson and Rogers [21] also observed in their past-dependent model that the resulting implied volatility is U-shaped as a function of strike price. However, they dealt with only a special case where the model can be reduced to a system of SDDEs. Unfortunately, not every past-dependent model can be reduced to a system of SDDEs, and a more sophisticated approach, as developed in this paper, is needed.

More precisely, we consider the model of $S(t)$ with volatility σ depending on t and the path $S_t = \{S(t+\theta),\ \theta \in [-\tau, 0]\}$. We call it a *level-and-past-dependent volatility* $\sigma \equiv \sigma(t, S_t)$, contrary to the level-dependent volatility (that is clearly a special case of the former one when the time delay parameter $\tau = 0$). Our model of stochastic volatility exhibits past-dependence: the behavior of a stock price right after a given time t not only depends on the situation at t, but also on the whole past (history) of the process $S(t)$ up to time t. This draws some similarities with fractional Brownian motion models due to a long-range dependence property. Our work is also based on the GARCH(1,1) model (see [5]) and the celebrated work of Duan [12] where he showed that it is possible to use the GARCH model as the basis for an internally consistent option pricing model. We should mention that in the work of Kind et al. [27], a past-dependent model was defined by diffusion approximation. In their model, the volatility depends on the quadratic variation of the process, while our model deals with more general dependence of the volatility on the history of the process over a finite interval.

In future work, we wish to derive the continuous-time GARCH model for stochastic volatility with delayed response which incorporates conditional expectation of log-returns, and we also expect to develop a method of estimation of the time delay parameter (as well as all the other parameters).

15.3 Stochastic delay differential equations

For any path $x : [-\tau, \infty) \to R^d$ at each $t \geq 0$ define the segment $x_t : [-\tau, 0] \to R^d$ by

$$x_t(s) := x(t+s) \quad a.s.,\ t \geq 0,\ s \in [-\tau, 0].$$

Denote by $C := C([-\tau,0],R^d)$ the Banach space of all continuous paths $\eta : [-\tau,0] \to R^d$ equipped with the supremum norm

$$||\eta||_C := \sup_{s\in[-\tau,0]} |\eta(s)|, \quad \eta \in C.$$

Consider the following *stochastic delay differential equation* (SDDE) (see [29])

$$\begin{cases} dx(t) = H(t,x_t)dt + G(t,x_t)dW(t), & t \geq 0 \\ x_0 = \phi \in C \end{cases} \tag{15.3.1}$$

on a filtered probability space $(\Omega, \mathcal{F}, (\mathcal{F}_t)_{t\geq 0}, P)$ satisfying *the usual conditions*; that is, the filtration $(\mathcal{F}_t)_{t\geq 0}$ is right-continuous and each $\mathcal{F}_t$, $t \geq 0$, contains all P-null sets in $\mathcal{F}$. $W(t)$ represents the m-dimensional Brownian motion.

The SDDE 15.3.1 has a *drift coefficient* function $H : [0,T] \times C \to R^d$ and a *diffusion coefficient* function $G : [0,T] \times C \to R^{d\times m}$ satisfying the following.

(i) H and G are Lipschitz on bounded sets of C uniformly with respect to the first variable, i.e., for each integer $n \geq 1$, there exists a constant $L_n > 0$ (independent of $t \in [0,T]$) such that

$$|H(t,\eta_1) - H(t,\eta_2)| + ||G(t,\eta_1) - G(t,\eta_2)|| \leq L_n ||\eta_1 - \eta_2||_C$$

for all $t \in [0,T]$ and $\eta_1, \eta_2 \in C$ with $||\eta_1||_C \leq n$, $||\eta_2||_C \leq n$.

(ii) There is a constant $K > 0$ such that

$$|H(t,\eta)| + ||G(t,\eta)|| \leq K(1 + ||\eta||_C)$$

for all $t \in [0,T]$ and $\eta \in C$.

A *solution* of 15.3.1 is a measurable, sample-continuous process $x : [-\tau,T] \times \Omega \to R^d$ such that $x|_{[0,T]}$ is $(\mathcal{F}_t)_{0\leq t\leq T}$-adapted and x satisfies 15.3.1 almost surely.

In Mohammed [29] it was shown that if hypotheses i)-ii) hold then for each $\phi \in C$ the SDDE 15.3.1 has a unique solution $x^\phi : [-\tau,\infty) \times \Omega \to R^d$ with $x^\phi|_{[-\tau,0]} = \phi \in C$ and $[0,T] \ni t \to x_t^\phi \in C$ being sample-continuous.

15.4 General formulation

The stock price value is assumed to satisfy the following SDDE:

$$dS(t) = rS(t)dt + \sigma(t,S_t)S(t)dW(t) \tag{15.4.1}$$

with continuous deterministic initial data $S_0 = \varphi \in C := C([-\tau,0],R)$, here σ represents a *volatility* which is a continuous function of time and the elements of C.

As mentioned in last section, the existence and uniqueness of solution of 15.4.1

are guaranteed if the coefficients in 15.4.1 satisfy the following local Lipschitz and growth conditions:

$$|\sigma(t,\eta_1)\,\eta_1(t) - \sigma(t,\eta_2)\,\eta_2(t)| \leq L_n||\eta_1 - \eta_2||_C \tag{15.4.2}$$

$\forall n \geq 1 \quad \exists L_n > 0 \quad \forall t \in [0,T] \quad \forall \eta_1, \eta_2 \in C,\ ||\eta_1||_C \leq n,\ ||\eta_2||_C \leq n$, and $\exists K > 0 \quad \forall t \in [0,T],\ \eta \in C:$

$$|\sigma(t,\eta)\,\eta(t)| \leq K(1 + ||\eta||_C). \tag{15.4.3}$$

Note that the stock price values are positive with probability 1 if the initial data is positive; that is, $\varphi(\theta) > 0$ for all $\theta \in [-\tau, 0]$.

We are primarily interested in an option price value, which is assumed to depend on the current and previous stock price values in the following way:

$$F(t,S_t) = \int_{-\tau}^{0} e^{-r\theta} H(S(t+\theta), S(t), t)\,d\theta, \tag{15.4.4}$$

where $H \in C^{0,2,1}(R \times R \times R_+)$. Such a representation is chosen since it includes sufficiently general functionals for which an analogue of Ito's lemma can be derived.

Lemma. Suppose a functional $F : R_+ \times C \to R$ has the following form

$$F(t,S_t) = \int_{-\tau}^{0} h(\theta) H(S_t(\theta), S_t(0), t)\,d\theta,$$

for $H \in C^{0,2,1}(R \times R \times R_+)$ and $h \in C^1([-\tau,0],R)$. Then

$$F(t,S_t) = F(0,\varphi) + \int_0^t \mathcal{A}F(s,S_s)ds + \int_0^t \sigma(s,S_s)S(s)\mathcal{B}F(s,S_s)dW(s), \tag{15.4.5}$$

where for $(t,x) \in R_+ \times C$,

$$\begin{aligned} \mathcal{A}F(t,x) &= h(0)H(x(0),x(0),t) - h(-\tau)H(x(-\tau),x(0),t) \\ &- \textstyle\int_{-\tau}^{0} h'(\theta)H(x(\theta),x(0),t)d\theta + \int_{-\tau}^{0} h(\theta)LH(x(\theta),x(0),t)d\theta, \\ \mathcal{B}F(t,x) &= \textstyle\int_{-\tau}^{0} h(\theta)H_2'(x(\theta),x(0),t)d\theta, \end{aligned}$$

and

$$\begin{aligned} LH(x(\theta),x(0),t) &= rx(0)H_2'(x(\theta),x(0),t) + \frac{\sigma^2(t,x)x^2(0)}{2}H_{22}''(x(\theta),x(0),t) \\ &\quad + H_3'(x(\theta),x(0),t), \end{aligned}$$

where $H_i', i = 1,2,3$, represents the derivative of $H(x(\theta),x(0),t)$ with respect to i-th argument. *Proof:* We defer the detailed proof to the Appendix.

In what follows, we assume that a riskless portfolio consisting of a position in the option and a position in the underlying stock is set up. In the absence of arbitrage opportunities, the return from the portfolio must be risk-free with the interest rate r. The portfolio Π has to be riskless during the time interval $[t, t+dt]$ and must instantaneously earn the same rate of return as other short-term risk-free securities. It follows that $d\Pi(t) = r\Pi(t)dt$ and this will be used in the proof of the following theorem.

Theorem 15.1 *Suppose the functional F is given by 15.4.4 with $S(t)$ satisfying 15.4.1 and $H \in C^{0,2,1}(R \times R \times R_+)$. Then, $H(S(t+\theta), S(t), t)$ satisfies the following equation*

$$0 = H|_{\theta=0} - e^{-r\theta} H|_{\theta=-\tau} + \int_{-\tau}^{0} e^{-r\theta} \left(H_3' + rS(t)H_2' + \frac{1}{2}\sigma^2(t, S_t)S^2(t)H_{22}'' \right) d\theta \tag{15.4.6}$$

for all $t \in [0, T]$.

Proof: To construct an equation for F, we need to consider a portfolio which consists of -1 derivative and $\mathcal{B}F(t, S_t)$ shares. Then, the portfolio value $\Pi(t)$ is equal to

$$\Pi(t) = -F(t, S_t) + \mathcal{B}F(t, S_t)\, S(t),$$

and the associated infinitesimal change in the time interval $[t, t+dt]$ is

$$d\Pi = -dF + \mathcal{B}F\; dS.$$

We should point out here that in the last statement we suppose that $(\mathcal{B}F)$ is held constant during the time-step dt, and hence term $d(\mathcal{B}F)$ is equal to zero. If this were not the case then $d\Pi$ would contain term $d(\mathcal{B}F)$.

Using 15.4.5 and 15.4.1, we obtain

$$d\Pi = -\mathcal{A}F\; dt - \sigma S \mathcal{B}F\; dW + \mathcal{B}F(rS\, dt + \sigma S\, dW).$$

Consideration of risk-free during the time dt then leads to

$$d\Pi = r\Pi\, dt,$$

that is,

$$-\mathcal{A}F(t, S_t) + rS(t)\mathcal{B}F(t, S_t) = r(-F(t, S_t) + \mathcal{B}F(t, S_t)S(t)),$$

or

$$\mathcal{A}F(t, S_t) = rF(t, S_t).$$

Therefore, the equation for $H(S(t+\theta), S(t), t)$ becomes

$$0 = H|_{\theta=0} - e^{-r\theta} H|_{\theta=-\tau} + \int_{-\tau}^{0} e^{-r\theta} \left(H_3' + rS(t)H_2' + \frac{1}{2}\sigma^2(t, S_t)S^2(t)H_{22}'' \right) d\theta.$$

This completes the proof.

Consider a *European call option* with the final payoff $\max(S-K, 0)$ at the maturity time T [23]. Then problem 15.4.6 has the boundary condition at the time T, either

$$F(T, S_T) = \max(S(T) - K, 0) \tag{15.4.7}$$

or, induced by the functional nature of F,

$$F(T, S_T) = \frac{1}{\tau} \int_{-\tau}^{0} \max(e^{-r\theta} S_T(\theta) - K, 0) d\theta, \tag{15.4.8}$$

where $1/\tau$ is a normalizing factor such that $F(T, S_T) \to \max(S(T) - K, 0)$ as $\tau \to 0$.

15.5 A simplified problem

We now consider the simplified problem 15.4.6, assuming that $H(S(t+\theta),S(t),t)$ is a sum of two functions, one of which depends on the current value of stock price $S(t)$ and another depends on the previous values of stock price $\{S(t+\theta),\theta\in[-\tau,0)\}$. That is, our option price 15.4.4 takes the form:

$$F(t,S_t)=h_1(S(t),t)+\int_{-\tau}^{0}e^{-r\theta}h_2(S(t+\theta),t)d\theta, \tag{15.5.1}$$

where $h_1(S(t),t)$ is a classical Black-Scholes call option price(see [4]) with the variance assumed equal to a long-run variance rate V (it is known that the stock price variance rate exhibits the so called *mean reversion*, see [23])

$$h_1(S(t),t)=S(t)N(d_1)-Ke^{-r(T-t)}N(d_2), \tag{15.5.2}$$

where $N(x)=\frac{1}{\sqrt{2\pi}}\int_{-\infty}^{x}e^{-x^2/2}dx$ and d_1 and d_2 are defined as

$$\begin{aligned} d_1 &= \tfrac{\ln(S(t)/K)+(r+V/2)(T-t)}{\sqrt{V(T-t)}}, \\ d_2 &= d_1-\sqrt{V(T-t)}. \end{aligned}$$

Note that the functional 15.5.1 seems to be a natural choice since we are interested in the difference between the original Black–Scholes option price and the one implied by the market with delayed response.

Theorem 15.2 *Assume the functional F is given by 15.5.1 with h_1 given by 15.5.2. Then,*

$$F(t,S_t)=h_1(S(t),t)+\frac{1}{2}\int_{t}^{T}e^{r(t-\xi)}\left[\sigma^2(\xi,S_t)-V\right]S^2(t)\frac{\partial^2 h_1}{\partial S^2}(S(t),\xi)\,d\xi. \tag{15.5.3}$$

Proof: Substituting 15.5.1 into 15.4.6 yields the following equation for h_2.

$$h_2(S(t),t)-e^{r\tau}h_2(S(t-\tau),t)+\int_{-\tau}^{0}e^{-r\theta}\frac{\partial h_2}{\partial t}d\theta=\frac{1}{2}(V-\sigma^2(t,S_t))S^2(t)\frac{\partial^2 h_1}{\partial S^2}.$$

Also from 15.5.1 we derive that

$$\begin{aligned} \tfrac{dF}{dt} &= \tfrac{dh_1}{dt}+\tfrac{d}{dt}\left[\int_{t-\tau}^{t}e^{-r(s-t)}h_2(S(s),t)ds\right] \\ &= \tfrac{dh_1}{dt}+h_2(S(t),t)-e^{r\tau}h_2(S(t-\tau),t) \\ &+ \int_{t-\tau}^{t}re^{-r(s-t)}h_2(S(s),t)ds+\int_{t-\tau}^{t}e^{-r(s-t)}\tfrac{\partial h_2}{\partial t}ds. \end{aligned}$$

Combining this expression with the one considered above we get the following equation for F:

$$\frac{\partial F}{\partial t}+rS(t)\frac{\partial h_1}{\partial S}+\frac{1}{2}\sigma^2(t,S_t)S^2(t)\frac{\partial^2 h_1}{\partial S^2}=rF. \tag{15.5.4}$$

We remark that the above equation is very similar to the well-known Black–Scholes PDE. Observing that h_1 satisfies the following PDE:

$$\frac{\partial h_1}{\partial t} + rS(t)\frac{\partial h_1}{\partial S} + \frac{1}{2}VS^2(t)\frac{\partial^2 h_1}{\partial S^2} = rh_1,$$

we have a new PDE for $f(t,S_t) := F(t,S_t) - h_1(S(t),t)$ as follows

$$\frac{\partial f}{\partial t} = rf + \frac{1}{2}\left[V - \sigma^2(t,S_t)\right] S^2(t)\frac{\partial^2 h_1}{\partial S^2}(S(t),t).$$

We can easily solve the above equation by using the variation-of-constants formula to obtain 15.5.3.

15.5.1 Continuous time version of GARCH model

Assume the $\sigma^2(t)$ satisfies the following equation

$$\frac{d\sigma^2(t)}{dt} = \gamma V + \frac{\alpha}{\tau}\ln^2\left(\frac{S(t)}{S(t-\tau)}\right) - (\alpha+\gamma)\sigma^2(t), \tag{15.5.5}$$

where V is a long-run average variance rate, α and γ are positive constants such that $\alpha+\gamma<1$. Here, $S(t)$ is a solution of the SDDE 15.4.1 with positive initial data $\varphi \in C$.

Consider a grid $-\tau = t_{-l} < t_{-l+1} < \cdots < t_0 = 0 < t_1 < \cdots < t_N = T$ with the time step size Δ_l of the form

$$\Delta_l = \frac{\tau}{l},$$

where $l \geq 2$. Then a discrete time analogue of 15.5.5 is

$$\sigma_n^2 = \gamma V + \frac{\alpha}{l}\ln^2(S_{n-1}/S_{n-1-l}) + (1-\alpha-\gamma)\sigma_{n-1}^2,$$

where $\sigma_n^2 = \sigma^2(t_n)$ and $S_n = S(t_n)$. Note that the process described by this difference equation is very similar to the GARCH(1,1) model proposed by Bollerslev in 1986 [5] (with returns assumed to have mean zero), which seems to provide a good explanation of stock price data,

$$\sigma_n^2 = \gamma V + \alpha\ln^2(S_{n-1}/S_{n-2}) + (1-\alpha-\gamma)\sigma_{n-1}^2. \tag{15.5.6}$$

Now, using a variation-of-constants formula for 15.5.5 we obtain

$$\begin{aligned}\sigma^2(t) &= \frac{\gamma V}{\alpha+\gamma} + \left(\sigma^2(t_0) - \frac{\gamma V}{\alpha+\gamma}\right)e^{-(\alpha+\gamma)(t-t_0)} \\ &\quad + \frac{\alpha}{\tau}\int_{t_0}^{t} e^{(\alpha+\gamma)(\xi-t)}\ln^2\left(\frac{S(\xi)}{S(\xi-\tau)}\right)d\xi\end{aligned} \tag{15.5.7}$$

for $t_0 \geq 0$. It is then natural that we consider the following expression for variance:

$$\bar{\sigma}^2(t,S_t) = \sigma^2(t_0)e^{-(\alpha+\gamma)(t-t_0)} + \left[\gamma V + \frac{\alpha}{\tau}\ln^2\left(\frac{S(t)}{S(t-\tau)}\right)\right]\frac{1-e^{-(\alpha+\gamma)(t-t_0)}}{\alpha+\gamma}, \tag{15.5.8}$$

since functions $\bar{\sigma}^2$ and σ^2 are close to each other in the following sense:

$$\sigma^2(t) = \bar{\sigma}^2(t) + o(|t-t_0|), \quad as\, t \to t_0.$$

Expression 15.5.8 for the volatility allows us to obtain a closed form for a call option price involving delayed market response.

Theorem 15.3 *Assume that the stock price satisfies SDDE 15.4.1 with the initial data $\varphi \in C$ and assume the volatility $\bar{\sigma}$ is given by 15.5.8. Then the European call option price with the strike price K and maturity T at the time t is given by*

$$F(t,S_t) = h_1(S(t),t) + (\Sigma(S_t)-V)\mathcal{I}(r,t,S(t)) + (\sigma^2(t)-\Sigma(S_t))\mathcal{I}(r+\alpha+\gamma,t,S(t)), \tag{15.5.9}$$

where $h_1(S(t),t)$ is given by 15.5.2 and

$$\Sigma(S_t) = \frac{\alpha}{\tau(\alpha+\gamma)}\ln^2\left(\frac{S(t)}{S(t-\tau)}\right) + \frac{\gamma V}{\alpha+\gamma},$$

$$\mathcal{I}(p,t,S(t)) = \frac{1}{2}S^2(t)\int_t^T e^{p(t-\xi)}\frac{\partial^2 h_1}{\partial S^2}(S(t),\xi)d\xi \quad for \quad p \geq 0.$$

Proof: Substituting the expression 15.5.8 for σ^2 in 15.5.3, we obtain

$$\begin{aligned} &F(t,S_t) = h_1(S(t),t) \\ +\;& \tfrac{1}{2}S^2(t)\textstyle\int_t^T e^{r(t-\xi)}\left[\sigma^2(t_0)e^{-(\alpha+\gamma)(\xi-t_0)} + \Sigma(S_t)(1-e^{-(\alpha+\gamma)(\xi-t_0)}) - V\right] \\ & \tfrac{\partial^2 h_1}{\partial S^2}(S(t),\xi)\,d\xi, \end{aligned}$$

which can be rewritten as

$$\begin{aligned} &F(t,S_t) = h_1(S(t),t) + (\Sigma(S_t) \\ -\;& V)\cdot\tfrac{1}{2}S^2(t)\textstyle\int_t^T e^{r(t-\xi)}\tfrac{\partial^2 h_1}{\partial S^2}(S(t),\xi)\,d\xi \\ +\;& (\sigma^2(t_0)-\Sigma(S_t))e^{-(\alpha+\gamma)(t-t_0)}\cdot\tfrac{1}{2}S^2(t)\textstyle\int_t^T e^{(r+\alpha+\gamma)(t-\xi)}\tfrac{\partial^2 h_1}{\partial S^2}(S(t),\xi)\,d\xi, \end{aligned}$$

which, in the case $t = t_0$, is

$$\begin{aligned} F(t,S_t) &= h_1(S(t),t) + (\Sigma(S_t) \\ &- V)\cdot\tfrac{1}{2}S^2(t)\textstyle\int_t^T e^{r(t-\xi)}\tfrac{\partial^2 h_1}{\partial S^2}(S(t),\xi)\,d\xi \\ &+ (\sigma^2(t)-\Sigma(S_t))\cdot\tfrac{1}{2}S^2(t)\textstyle\int_t^T e^{(r+\alpha+\gamma)(t-\xi)}\tfrac{\partial^2 h_1}{\partial S^2}(S(t),\xi)\,d\xi. \end{aligned}$$

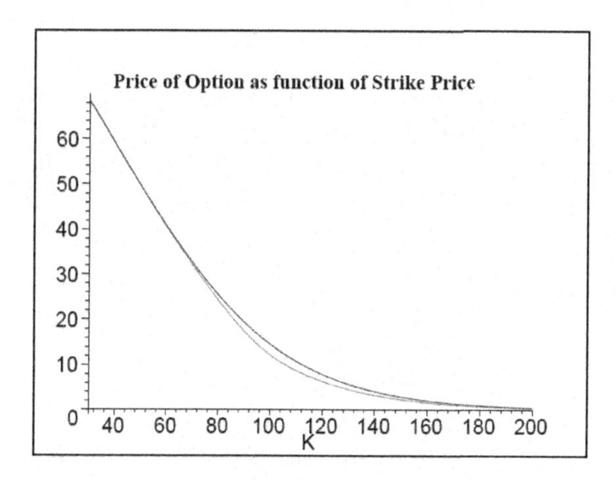

Figure 15.1 *The upper curve is the original Black–Scholes price and the lower curve is the option price given by the formula 15.5.9; here* $S(0) = 100$, $r = 0.05$, $\sigma(0) = 0.316$, $T = 1$, $V = 0.127$, $\alpha = 0.0626$, $\gamma = 0.0428$, $\tau = 0.002$.

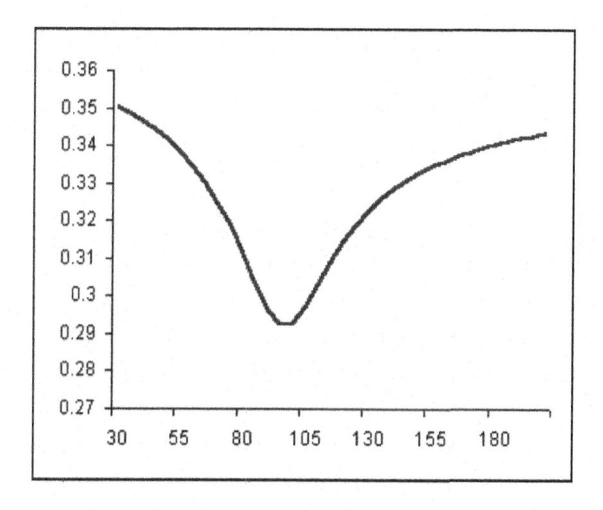

Figure 15.2 *Implied volatility of the call option price computed by 15.5.9 vs. strike price; the set of parameters is the same as for Figure 15.1.*

15.6 Appendix

Here we give a proof of Lemma 15.4. Fix $t > 0$ and denote $C \ni x = S_t$ with $S(t)$ satisfying 15.4.1. Then for a sufficiently small s

$$[F(t+s, x_s) - F(t, x)] = I_1 + I_2 + I_3 + I_4 + I_5,$$

where

$$
\begin{aligned}
I_1 &= \int_{-\tau}^{0} h(\theta - s)\,[H(x(\theta),x(s),t+s) - H(x(\theta),x(s),t)]\,d\theta,\\
I_2 &= \int_{-\tau}^{0} (h(\theta - s) - h(\theta))H(x(\theta),x(s),t)d\theta,\\
I_3 &= \int_{-\tau}^{0} h(\theta)\,[H(x(\theta),x(s),t) - H(x(\theta),x(0),t)]\,d\theta,\\
I_4 &= \int_{0}^{s} h(\theta - s)H(x(\theta),x(s),t+s)d\theta,\\
I_5 &= -\int_{-\tau}^{-\tau+s} h(\theta - s)H(x(\theta),x(s),t+s)d\theta.
\end{aligned}
$$

Then, by letting $s \to 0$,

$$
\begin{aligned}
I_1 &\to \int_{-\tau}^{0} h(\theta)H_3'(x(\theta),x(0),t)d\theta dt,\\
I_2 &\to -\int_{-\tau}^{0} h'(\theta)H(x(\theta),x(0),t)d\theta dt,\\
I_3 &\to \int_{-\tau}^{0} h(\theta)TH(x(\theta),x(0),t)d\theta dt\\
&\quad + \int_{-\tau}^{0} h(\theta)\sigma(t,x)x(0)H_2'(x(\theta),x(0),t)d\theta\; dW(t),\\
I_4 &\to \quad h(0)H(x(0),x(0),t)dt,\\
I_5 &\to -h(-\tau)H(x(-\tau),x(0),t)dt,
\end{aligned}
$$

where

$$
TH(x(\theta),x(0),t) = rx(0)H_2'(x(\theta),x(0),t) + \frac{\sigma^2(t,x)x^2(0)}{2}H_{22}''(x(\theta),x(0),t).
$$

The limit for I_3 is obtained by using the Ito's lemma. Then expression 15.4.5 follows.

References

[1] Akgiray, V. (1989). Conditional heteroscedasticity in time series of stock returns: evidence and forecast, *J. Business* 62, no. 1, 55-80.

[2] Avelanda, M., Levy, A., and Parais, A. (1995). Pricing and hedging derivative securities in markets with uncertain volatility, *Appl. Math. Finance* 2, 73-88.

[3] Baxter, M., and Rennie, A. (1996). *Financial Calculus*. Cambridge: Cambridge Univ. Press.

[4] Black, F., and Scholes, M. (1973). The pricing of options and corporate liabilities, *J. Political Economy* 81, 637-654.

[5] Bollerslev, T. (1986). Generalized autoregressive conditional heteroscedasticity, *J. Economics* 31, 307-327.

[6] Buff, R. (2002). *Uncertain volatility model. Theory and Applications*. NY: Springer.

[7] Chang, M.H., and Yoree, R.K. (1999a). The European option with hereditary price structure: basic theory, *Appl. Math. and Comput.* 102, 279-296.

[8] Chang, M.H., and Yoree, R.K. (1999b). The European option with hereditary price structure: a generalized Black–Scholes formula. Preprint.

[9] Chesney, M., and Scott, L. (1989). Pricing european currency options: a comparison of modified Black–Scholes model and a random variance model, *J. Finan. Quantit. Anal.* 24, no.3, 267-284.

[10] Cox, J.C., and Ross, S.A. (1976). The valuation of options for alternative stochastic processes, *J. Financial Economics* 3, 146-166.

[11] Cox, J.C., and Rubinstein, M. (1985). *Options Markets*. NJ: Prentice Hall.

[12] Duan, J.C. (1995). The GARCH option pricing model, *Math. Finance* 5, 13-32.

[13] Duffie, D. (1996). *Dynamic Asset Pricing Theory*. NJ: Princeton Univ. Press.

[14] Dunkan, T., Hu, Y., and Pasik-Dunkan, B. (2000). Stochastic calculus for fractional Brownian motion. 1. Theory, *SIAM J. Control Optim.* 38, no. 2, 582-612.

[15] Elliott, R.J., and van der Hoek, H. (2003). A general fractional white noise theory and application to finance, *Math. Finance* 13, no. 2, 301-330.

[16] Elliott, R., and Swishchuk, A. (2002). Studies of completeness of Brownian and fractional (B,S,X)-securities markets. Working paper.

[17] Frey, R. (1997). Derivative asset analysis in models with level-dependent and stochastic volatility, *CWI Quarterly* 10, 1-34.

[18] Geske, R. (1979). The valuation of compound options, *J. Financial Economics* 7, 63-81.

[19] Griego, R., and Swishchuk, A. (2000). Black-Scholes formula for a market in a Markov environment, *Theory Probab. and Mathem. Statit.* 62, 9-18. (in Ukrainian)

[20] Heston, S.L. (1993). A closed-form solution for options with stochastic volatility with applications to bond and currency options, *Review Finan. Studies* 6, 327-343.

[21] Hobson, D., and Rogers, L.C. (1998). Complete models with stochastic volatility, *Math. Finance* 8, no.1, 27-48.

[22] Hu, Y., and Øksendal, B. (2000). Fractional white noise analysis and applications to finance. Preprint, University of Oslo.

[23] Hull, J.C. (2000). *Options, Futures and Other Derivatives*, Prentice Hall, 4th edition, 2000.

[24] Hull, J.C., and White, A. (1987). The pricing of options on assets with stochastic volatilities, *J. Finance* 42, 281-300.

[25] Johnson, H., and Shanno, D. (1987). Option pricing when the variance is changing, *J. Finan. Quantit. Anal.* 22, 143-151.

[26] Kazmerchuk, Y., Swishchuk, A., and Wu, J. (2002). Black-Scholes formula revisited: security markets with delayed response, *Bachelier Finance Society 2nd World Congress*, Crete, Greece. (See also: Kazmerchuk, Y., Swishchuk, A. and Wu, J.-H. (2006) The pricing of options for security markets with delayed response, *Mathematics and Computers in Simulation*)

[27] Kind, P., Liptser, R., and Runggaldier, W. (1991). Diffusion approximation in past-dependent models and applications to option pricing, *Ann. Probab.* 1, no. 3, 379-405.

[28] Merton, R.C. (1976). Option pricing when underlying stock returns are discon-

tinuous, *J. Financial Economics* 3, 125-44.

[29] Mohammed, S.E. (1998). Stochastic differential systems with memory: theory, examples and applications, In *Stochastic Analysis and Related Topics VI*, Birkhäuser Boston, 1-77.

[30] Mohammed, S.E., Arriojas, M. and Pap, Y. (2001). A delayed Black and Scholes formula. Preprint, Southern Illinois University.

[31] Øksendal, B. (1998). *Stochastic Differential Equations: An Introduction with Applications*. NY: Springer.

[32] Rogers, L.C. (1997). Arbitrage with fractional Brownian motion, *Math. Finance* 7, no. 1, 95-105.

[33] Scott, L.O. (1987). Option pricing when the variance changes randomly: theory, estimation and an application, *J. Fin. Quant. Anal.* 22, 419-438.

[34] Sheinkman, J., and LeBaron, B. (1989). Nonlinear dynamics and stock returns, *J. Business* 62, no. 3, 311-337.

[35] Stein, E.M., and Stein, J.C. (1991). Stock price distributions with stochastic volatility: an analytic approach, *Review Finan. Studies* 4, 727-752.

[36] Swishchuk, A. (1995). Hedging of options under mean-square criterion and with semi-Markov volatility, *Ukrain. Math. J.* 47, no. 7, 1119-1127.

[37] Swishchuk, A. (2000). *Theory of Random Evolutions. New Trends*. Dordrecht: Kluwer.

[38] Swishchuk, A., Zhuravitskii, D., and Kalemanova, A. (2000). An analogue of Black–Scholes formula for option prices of (B,S,X)-securities markets with jumps, *Ukrain. Mathem. J.* 52, no. 3, 489-497.

[39] Wiggins, J.B. (1987). Option values under stochastic volatility: theory and empirical estimates, *J. Finan. Econ.* 19, 351-372.

[40] Willmott, P., Howison, S., and Dewynne, J. (1995). *Option Pricing: Mathematical Models and Computations*. Oxford: Oxford Financial Press.

Chapter 16

Epilogue

Random dynamical systems (RDS) provide a useful framework for modeling and analyzing various physical, social, financial, and economic phenomena.

In this book, we presented many models of RDS and developed techniques which can be implemented in finance and economics.

The first three chapters of the book described deterministic and random dynamical systems, random maps and position dependent random maps, and their applications in finance with generalized binomial models for stocks, options, and interest rates.

In Chapter 5 we introduced another class of random dynamical systems, namely, the random evolutions, which are operator valued random dynamical systems, and we described their many properties that we use in the next four chapters for geometric Markov renewal processes (GMRPs).

The GMRPs are generalizations of the classical Cox–Ross–Rubinstein binomial model and Aase geometric compund Poisson model in finance. We presented the GMRP in series scheme and obtained a limit result in the case of averaging and merging principles (Chapter 6), diffusion approximation, including merging and double averaging schemes (Chapter 7), normal deviations, with merging and double averaging schemes (Chapter 8), and Poisson approximations cases (Chapter 9). We considered the particular case of two-state Markov switching for all above-mentioned cases and presented numerical results. The averaged and merged principle describes an idea to use GMRP for modeling of regime-switching interest rate. The diffusion and normal deviation principles give the idea to use the GMRP for modeling of stock price as a diffusion process with coefficients that depend on the states of a Markov chain. The Poisson approximation can be used to model a stock price with jumps and regime-switching coefficients. We also presented option pricing formulas in the case of diffusion approximation and normal deviations schemes, and discussed the hedging strategies in both cases.

The next four chapters (11–13) were devoted to the stability properties and optimal control conditions for RDS that we can find in finance and economics: various interest models with jumps; stock prices described by stochastic delayed differential equations with jumps and regime-switchings; vector stochastic delayed differential equations. We also presented some optimization examples, including the stochastic Ramsey model and optimal portfolio selection problem.

The last, but not least Chapter 15, was devoted to the option pricing formula for a model of a stock price that is described by stochastic delay differential equation. The option pricing formula can be used in case we would like to include in our analysis not only current stock price, but also path-dependent information, or, as we call it, delay.

We hope that this book will be useful for the many researchers and academics that work in RDSs, mathematical finance and economics, and also for practitioners working in the financial industry. We believe that it will also be useful for graduate students specializing in the areas of RDSs, mathematical finance, and economics.

Index

Printed in the United States
by Baker & Taylor Publisher Services